T0222287

Springer Undergraduate Mathematics Series

More information about this series at http://www.springer.com/series/3423

Jeremy Gray

The Real and the Complex: A History of Analysis in the 19th Century

Springer

Jeremy Gray
Department of Mathematics and Statistics
The Open University
Milton Keynes
UK

and

University of Warwick
Coventry
UK

ISSN 1615-2085 ISSN 2197-4144 (electronic)
Springer Undergraduate Mathematics Series
ISBN 978-3-319-23714-5 ISBN 978-3-319-23715-2 (eBook)
DOI 10.1007/978-3-319-23715-2

Library of Congress Control Number: 2015950876

Mathematics Subject Classification (2010): 01A55, 26-03, 30-03

Springer Cham Heidelberg New York Dordrecht London

Printed on acid-free paper

Springer International Publishing AG Switzerland is part of Springer Science+Business Media (www.springer.com)

Preface

It is the task of the historian to give back to the past its sense of the future.

—unattributed quote.

This book is based on a course on the history of mathematical analysis that I taught for four years to third- and fourth-year undergraduates at the University of Warwick from 2005 to 2008. It concentrates on the history of topics in analysis, starting with three topics that were active around 1800 and it stops just short of Lebesgue's introduction of measure theory.

It is a historical approach, concerned with how ideas came about, what problems people were addressing, why they thought these ideas would help, and what they made of what they discovered. Very often, the mathematicians were surprised and puzzled too. The thread to follow is what interested them, not to get lost in the details of mathematics, which can indeed be unclear. That said, many of these topics are the staple ingredients of courses in mathematical analysis. Much of this material in this book illuminates this classical analysis by showing what can go wrong and so motivates the need for carefully worded theorems. But such examples should not be regarded as unwelcome guests at the feast, traps for the unwary, and certainly not as merely providing 'catch questions' for problem sheets and exams. Rather, they are examples of what can happen, clues to what else is to be discovered, stages on the endless path to giving a correct and comprehensive account of a subject. And, as this book hopes to make clear, a certain degree of insight is needed to appreciate which are the important ones, the fruitful examples to pursue, and we shall see that mathematicians have legitimately differed about this.

While giving the course I was frequently reminded of one of the most striking problems involved in teaching mathematics: indicating how valuable new mathematics is possible and actually gets done. A student in any of the sciences may easily enter university with some idea of what is happening in research in their subject, but despite the growing numbers of very welcome popularisations of mathematics in recent years it remains almost impossible to bring, shall we say, any of the Clay Problems to an audience of beginners (with the possible exception of the P versus NP problem). The problem is especially acute with the purer and less utilitarian branches of mathematics.

It is my belief that working mathematicians assimilate a certain body of knowledge and make it their own. They know how to use it to tackle certain problems, but they also have an idea of what they would like to know but do not. With some further work they may form a programme for getting there, conjecture a useful interim result, discover a misunderstanding, even a mistake in what they thought they knew. The attempt to obtain new results shapes their sense of what they know and what they would like to discover: a lemma, an estimate, an example, a counter-example, perhaps even the long sought-for result. I have tried in this book to convey my sense of how the analysts of the nineteenth century found their ways, imperfectly, from what they knew to what they wanted to know.

I found it curiously helpful to use Maple to generate pictures of functions. These graphs are not proofs, they may well be misleading, but they are suggestive, and they provide a way of looking carefully at the functions themselves. I hope they form a bridge between the student's understanding and that of the well-educated mathematician of more than a hundred years ago.

I have not respected the modern distinction between real and complex analysis for the simple reason that it did not operate in the nineteenth century. Indeed, as will be explained, neglecting it proved to be one of the sources of one of the finest discoveries of real analysis: functions that are continuous on an interval and differentiable nowhere (see Sect. 20.4).

It is my hope that this book will be part of a four-volume series, the first volume of which, *Worlds out of Nothing* (2011), covers topics in the history of geometry in the nineteenth century, the third, now almost completed, will be on algebra in the same period, and the final volume of which will be on the history of differential equations. Like the first book, this was originally a course of 30 lectures, structured around three assessment points. I do not want to repeat the advice about writing essays that I gave in the earlier book, although I stand by it, but I have taken the opportunity to treat three chapters as an opportunity not only to look back at what has been discussed but to think about how it can be described and to address the student more directly. The point at issue is the difference between a list of facts and a good essay. A good essay is an argument; it expresses a point of view which it supports with evidence. It is attentive to other points of view, refuting them as necessary. It is an attempt to persuade.

There can be many different but equally good answers to an essay question. This depends partly on who forms the intended audience, partly on the author's sensibilities, partly on the space and time available. As a teacher I favour the short essay because it is hard: it is difficult to select the most important items, relatively easy to reply with 'everything'. It is, of course, difficult as a teacher not to reward students who uncritically but astutely follow the 'party line', especially when, as here, the author has strong opinions. To that end students should be encouraged to track down other books in their university library, and I have made some suggestions as this book proceeds.

A problem with teaching the history of the mathematics of the nineteenth century to an English-speaking audience is that so little was written in English. In an Appendix I have given my own translations of extracts from a paper by Dirichlet on Fourier series, papers by Riemann introducing complex function theory and the Riemann integral, and the paper by Schwarz where he mapped a circle analytically onto a square. An extract from the English translation of Fourier's account of Fourier series is also supplied. A small number of other sources are now available in English including the collected works of Riemann, which is very welcome.

The Structure of This Book

The first three chapters set the scene by describing Lagrange's attempt to provide rigorous foundations for the calculus, Fourier's introduction of the series now named after him, and Legendre's work on elliptic integrals. All these topics are sources for much of the subsequent development, although only the second is likely to seem familiar. But Lagrange's ideas set the scene for Cauchy's, and the topic of elliptic integrals is not hard, was obviously important by the standards of 1800, and has surprising later ramifications.

The scene being set we now meet Cauchy, with two chapters on his work on real analysis and one on the much more confused state of his work in the 1820s and 1830s on complex analysis. The marked difference in his understanding was mirrored by the circumstances of publication, with important implications for decades. Three chapters follow on how Abel, Jacobi, and Gauss made complex elliptic functions out of real elliptic integrals, which I argue was one of the most important sources of subsequent interest in complex analysis (and more visible than Cauchy's work). Then come two chapters bringing elliptic functions and complex functions together, and we can pause to take stock.

The next part of the book picks up the topic of potential theory and the contributions of Gauss, Green, and Dirichlet, before turning to the topic of Dirichlet's work on Fourier series and Riemann's work on trigonometric series. Then come several chapters on the rival developments of complex analysis by Riemann and his successors on the one hand and Weierstrass and his supporters on the other. The lecture course included a chapter on minimal surfaces at this point, as an illustration of how new subjects in mathematics sell themselves by the unsuspected but powerful connections they make with other branches of the subject, but I have decided that that story is better placed in the planned history of differential equations.[1]

After a further pause for taking stock the book turns to topics in the real analysis of the later nineteenth century: uniform convergence and what lies beyond, what

[1]For a much fuller history of the study of minimal surfaces, see the forthcoming (Gray and Micallef).

were called 'assumptionless functions' and their strange properties, with the questions they raise about the nature of point sets and the properties of the integral, culminating in the collapse of the fundamental theorem of the calculus. Then comes the explicit construction of the real numbers and a look at the first proofs of the implicit function theorem (this replaces a chapter on the first rigorous methods in potential theory, which I have moved to the projected volume on the history of differential equations). The final three chapters before the last revision chapter look forward and deliberately look at topics that were not by any means resolved by 1900: paradoxes to do with length and area and the behaviour of the integral; the Cantorian theory of infinite sets; and some topics where topology finally surfaces.

Acknowledgements

In preparing this course and writing this book I made considerable use of three books: Bottazzini's *The Higher Calculus* (1986); Bressoud's *A radical approach to real analysis* (1994); and Hawkins' *Lebesgue's Theory of Integration; its Origins and Development* (1999). This blanket acknowledgement is intended to obviate the need for a dense set of footnotes throughout this book. Anyone wanting to know more about the history of analysis should take up these books as a matter of course. I had intended to augment the course by including measure theory, basing my account on the two long papers that Lebesgue himself wrote that provide a very sympathetic, and historical, introduction to the subject. Happily, that task has been carried out admirably by Bressoud in his book *A radical approach to Lebesgue's theory of integration* (2008).

The material on the history of complex function theory largely comes from joint work with Umberto Bottazzini, *Hidden Harmony—Geometric Fantasies; The rise of complex function theory* (2013). But to put it that way understates the debt I owe to him. In the course of the many years we worked on that book I learned a great deal from him, not only, but obviously, about Cauchy, Weierstrass, and every aspect of the history of complex function theory, but about doing the history of mathematics . His influence is visible, to me, on every page—Umberto, thank you!

I also thank David Rowe for his many very helpful comments on the final version of this book, and the anonymous referees for their criticisms both small and large.

Finally, my thanks go to the people at the Open University and the University of Warwick for making it possible for me to teach the course, and to the students who did it and enjoyed it.

Contents

List of Figures

Chapter 1
Lagrange and Foundations for the Calculus

1.1 Introduction

Mathematical analysis has many beginnings, and in the 19th century its history has a lot to do with the reformulations of the calculus that gradually led to the creation of adequate foundations and standards of rigorous argument. This does not mean that these were the only issues occupying the minds of mathematicians, or that they succeeded completely, but in important ways the subject was greatly transformed as the century progressed.

This chapter and the next two are intended to set the scene, and to give a glimpse of what was happening between, roughly, 1780 and 1820. This covers the period of the French Revolution and the Napoleonic wars, and it is a largely French, even Parisian, story. In this chapter we look briefly at various 18th-century attempts to say why the calculus worked, concluding with what can be regarded as the last of the 18th-century attempts. It is due to Joseph-Louis Lagrange, and its failure opened the way for the radically different accounts that followed.[1]

1.2 The Calculus c. 1800

The calculus was the independent creation of Isaac Newton in the 1660s and Gottfried Wilhelm Leibniz in the 1660s and 1670s. In the 18th Century it ground slowly to a halt in Britain, but it developed with remarkable vigour on Continental Europe, in the hands of the brothers Jakob and Johann Bernoulli, who were in direct contact with Leibniz, and with Johann Bernoulli's one-time pupil, the remarkable Leonhard Euler. In the course of a long life spend first as a student in Basel and then divided

[1] For a rich and stimulating account of rigour in 18th-century mathematics, and of what was involved in the transition to the 19th century, I particularly recommend Schubring (2005).

J. Gray, *The Real and the Complex: A History of Analysis in the 19th Century*,
Springer Undergraduate Mathematics Series, DOI 10.1007/978-3-319-23715-2_1

between the Academies of St. Petersburg, Berlin, and St. Petersburg again Euler wrote so much that the still-incomplete edition of his collected works has run to 78 volumes.[2]

The 18th century calculus dealt in both single and many variable problems, it embraced differential equations, first ordinary and then partial, it had a theory of differentiation and integration, it was full of clever techniques. In Euler's hands particularly, it was a theory of functions: dependent variables were functions of independent variables. (The calculus of Newton and Leibniz had dealt in variables mutually varying.) Above all, the calculus became the core domain of mathematics, it illuminated geometry and it seemed to be the only way to the mathematical study of nature.

At its core stood a painful paradox. The simple and invariably correct rules for differentiation and integration were established by arguments that invoked: the vanishing of negligible quantities; arguments about infinitesimal quantities; plausible limit arguments that nonetheless seemed close to giving rules for evaluating 0/0. In short, the calculus worked—but no-one could adequately say why.

This was not for the want of trying. Euler wrote at length on this, as on everything else, but his view was that the naïve intuitions could be trusted if they were stated as clearly as they could be. However, neither he, nor anyone else, achieved that goal. For example, in his *Institutiones calculi differentialis* (1755), one of his major treatises that defined the calculus for the later 18th Century, Euler spelled out his opinion in this frequently quoted passage from his *Institutiones calculi differentialis* (1755).[3]

> 83. This doctrine of the infinite, however, will be better explained when we explain what the infinitely small of the mathematicians is. There is no doubt that any quantity can be diminished until it vanishes and is transformed, into nothing. But an infinitely small quantity is nothing else but a vanishing quantity and, therefore, actually will be $= 0$. This definition of the infinitely small is in agreement with the other which states that it is smaller than any given quantity. Indeed, if a quantity were so small that it is smaller than any given one, then it certainly could not be anything else but zero; for if it were not $= 0$, then a quantity equal to it could be shown, which is against the hypothesis. To those who ask what the infinitely small quantity in mathematics is, we answer that it is actually $= 0$. Hence there are not so many mysteries hidden in this concept as there are usually believed to be. These supposed mysteries have rendered the calculus of the infinitely small quite suspect to many people. Those doubts that remain we shall thoroughly remove in the following pages, where we shall explain this calculus.

After noting that it may make sense to write $2 : 1 = 0 : 0$ because the first term is twice the second and the third time twice the fourth, Euler went on

> §85 …It is known that …$n.0 = 0$ as well as $n : 1 = 0 : 0$. From this it seems possible that two quantities, whatever their geometric ratio may be, will always be equal if we look at them from the arithmetic point of view. Hence if two zeros can have an arbitrary ratio, then I judge that different signs should be applied, especially when we have to consider a geometric ratio of different zeros. The calculus of the infinitely small is therefore nothing but the investigation of the geometric ratio of different infinitely small quantities. This enterprise will be thrown

[2]For a growing and invaluable guide, consult the Euler Archive on the web.

[3]The translation is taken from Struik (1969, 384).

> into the greatest confusion unless we use different signs to indicate these infinitely small quantities ....
>
> §86 Hence, if we introduce into the infinitesimal calculus a symbolism in which we denote dx an infinitely small quantity, then $dx = 0$ as well as $adx = 0$ (a an arbitrary finite quantity). Notwithstanding this, the geometric ratio $adx : dx$ will be finite, namely $a : 1$, and this is the reason that these two infinitely small quantities dx and adx (though both $= 0$) cannot be confused with each other when their ratio is investigated. Similarly, when different infinitely small quantities dx and dy occur, their ratio is not fixed though each of them $= 0$.

Whatever this may mean, it cannot be said to do more than gesture at what might be involved in rigorising the calculus,

Euler's only rival, for most of his life, was the French mathematician Jean le Rond d'Alembert, who concentrated his efforts on problems in applied mathematics and referred to the 18th century as the century of mechanics.[4] He owes his name to the fact that he had been abandoned as an infant on the steps of the church of St Jean le Rond in 1717, which is near Notre Dame in Paris, and he was brought up by the glazier's wife, to whom he remained devoted. He had an excellent memory and grew up to be a brilliant conversationalist, which enabled him to become prominent in the competitive world of the Parisian salons that were frequented by influential people who were neither as rich and powerful as the nobility nor as conservative as the Church. In this milieu he rose to become a leading member of the group around Denis Diderot who produced the *Encyclopédie (or, Dictionnaire raisonné des sciences, des arts, et des métiers*, (1751–1772). This remarkable series of volumes was animated by the philosophy of rationalism, and is one of the major documents of the Enlightenment, and this led to heated conflict with the Jesuits and other forces of Catholic conservatism. In the ensuing struggles Diderot and his colleagues were supported by Voltaire and later by Condorcet, and eventually they succeeded in taking control of the Académie des Sciences.

His interest in mechanics led d'Alembert to take a Newtonian view of the foundations of the calculus. Newton had indicated the foundations of his calculus in these terms in a series of Lemmas on pages 38–39 of the *Principia*

> These Lemmas are premised to avoid the tediousness of deducing involved demonstrations ad absurdum, according to the method of the ancient geometers. For demonstrations are shorter by the method of indivisibles; but because the hypothesis of indivisibles seems somewhat harsh, and therefore that method is reckoned less geometrical, I chose rather to reduce the demonstrations of the following Propositions to the first and last sums and ratios of nascent and evanescent quantities, that is, to the limits of those sums and ratios, and so to premise, as short as I could, the demonstrations of those limits.

These Lemmas have been much discussed. D.T. Whiteside, the editor of *The Mathematical Papers of Isaac Newton*, brushed them aside as not, after all, central to the arguments in the *Principia*, and called them "undeniably a retrospective gloss".[5] Indeed, it is clear that Newton did not derive the results of the *Principia* by his calculus

[4]On d'Alembert's life and times see Hankins (1970). A vigorous new edition of all of d'Alembert's papers is also underway.

[5]See Whiteside, *The Mathematical Papers of Isaac Newton*, Vol. 6, p. 108.

and then cover his tracks with geometry, as was once suggested. However, Pourciau (1998, 281) argued forcefully that "each lemma *is itself* (a natural replacement for) an elementary and basic definition, property, or theorem of the calculus" [Pourciau's italics]. The point to note in the context of 18th-century explorations of these issues is that an approach to the calculus in terms of some, not entirely precise, idea of limits was advocated by some authors and carried a Newtonian pedigree.

D'Alembert addressed directly the question of how to defend a limiting argument that appears to make sense of an expression of the form 0/0.[6] Unfortunately his answer is confused at crucial points—a familiar feature of his written work, which lacked the elegance of his conversations.

He first suggested that 0/0 can have any value, including the correct one, and then argued that the limit is not the ratio 0/0 because this ratio is not defined. This may be the source of another problem with his concept of a limit, which he defined this way.

> One magnitude is said to be the limit of another magnitude when the second may approach the first within any given magnitude, however small, though the first may never exceed the magnitude it approaches; so that the difference of such a quantity to its limit is absolutely unassignable.

It seems that, for d'Alembert, a limit may be approached, but only from one side, and never attained.

1.3 Joseph Louis Lagrange

Joseph Louis Lagrange (1736–1813)

[6]See 'Limite', *Encyclopédie*, vol. 9, 1765. There is an extract from d'Alembert's essay on differentials, *Encyclopédie*, vol. 4, (1754) in Struik (1969, 342–345) and also in Fauvel and Gray (1987, 18.A3).

There matters rested until Joseph Louis Lagrange took up the question. Lagrange was born in Turin in 1736. He was not an Italian, for a unified Italy was over a century away, but rather someone who was in some ways Italian and in others French. From his youth onwards he was devoted to the work of Euler, his earliest interests were in the calculus of variations and in mechanics, notably planetary astronomy, branches of the calculus in which Euler was the master. The younger man then followed him into researches in algebra and number theory. In all of these areas he was eventually to surpass Euler, for where Euler was a brilliant 'experimenter' if you like, a discoverer of remarkable facts, Lagrange was the one who came up with the proofs; proofs that elude Euler are of a high calibre, and Euler was impressed. Yet Lagrange was a very shy man, as diffident as he was gifted; Clairaut described him in 1764 as "a young man as remarkable for his talents as for his modesty; his character is mild and melancholy; he knows no other pleasure than to study".[7] Remarkably, although Euler and Lagrange corresponded frequently, they never met. Indeed, Lagrange would only go to the Berlin Academy of Sciences after Euler had left in 1766 to resume directorship of the St Petersburg Academy; it seems Lagrange believed Euler to be too much his superior—even though Lagrange was succeeding Euler as director in Berlin. He stayed in Berlin until he moved to Paris in 1787, and over the next twenty years—the years of the Revolution and of Napoleon—Lagrange contributed to establishing Paris as the new world centre for mathematics in France.

Higher Education in the French Revolution

By the mid-1790s the French educational system had totally collapsed, and out of the chaos came two vigorous new institutions. One, founded in 1794, was the École Normale de l'An III—that is, Normal School of the Year Three (of the Republic, the revolutionaries redefined the calendar and in that way sought to redefine time itself) where teachers were to be trained. Potential student-teachers from throughout France were assembled in Paris and lectured to, in an endeavour to establish educational norms for the whole country, by the leading mathematicians of France: Lagrange, Pierre-Simon Laplace and Gaspard Monge. In this form the school lasted only a few months (it was later refounded as the École Normale Supérieure, originally intended to provide training for teachers), but its influence lasted longer, because the lectures were copied down (by stenographers in the audience) and printed. The educational reformers were clear that the educational system for all France should be determined and influenced by decisions and practices in Paris, and further, that there should be a strong mathematical component in the curriculum at all levels.

The École Polytechnique was also founded in 1794, with the task of training large numbers of students in physics, mathematics and engineering.[8] It was the first institution in the world of this kind, and every year students competed across the whole of France to become one of the two hundred or so entrants. Monge and

---

[7] See the letter to D. Bernoulli, quoted in Euler, *Opera Omnia*, IV, A5, p. 330, n. 2.

[8] It was originally called the École Centrale des Travaux Publics (the Central School for Public Works) thus for civil and military engineering, and given its present name the in the following year.

Lagrange taught there, and Pierre Simon Laplace held the even more influential position of examiner—he determined what would be examined as well as what level of attainment the students ought to display. The founding of the École Polytechnique is a significant event in the rise of the modern university.

In 1799, the École Polytechnique was re-organised. It now provided two years of courses that were preparatory to the specialised training that graduates of the school could go on to obtain in one of the higher Écoles (such as the School of Mining, the School of Artillery, the School of Military Engineering, and the School of Bridges and Roads). In 1804 Napoleon ordered that the École Polytechnique be militarised, and the students treated like cadet soldiers. It has been suggested that a factor in his wish for a more disciplined regime at the school was the students' reluctance to welcome Napoleon's appointing himself Emperor.

1.4 Lagrange on the Foundations of the Calculus

To return to the story of the calculus, in 1784 Lagrange, then still at the Berlin Academy, proposed that the Academy award a prize for

> a clear and precise theory of what is called Infinity in mathematics [because] It is well known that higher mathematics continually uses infinitely large and infinitely small quantities. Nevertheless, geometers, and even the ancient analysts, have carefully avoided everything which approaches the infinite.

The matter was to be "treated with all possible rigour, clarity, and simplicity", so it is clear that the judges were not asking for a few well-phrased definitions but for a new and substantial effort to vindicate the calculus.

In this they were unsuccessful.[9] No major mathematician entered, and the judges reported that the entries lacked rigour, they failed to deal with the problem of deduction from contradictory assumptions (for example, that a quantity is both zero and non-zero), and they did not explain "how so many true theorems have been deduced".

So when Lagrange began teaching analysis at the École Polytechnique in 1797 the pressure was on him to meet the challenge he had earlier set, and he published his *Théorie des fonctions analytiques*. All his mathematics has an algebraic cast, and the full title of this work makes clear that this is no exception. It runs

> Théorie des fonctions analytiques contenant les principes du calcul différentiel, dégagé de toute considération d'infiniment petits ou d'évanouissans, de limites ou de fluxions, et réduits à l'analyse algébrique des quantités finis.

That is: *The theory of analytic functions containing the principles of the differential calculus disengaged from any consideration of the infinitely small or vanishing quantities, limits and fluxions, and reduced to the algebraic analysis of finite quantities.*

---

[9]The prize was awarded to Simon Antoine Jean L'Huillier for his *Exposition élémentaire* (1787).

He argued that the basic principles of the calculus had been neglected by earlier mathematicians who had been concerned only with obtaining solutions to problems, and had therefore entertained unrigorous concepts such as infinitesimals. However unfair—the method of infinitesimals proved to contain more of the kernel of the eventual solution than Lagrange's own methods—his criticisms were valid. He rejected Newton's methods on two grounds: they depended the "foreign idea" of motion that belonged to physics, not mathematics; and there was no adequate definition of 'instantaneous variable velocity'. He rejected d'Alembert's limit concept as being too vaguely geometric, and geometry, like motion, was "foreign" to the very spirit of analysis. And once he had rejected all previous attempts, he set about placing all previous results in the calculus within a new, rigorous framework.

Lagrange showed, to his satisfaction, that every function f can almost always be expanded as a power series of the form

$$f(x+i) = f(x) + a_1 i + a_2 i^2 + \cdots,$$

where i stands for an arbitrary increment, provided that one could somehow derive the coefficient of the first power of i from f itself.[10] He made a strong case that the meaning of $f'(x)$ does not have to depend upon limits, infinitesimals, or prime and ultimate ratios: it is simply the coefficient of i in the series expansion of $f(x+i)$. His line of reasoning is thoroughly algebraic, and such a claim, if valid, would go a long way towards making the calculus rigorous. However, his ideas were not completely convincing and, as we shall see in the next few chapters, they were soon replaced. But the force of his example, and his eminence as a mathematician, generated much debate over the foundations of the calculus: those mathematicians were not satisfied with merely obtaining results would now find it necessary to justify their methods much more rigorously.

One might object that Lagrange's enterprise is hopeless, and point out that the function defined by

$$y = \begin{cases} x & \text{if } x \geq 0 \\ -x & \text{if } x \leq 0 \end{cases}$$

cannot be differentiable at the origin. But on such occasions Lagrange was known to observe that[11] "This principle is generally valid, but I have remarked that it can be subject to exceptions that can make the preceding demonstration defective." One interpretation of this remark might be that a claim with a convincing proof that was surely true in general but occasionally false might be viewed as true in the way that we today say that "Everyone can vote in this election" and mean something more like "Everyone can vote in this election except those barred by specific, well-known, pieces of legislation". We shall meet this viewpoint that theorems may admit exceptions in a more sophisticated context below in Sect. 4.4.

In a little more detail, Lagrange argued as follows.

---

[10] See the extract in Fauvel and Gray (1987, 18.A4) and below.

[11] In Lagrange (1826, 182) quoted in Sørensen (2005, 470).

1. In the expansion of $f(x+i)$, where i is a small increment, no fractional powers of i can occur except for particular values of x.
2. So $f(x+i) = f(x) + f'(x)i + f''(x)i^2 + \cdots$; this expression defines $f'(x), f''(x), \ldots$.
3. In the expansion $f(x+i) = f(x) + pi + qi^2 + ri^3 + \cdots$ replace x by $x+o$ where o is independent of x, and by expanding $f(x+i+o)$ in two ways (first, replace i by $i+o$ in the above expansion, second replace x by $x+o$) deduce that the process which determines p from f, say $p = f'$ also determines q from p by $2q = p'$, r from q by $3r = q'$, and so on.
4. By considering the well-known cases from the elementary calculus, deduce that $f'(x) = \frac{df}{dx}$ and hence the familiar formula for the Taylor series.

The best account of Lagrange's approach to the foundations of the calculus, and the only one to respect its depth and novelty and so grasp its ultimate failure without anachronism is Ferraro and Panza (2012). They argue (p. 99) that

> numbers and geometric and mechanical magnitudes are, for Lagrange, quantities of a particular sort, whereas his theory of analytical functions is intended to deal with quantities in general, or better *in abstracto*.

So Lagrange took his algebra to apply to entirely abstract quantities, which he often called 'algebraic quantities', and (p. 100)

> Lagrange's theory is algebraic insofar as it deals with algebraic quantities, [and] formal, insofar as these quantities are identified through the relations they have with each other, which are in turn displayed by appropriate formulas.

But this saddled Lagrange with a problem, which was why his 'algebraic' quantities obey the usual rules for quantities, such as being ordered and varying continuously. He did not want to reduce his general quantities to the special case of geometrical ones, for that would prevent him explaining why the calculus could be applied to geometry, and Ferraro and Panza argue that this is where and how he failed. Their conclusion is (p. 100)

> Our basic point in explaining Lagrange's failure is precisely that his notion of algebraic quantity does not guarantee that algebraic quantities meet this crucial requirement, so much so that he cannot but surreptitiously suppose that they do meet it. Hence, this notion appeared too weak to bear the weight of Lagrange's reductionist purpose. But once reinforced with this surreptitious assumption, it became too strong to play the role of a starting point of his foundational program.

1.5 The Infinitely Small and the Infinitely Large

It will become clear as this book unfolds that a number of mathematicians, Cauchy among them, were able to make great progress in developing and rigorising the calculus without always being clear what they were talking about. This is not the paradox it may seem; they were dealing with exceptionally difficult topics where being productive is the best that can be hoped for. The historians' job, on such occasions, is to make sense of the texts they wrote on their own terms, giving good reasons for their sense of what the phrase "on their own terms" means.

To do this, careful reading of each specific text is required. Other texts may be consulted: those by the same author, those by contemporaries and influential predecessors. Definitions, when they were given, can be usefully compared with the use of the terms in practice, and may not agree. Meanings may be elucidated by establishing a set of ideas that the original author shared with his or her contemporaries. When the original text had an influence, the historian can gain a feeling for what the text may say by finding out how it was read by the author's contemporaries and successors. If the author's ideas were taken up, then on what grounds? How were they altered in subsequent use? If they were not taken up, then why not? Were they neglected, or explicitly rejected? Were they understood or misunderstood?

Specifically, there was a profusion of methods in the early calculus that condensed in the hands of Newton and Leibniz to talk of limits and infinitesimals. In keeping with the procedures just outlined, these concepts are often glossed by historians of mathematics as something like "very very small" or "arbitrarily small" or "tending to zero", in a loose chain of influences from Newton and Leibniz via the major names to Weierstrass and the language of standard analysis ($\varepsilon - \delta - N$ analysis). Most historians of mathematics give old authors the benefit of the doubt, while noting that more needs to be said by modern standards. It is indeed tiresome and ultimately ahistorical to read an old text as if marking an undergraduate essay. But interpretation can be difficult when subtleties may have led mathematicians into foundational problems, or even, as we shall see with Cauchy in Chap. 4, into error.

For example, Leibniz introduced his differential calculus in 1684 in terms of an axis *AX*, a curve *VV*, the perpendicular (or ordinate) *VX* to the axis, and a tangent *VB* intersecting the axis at B. Then he wrote[12]

> Now some straight line selected arbitrarily is called dx, and the line which is to d as v is to *XB* is called dv, or the difference of these v. Under these assumptions we have the following rules of the calculus.
>
> If a is a given constant, then $da = 0$, and $d(ax) = adx$. If $y = v$ (that is, if the ordinate of any curve *YY* is equal to any corresponding ordinate of the curve *VV*), then $dy = dv$. Now *addition* and *subtraction*: if $z - y + w + x = v$, then $d(z - y + w + x) = dv = dz - dy + dw + dx$. *Multiplication*: $d(xv) = xdv + vdx$, or, setting $y = xv$, $dy = xdv + vdx$.

No matter that the rules are not explained—what is this difference dv? It would seem to be some, presumably small, segment of a straight line. But less than a page later, Leibniz wrote:

> We have only to keep in mind that to find a *tangent* means to draw a line that connects two points of the curve at an infinitely small distance, or the continued side of a polygon with an infinite number of angles, which for us takes the place of the curve. This infinitely small distance can always be expressed by a known differential like dv, or by a relation to it, that is, by some known tangent.

---

[12]See Struik *Sourcebook*, 272–280, F&G 13.A3. I have removed some repetitions from the text.

Now it is presented as an infinitely small distance. Could it be that Leibniz did indeed think of there being infinitely small distances, or was that more a way of speaking, a useful fiction? It is already clear that they have contradictory properties, and why should $d(xv)$ not be written as $(x + dx)(v + dv) - xv = xdv + vdx + dxdv$?

In 1701 Leibniz published an article in which he said that the infinite need not be taken rigorously, "For in place of the infinite or the infinitely small we can take quantities as great or as small as necessary in order that the error be less than any given error." Several people who had learned the calculus from him asked for a clarification, and he wrote a letter in 1702 saying that although infinitely small magnitudes cannot be real nonetheless geometry proceeds as if they are. However, he went on, actual infinite and infinitesimal quantities are only fictions that can be used for the sake of brevity.[13] To quote a recent authority (Arthur 2013, 553–554):

> …the idea that Leibniz was committed to infinitesimals as actually infinitely small entities is a misreading: his mature interpretation of the calculus was fully in accord with the Archimedean Axiom. Leibniz's interpretation is (to use the medieval term) syncategorematic: Infinitesimals are fictions in the sense that the terms designating them can be treated as if they refer to entities incomparably smaller than finite quantities, but really stand for variable finite quantities that can be taken as small as desired.

Euler's use of the concepts of infinitely large and infinitely small numbers also indicates the issues. The exponential function a^z, $a > 1$ was presented in his *Introductio in analysin infinitorum* §98 as a function that, among other properties, goes to infinity as z goes to infinity. Then, when explaining how to write the logarithm and exponential functions as power series, he considered what is meant by the expression a^ω, $a > 1$, and wrote (§114):

> Let ω be an infinitely small number, or a fraction so small that, although not equal to zero, still $a^\omega = 1 + \psi$, where ψ is also an infinitely small number ….

From this he deduced that one can write $a^\omega = 1 + k\omega$.

A few pages later (in §134) when explaining how to write the sine and cosine functions as power series, he wrote

> Let the arc z be infinitely small, then $\sin z = z$ and $\cos z = 1$. If n is an infinitely large number, so that nz is a finite number, say $nz = v$, then ….

How might these and many other remarks have been understood by him or his readers? The use of infinites and infinitesimals was not explained by him in the *Introductio*, so we must start by taking his words at face value. For example, there are infinitely small and infinitely large numbers, and they are not, apparently, the result of a limiting argument. This does not mean that Euler fully understood what has to be said to make the concept consistent with other concepts he used. It does not mean that his use of other terms, such as infinite series, was in any way different from the usual spread of ideas from Newton to Cauchy and Weierstrass. In the *Introductio*, Euler wrote an infinite series in the usual way, e.g. as $A + Bz + Cz^2 + Dz^3 + \cdots$,

[13] See Jesseph (2015, 194–195); the volume contains a number of important articles on Leibniz's mathematics and philosophy.

and indeed insisted (§62) that "There is a law by which any coefficient is determined by a certain number of its predecessors".

The same is true of Euler's use of the concept of the infinitesimal, as the passage from his *Institutiones calculi differentialis* showed. Taken at face value, Euler said that an infinitely small quantity is really equal to zero. Moreover, his handling of terms like dx in his discussion of foundations is open to obvious misuse; we can note that he does not explain how to avoid obvious contradictions, and he even, on occasion and presumably for pedagogic purposes, regarded one millionth as the "smallest possible amount" (*Introductio* §148).

The most obvious interpretation of all this is that Euler was no more able to explain the well-known problems with the calculus than anyone else, but was comfortable with naive arguments that allowed him to get on with the job: infinite numbers are very, very big, bigger than any finite number; infinitesimals are very, very small, even vanishing. Implicit limiting arguments could be made explicit, but arguments with infinites and infinitesimals are quicker when you have learned how to use them wisely. Euler is intelligible, and of his time, when read in that way.

Recently there have been attempts to argue that Leibniz, Euler, and even Cauchy could have been thinking in some informal version of rigorous modern non-standard analysis, in which infinite and infinitesimal quantities do exist. However, a historical interpretation such as the one sketched above that aims to understand Leibniz on his own terms, and that confers upon him both insight and consistency, has a lot to recommend it over an interpretation that has only been possible to defend in the last few decades. It is parsimonious and requires no expert defence for which modern concepts seem essential and therefore create more problems than they solve (e.g. with infinite series). The same can be said of non-standard readings of Euler; for a detailed discussion of Euler's ideas in this connection, see Schubring (2005). This is not to say that Leibniz, Euler, or even Cauchy in certain ways, almost had the modern, classical, Weierstrassian kind of real analysis in mind. Plainly they did not. But it is to say that the foundations of the calculus were for at least two centuries the subject of shifting, partial, and largely coherent speculations that form the opening chapters of the history of analysis.

Chapter 2
Joseph Fourier

Joseph Fourier (1768–1830)

2.1 Introduction

Fourier series are trigonometric series used to represent a function, and they are widely used throughout pure and applied mathematics. Fourier was not the first to use them, but his name is rightly attached to them because he was the first to use them in the study of heat diffusion, to display their use in the solution of a partial differential equation, and to argue successfully for their generality.

There had been a long 18th-century debate about trigonometric series in connection with solutions to the wave equation and the shape of a vibrating string. On the one hand it seemed reasonable that a string could have any continuous initial shape—that was Euler's view—on the other hand the equation could only be solved by functions to which the calculus applied (we would say that the solutions had to

J. Gray, *The Real and the Complex: A History of Analysis in the 19th Century*,
Springer Undergraduate Mathematics Series, DOI 10.1007/978-3-319-23715-2_2

be twice differentiable) and this made them what d'Alembert called analytic. Convergence questions were not central to this debate, which was left unresolved in a number of ways.

Fourier proposed to reopen the debate by boldly asserting that any solution to the heat equation, which he was the first to derive, could be written as an infinite sum of sines and cosines for the simple reason that any function could be written that way. This is a dramatic claim, and it was still more so in his day, because the consensus was that however broadly a function might be defined all the functions that arise in practice are finite sums of familiar ones: polynomials, sines, cosines, exponentials and logarithms, nth roots, and the like. They could also be infinite power series, and indeed infinite trigonometric series, but nonetheless they had the usual sorts of properties, such as smoothly varying graphs. No-one said so in so many words, but it is clear that the expectation was there, and Fourier in particular simply assumed that every function is continuous, as is clear from his account of the coefficients of a Fourier series in his (1822, §423).

One of the dramas introduced by Fourier's series was that they readily flout all these expectations. As we shall see, at various stages in the 19th century they provided fresh, and disturbing, examples of just what functions could do. Contrary to what Fourier himself believed, if Cauchy's work began the exploration of what rigorous mathematics can do, Fourier series can indicate just what theory is up against.

2.2 Fourier's Career

Joseph Fourier was born in Auxerre, France in 1768. He was orphaned at the age of 9 and placed in the town's military school where he learned mathematics and a sense of civic responsibility. He was nearly guillotined at the height of the Terror in 1794, but the sentence was withdrawn and Fourier was able to go to the École Normale. In 1795 he was appointed an assistant lecturer at the École Polytechnique, working under Lagrange and Monge, and in 1798 Monge, a prominent supporter of Napoleon, selected Fourier to go on the French expedition to Egypt. After the British defeated them there, Fourier returned to France in 1801, hoping to resume his work at the École Polytechnique, but Napoleon had been impressed by his organisational talents and sent him instead to be the prefect of Governor of the Department of Isère.[1] He was so successful here that Napoleon made him a Baron in 1808, and in 1809 he finished his contribution to the *Description d'Egypt*, a massive account and glorification of ancient Egypt based on the surveys that French engineers had made of Egyptian pyramids and other remains.[2]

---

[1]An administrative region of France that extended from Grenoble to the French border.

[2]This period is the start of the celebration of ancient Egypt in the modern world, from Cleopatra's and other needles to fanciful statements about ancient wisdom, secret knowledge, and so forth, none of which can be held against Fourier. See Buchwald and Feingold (2012).

The eventual defeat of Napoleon was the lowest point of Fourier's life, but in 1816 he obtained a position as Director of the Bureau of Statistics for the Department of the Seine, a position which left him good time for research. His political enemies now in power delayed his appointment to the reformed Academy of Sciences for a year but he eventually rose to become the permanent secretary of the Academy in 1822 and to be elected to the Académie Française in 1827. He died in 1830 as the result of complications from an illness caught in Egypt.

2.3 Fourier and Series of Sines and Cosines

The book *Théorie analytique de la chaleur*, in which Fourier presented his ideas, was written work in several stages. He submitted a version to the Paris Academy of Sciences in 1807, but although Laplace, Lacroix and Monge were in favour of publishing it, Lagrange blocked publication, apparently because its treatment of trigonometric series differed markedly from the way he, Lagrange, had stipulated in the 1750s. Another chance came in 1810, when the Academy of Sciences announced a prize competition on heat diffusion. Fourier submitted a revised memoir, which won, but was criticised for a lack of rigour and generality. Fourier thought the criticism unfair, but revised it again, and the resulting book came out in 1822 (after Lagrange's death and when Fourier's standing was rising in the Academy).

Heat Diffusion

Fourier was interested in finding the temperature at every point of a solid body, perhaps as a function of time, when the shape of the body, its physical properties, and the temperature on some or all of its boundary is given. He made no assumptions about the nature of heat and concentrated on how it flowed.

He considered that any solid body could be regarded as made of infinitesimal cubes, and argued, on the basis of some observational evidence, that the amount of heat that passes from the hotter part of the body to an adjacent colder part in an instant of time is proportional to the duration of the instant, the infinitesimal temperature difference between opposite faces of the cubes and a certain function of the distance between the particles that depends on the nature of the body. So each body determines some constants that characterise how heat flows in them, such as its conductivity and its specific heat. In what follows all these physical matters will be consumed in the single letter K.

He considered what happens as the heat flows through one of these infinitesimal cubes, where temperature v is a function of x, y, z and t, the time. What enters the face with sides dx and dy is $Kdxdy\frac{\partial v}{\partial z}$ evaluated at that face.[3] What leaves the opposite face is $Kdxdy\frac{\partial v}{\partial z}$ evaluated at that face. This amount Fourier evaluated by saying that the faces are a distance dz apart, so he replaced z by $z + dz$, and the difference in the amount of heat between what enters and what leaves is given by

[3] I have modernised Fourier's 'd' notation by writing ∂v where he wrote dv.

$$Kdydxd\left(\frac{\partial v}{\partial z}\right) = Kdxdydz\frac{\partial^2 v}{\partial z^2}.$$

Should the temperature be in a steady state the sum of these quantities taken over the three pairs of opposite faces of a cube is zero and the resulting equation is

$$\frac{\partial^2 v}{\partial x^2} + \frac{\partial^2 v}{\partial y^2} + \frac{\partial^2 v}{\partial z^2} = 0. \tag{2.1}$$

The more important situation is when the temperature is changing. Fourier now argued that the amount of heat leaving a cube in the z direction is once again $Kdxdydz\frac{\partial^2 v}{\partial z^2}$, but now the sum over the pairs of opposite faces equals the rate of change of temperature, which is given by $\frac{dv}{dt}$. The result (see *Théorie* §128, p. 102) is the heat equation:

$$\frac{\partial v}{\partial t} = K\left(\frac{\partial^2 v}{\partial x^2} + \frac{\partial^2 v}{\partial y^2} + \frac{\partial^2 v}{\partial z^2}\right). \tag{2.2}$$

In each case, solutions of this partial differential equation are required that satisfy the given boundary conditions, and Fourier confined his attention to bodies with simple shapes, such as a cuboid, or one equivalent to this by a suitable coordinate transformation.

For example, in (§166, p. 133) Fourier considered a semi-infinite strip of a given width, π in suitable units. He supposed that the temperature at the base is kept constant at 1 in some units and that the temperature of the infinite sides is kept at at 0, and looked for the corresponding steady state distribution of temperature. Let y measure the height above the base and x the horizontal distance of a point from the mid-line of the strip (I have relabelled his coordinates) so the differential equation, which now involves only two variables, is

$$K\left(\frac{\partial^2 v}{\partial x^2} + \frac{\partial^2 v}{\partial y^2}\right) = 0, \tag{2.3}$$

Fourier looked for a solution of the form

$$v(x, y) = f(x)g(y),$$

which leads to the equation $g''(y)/g(y) = -f''(x)/f(x)$ in which both sides must be constant, say m, so the solutions are of the form

$$f(x) = \cos(mx), \quad g(y) = e^{-my} \tag{2.4}$$

The temperature in the bar surely does not become infinite, so the exponential term must decrease and so m must be positive. Also, m must be odd so that the solution vanishes for $x = \pm\pi$ for all y, as required.

It was clear to Fourier that a sum of solutions of this form is also a solution and he proceeded at once to consider infinite sums, solutions of the form, as he wrote (§169, p. 135),

$$ae^{-y}\cos x + be^{-3y}\cos 3x + ce^{-5y}\cos 5x + de^{-7y}\cos 7x + etc. \tag{2.5}$$

subject to the boundary condition at the base that

$$1 = a\cos x + b\cos 3x + c\cos 5x + d\cos 7x + etc. \tag{2.6}$$

The arbitrary constants had now to be determined. Fourier first gave a marvellous argument that involved him in solving the infinitely many equations he could obtain for his infinitely many unknowns by differentiating equation (2.6) arbitrarily often (see §§171–176 of the *Théorie*). Only then did he give the simpler and more general way that has become standard, and start to claim (§220, see §A.1) that every function can be written as one of these series. By this he meant that every function is equal to its corresponding series, and that there is a simple rule for writing down the coefficients of the series.

He noted (§221) that the integral

$$\int_0^{\pi} \sin jx \sin kx dx = \frac{1}{2}\left(\frac{1}{k-j}\sin(k-j)x - \frac{1}{k+j}\sin(k+j)x\right)\Big|_0^{\pi} \tag{2.7}$$

vanishes when $j \neq k$ and is $\pi/2$ when $k = j$, and claimed that the coefficients of a series such as his can be found by integrating the product of the series with $\sin jx$ for each value of $j > 0$. Similar results apply to series of cosines, to series of sines and cosines, and to series obtained when the period is different (as it might be, 2π or 1).

He went on to claim that any function f defined on the interval $[-\pi, \pi]$ can be written as an infinite series of sines and cosines in any of these forms (called the mixed series, the cosine series and the sine series, respectively):

$$f(x) = \frac{1}{2}a_0 + \sum_{n=1}^{\infty} a_n \cos nx + b_n \sin nx.$$

$$f(x) = \frac{1}{2}a_0 + \sum_{n=1}^{\infty} a_n \cos nx.$$

$$f(x) = \frac{1}{2}a_0 + \sum_{n=1}^{\infty} b_n \sin nx.$$

The coefficients of the mixed series are given by the formulae $a_0 = \frac{1}{\pi}\int_{-\pi}^{\pi} f(x)dx$ and

$$a_n = \frac{1}{\pi}\int_{-\pi}^{\pi} f(x)\cos kx dx, \quad b_n = \frac{1}{\pi}\int_{-\pi}^{\pi} f(x)\sin kx dx.$$

Examples of these series had already been studied by Euler and Daniel Bernoulli.

He then gave several simple examples of his series, some of which such a function constant on a given interval, or equal to x on a given interval. He showed how to obtain the function $\cos x$ as an infinite series of sines, dealt with cosine series as well as sine series, and

Fourier was very proud of his series for the function $F(x) = \pm\pi/4$:

$$\cos(x) - \frac{1}{3}\cos(3x) + \frac{1}{5}\cos(5x) - \frac{1}{7}\cos(7x) + \cdots,$$

as well he might be, once you see what it looks like. Here are three graphs of it: Fig. 2.1 shows the sum of only the first 5 terms in the series, Fig. 2.2 the first 25, and Fig. 2.3 the first 105.

Note that the difference between the sum of the series and the sum of its first 105 terms is certainly less than 1/209 and generally much less.

It is clear that the infinite series represents a function that is $+\pi/4$ on the range $(-\pi/2, \pi/2)$ and that is $-\pi/4$ on the range $(+\pi/2, 3\pi/2)$. Indeed, it represents a function that is $+\pi/4$ on the range $((4n-1)\pi/2, (4n+1)\pi/2)$, that is $-\pi/4$ on the range $((4n+1)\pi, (4n+3)\pi/2)$, and that is zero at the points $x = (4n \pm 1)\pi/2$.

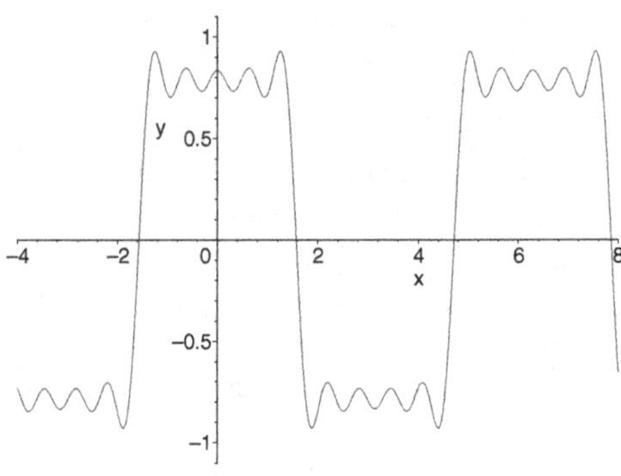

Fig. 2.1 The first 5 terms of a Fourier series

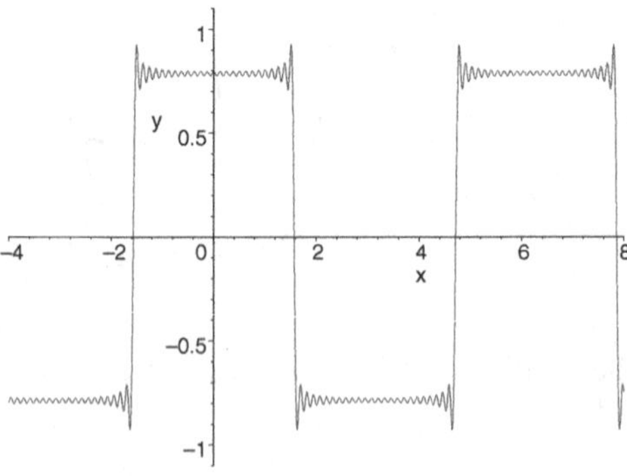

Fig. 2.2 The first 25 terms of a Fourier series

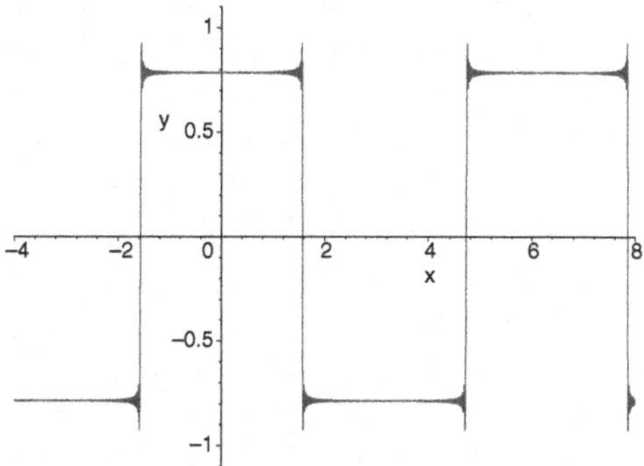

Fig. 2.3 The first 105 terms of a Fourier series

Note that it is plainly not the case that a series of analytic (and therefore in particular continuous) functions is itself continuous. We shall see that Fourier's confident remark (§235) that all these series converge everywhere to the function that they represent, and his proof of the claim in §423 (see Appendix A.1) that the coefficients in a 'Fourier' series can be evaluated as he indicated, were to be the occasion for much significant later work.

Chapter 3
Legendre and Elliptic Integrals

3.1 Introduction

As their name suggests, by 1800 elliptic integrals had an established place in contemporary astronomy. Kepler's 2nd law implies that a planet in an elliptical orbit sweeps out equal areas in equal times, so mathematicians had long grappled with questions involving the rectification of an ellipse, which is given by an elliptic integral. The topic acquired a new urgency in the early 19th century with the discovery of the asteroids. Adrien-Marie Legendre's computation of the values of elliptical integrals therefore fitted very comfortably into contemporary astronomy.

However, elliptic integrals had long perplexed mathematicians because they seemed impossible to evaluate. In the early 18th century an Italian mathematician, Count Giulio Fagnano, showed that there were similarities between elliptic and trigonometric integrals, and when his discoveries were reprinted in 1750 they came to the attention of Euler, who drew further results along the same lines.

Legendre's contribution was to create a theory of elliptic integrals. He showed that they could be reduced to canonical forms, in which they depend on a single parameter, that they were intimately connected to two linear ordinary differential equations, and that they were involved in several important problems in mechanics. Equally remarkably, he calculated tables of values of these integrals that made them as useful as any other functions in the mathematics of the day. These tables continued to be used until the advent of modern personal computers, and it has been estimated that producing them on his own took Legendre three years.[1]

[1]Dominique Tournes, private communication.

J. Gray, *The Real and the Complex: A History of Analysis in the 19th Century*,
Springer Undergraduate Mathematics Series, DOI 10.1007/978-3-319-23715-2_3

3.2 Legendre's Career

Adrien-Marie Legendre was born into a well-to-do family in Paris in 1752 and was able to live on private means and devote himself entirely to research, until the French Revolution wiped out his savings. He was by then well-known as a mathematician, and he briefly became one of the three commissioners involved in the determination of the standard meter. He then wrote an introduction to Euclidean geometry, his *Éléments de géométrie*, which ran to many editions and became the work every aspiring applicant to the École Polytechnique had to read, thus solving his financial problems.[2] After this he became involved in implementing the decimal system for angles (a reform that did not last) and peripherally involved in the École Polytechnique, examining students applying to the artillery school. In 1813 he succeeded Lagrange, who had just died, at the Bureau des Longitudes, where he remained until his death in 1833 at the age of 80.

His scientific work was concentrated on a number of Eulerian themes: number theory, and a study of the new functions Euler had introduced into analysis. It brought him into painful collision with Carl Friedrich Gauss on a number of occasions. Legendre was the first to publish the method of least squares in statistics, but Gauss claimed, correctly that he had been the first to use it. Gauss also found fault with Legendre's attempts to establish the parallel postulate as a theorem in Euclidean geometry in his *Éléments de géométrie*; the attempts were indeed flawed. However, his main contribution to mathematics is probably his elaborate theory of elliptic integrals, even though, as we shall see and as Legendre was please to acknowledge, it was to be eclipsed before the end of his life.

3.3 Elliptic Integrals

The paradigm elliptic integral is

$$u = \int_0^v \frac{dt}{\sqrt{1-t^4}}. \tag{3.1}$$

It measures arc-length along the lemniscate $r^2 = \cos 2\theta$, which is a curve in the shape of a figure eight. This is easy to prove, for differentiating the equation for the lemniscate gives

$$rdr = -\sin(2\theta)\, d\theta.$$

[2]See Dhombres (1985) for a discussion of mathematical publications in France after the revolution. He gives these sales figures for 1812: 3000 copies of the ninth editions of textbooks on geometry by Lacroix and Legendre, and 2000 copies of Monge's work on descriptive geometry. In 1813 he estimated that 1500 copies of Lagrange's book on the theory of functions and also of Lacroix's textbook on the calculus were published.

The element of arc length in polar coordinates is given by

$$ds^2 = dr^2 + r^2 d\theta^2,$$

and the claim follows on eliminating θ and obtaining this expression for arc-length:

$$ds^2 = \frac{dr^2}{1 - r^4}.$$

The total arc length of the lemniscate is denoted 2ω.

The integral (3.1) can be regarded either as defining u as a function of v, or v as a function of u. In either case, one is led to look for analogues of the behaviour of the trigonometric functions, because the integral

$$u = \int_0^v \frac{dt}{\sqrt{1 - t^2}}$$

may be taken to define the function $v = \sin u$.

Fagnano and Euler

In 1750 Count Fagnano, an Italian mathematician, submitted a book containing his life's work to the Berlin Academy.[3] The *Produzioni matematiche*, as it was called, contained a remarkable analogy with the trigonometric integrals, that the Count had discovered in 1714.

Arc-length along the lemniscate is a transcendental function of the parameter, t, but Fagnano showed that the values of the parameter for an arc and an arc of twice the length are algebraically related. More precisely, if

$$\int_0^v \frac{dt}{\sqrt{1 - t^4}} = 2 \int_0^w \frac{dt}{\sqrt{1 - t^4}}$$

then v and w are algebraically related (we shall see this more generally below). Moreover, he showed that the values of t that divide the lemniscatic arc into 2.2^m, 3.2^m, or 5.2^m pieces are also algebraically related.

The book was naturally sent to Euler, who had been interested in elliptic integrals for many years, and he immediately generalized Fagnano's results.[4] He regarded the integrals as solutions to the differential equation

$$\frac{dx}{\sqrt{1 - x^4}} = \pm \frac{dy}{\sqrt{1 - y^4}} \tag{3.2}$$

[3] See Enneper (1890, Note III).

[4] See his paper E252 of 1752 and his later, more extensive treatment E251 (1760) that we follow here. There is an English translation of E251 by Stacy Langton on the Euler Archive.

and using the analogy with the trigonometric case

$$\frac{dx}{\sqrt{1-x^2}} = \pm\frac{dy}{\sqrt{1-y^2}}$$

Euler guessed correctly and then verified (E251, §9) that the solution of the differential equation (3.2) is

$$x^2 + y^2 + c^2x^2y^2 = c^2 + 2xy(1+c^4)^{1/2} \tag{3.3}$$

where c is an arbitrary constant. This yields what is now called Euler's algebraic addition theorem for elliptic integrals:

$$\int_0^x \frac{dt}{\sqrt{1-t^4}} + \int_0^y \frac{dt}{\sqrt{1-t^4}} = \int_0^c \frac{dt}{\sqrt{1-t^4}} \tag{3.4}$$

which says that the sum of any two arcs of a lemniscate is an arc whose parameter, c, depends only algebraically on the parameter values of the original arcs. Equation (3.4) and Euler's equation (3.3) confirm what Fagnano had already noticed, that doubling the lemniscatic arc yields a quartic equation, whereas doubling of a circular arc yields only a quadratic equation.

At this point the analogy seems to have deserted Euler, and he did not proceed to create a theory of 'lemniscatic integrals'

$$u = u\,(v) = \int_0^v \frac{dt}{\sqrt{1-t^4}}.$$

Krazer, in the preface to Euler's *Opera Omnia* (1), vol. 20, p. 21, argued that on this occasion Euler remained too close to the geometry of conics, and before Legendre, as Enneper pointed out (Enneper 1890, 542), "the writings of Fagnano seem to have been very little known, and the writings of the great Euler too little read".[5]

Legendre's Elliptic Integrals

Legendre was the first to study the lemniscatic integral from a purely functional point of view. After two memoirs occupied with geometrical questions, his (1788a) and (1788b), his 'Mémoire sur les transcendantes elliptiques' considered the new functions that the integrals define, as the sub-title makes clear: 'Containing easy methods of comparing and valuing these transcendentals, which include elliptic arches, and which frequently occur in the application of the Integral Calculus'.[6]

[5] In E251 Euler did go on to study more general integrands that involve square roots of quartics, but nothing that amount to a theory.

[6] Read to the Paris Academy in 1792.

Legendre began with a crucial simplification and three-fold classification of the way in which elliptic integrals can arise that has been employed by everyone ever since, and that Legendre rightly said in his *Traité* 33 years later was at the basis of his method. He first showed that any integral of the form $\int \frac{Pdx}{R}$, where P is a rational function in x and R is the square root of a quartic (with real coefficients), can be simplified to one of the form

$$\int \frac{Qdt}{\sqrt{(1-t^2)\left(1-c^2t^2\right)}},$$

where Q is another rational function. The substitution $t = \sin\phi$ reduces this to $\int \frac{Qd\phi}{\sqrt{(1-c^2\sin^2\phi)}}$.

Legendre called the variable ϕ the amplitude of the elliptic integral, the parameter c the modulus, and quantity $b = \sqrt{1-c^2}$ the complementary modulus. He denoted the square root $\sqrt{(1-c^2\sin^2\phi)}$ by Δ or $\Delta(\phi)$, and showed how this reduction led to one of three distinct kinds,

1. elliptic integrals $\int\limits_0^x \frac{d\phi}{\Delta}$ are of the first kind;
2. elliptic integrals $\int\limits_0^x \Delta d\phi$ are of the second kind[7]; and
3. elliptic integrals $\int\limits_0^x \frac{d\phi}{(1+n\sin^2\phi)\Delta}$ are of the third kind.

Only in the third case did Legendre allow that the coefficient n could be imaginary, in which case, he observed, the integral would be accompanied by another multiplied by $\sqrt{-1}$, and the sum of two will always be real. This is typical of the way complex considerations could enter mathematical arguments at the time, but only on condition that the final expressions are purely real. He insisted that the modulus c and the amplitude ϕ were real, and that $c < 1$.

Legendre regarded all these integrals as functions of their upper end point x. He denoted by F a typical indefinite integral of the first kind, and a typical indefinite integral of the second kind by E (or sometimes G). The corresponding complete integrals, as they are called, he wrote as follows:

$$F^1 = \int\limits_0^{\pi/2} \frac{d\phi}{\Delta} \text{ and } E^1 = \int\limits_0^{\pi/2} \Delta d\phi. \tag{3.5}$$

He wrote $F^1(c)$ and $E^1(c)$ when he regarded them as functions of the modulus c.

[7]The arc length of an ellipse is given by an integral of this kind.

Among other results, he now rederived Euler's differential equation (3.2), in the form

$$\frac{d\phi}{\sqrt{1-c^2\sin^2\phi}} = \frac{d\psi}{\sqrt{1-c^2\sin^2\psi}}.$$

He deduced that its integrals satisfy

$$F(\phi) + F(\psi) = F(\mu), \tag{3.6}$$

where μ is an arbitrary constant. But when $\phi = 0, \psi = \mu$, so a comparison with Euler's result shows that

$$\cos\phi\cos\psi - \sin\phi\sin\psi\Delta(\mu) = \cos\mu. \tag{3.7}$$

He concluded the paper by indicating how tables of the values of $F(\phi)$ could be calculated for specified values of the modulus c.

It is clear that Legendre thought of himself as enriching mathematics in the manner of Euler with whole families of new, interesting functions. Legendre took the function-theoretic point of view even further in his *Exercises de Calcul Intégrale* (3 vols, 1811–1817). The books begin with his study of what he called the elliptic functions —meaning $F(\phi)$, $E(\phi)$, and related functions—but to avoid confusion with their inverses, which are called elliptic functions today, I shall call Legendre's functions elliptic integrals.

For a new function to be accepted into mathematics it has to be shown to be useful, and if it is to be applied then its values have to be known numerically. Legendre devoted Book III (1816) to the production of several sets of tables (which we shall consider later), and then put the new functions to work solving geometrical and mechanical problems.[8]

In the *Exercises* Legendre, drawing on results in his study of elliptic integrals (1792), deduced by differentiating under the integral sign that the complete elliptic integrals satisfy two linear differential equations. The first one,

$$\left(1-c^2\right)\frac{d^2F^1}{dc^2} + \frac{1-3c^2}{c}\frac{dF^1}{dc} - F^1 = 0 \tag{3.8}$$

had already been given by Euler in 1750.[9] The second one is

$$\left(1-c^2\right)\frac{d^2E^1}{dc^2} + \frac{1-c^2}{c}\frac{dE^1}{dc} - E^1 = 0. \tag{3.9}$$

[8] He also wrote extensively on Eulerian integrals, and computed tables for the Beta and Gamma functions. These had been introduced by Euler, who showed they were closely connected to elliptic integrals.

[9] See Euler, *Opera Omnia* (1) vol. 20, p. 40.

Legendre solved these equations and exhibited power series expansions for the complete integrals $F^1(c)$ and $E^1(c)$. He also obtained a strikingly attractive result connecting complete integrals of the first two kinds with complementary moduli (c and b):

$$\frac{\pi}{2} = F^1(c)\,E^1(b) + F^1(b)\,E^1(b) - F^1(b)\,F^1(c) \tag{3.10}$$

The Pendulum and the Elliptic Integral

In his (1792) Legendre had discussed the oscillations of a simple pendulum of length ℓ. He set the constant $c^2 = 2\ell/h$ where h is the maximum height of the pendulum (in units where the acceleration due to gravity = 1). He defined the angle ψ as a function of the angle of displacement from the vertical, θ, by the formula $c \sin\psi = \sin(\theta/2)$, and deduced that the motion of the pendulum is given by the equation

$$dt = \frac{\ell d\psi}{\sqrt{1 - c^2 \sin^2 \psi}}$$

It follows that Legendre's equation implies that the familiar approximate equation for small arcs under-estimates the time needed to make a complete swing. (To see this, compare dt and $dt' = \frac{\sqrt{l}d\psi}{\sqrt{1-c^2\psi^2}}$, and see that $dt' > dt$.) Legendre was surprised to notice that he was the first to show that there were algebraic relations connecting the times of swings of a circular pendulum, just as there were for divisions of circular arcs.

To see how Legendre obtained his equation, suppose that the pendulum (in the form of a weightless rigid rod of length ℓ with a bob of mass m at the end) is set in motion from the vertical position with an initial velocity v. When it is at an angle θ with the vertical, the tangential force on the bob is $mg \sin\theta$, and so the equation for the rate of change of angular momentum is $m\ell\ddot{\theta} = -mg\sin\theta$. Divide through by m, multiply both sides by $\dot{\theta}$, integrate, and obtain

$$\ell\frac{\dot{\theta}^2}{2} = g\cos\theta + a,$$

where a is a constant determined by the initial conditions. If we take the initial condition to be $\ell\dot{\theta} = v$ when $\theta = 0$ then we find that

$$a = \frac{v^2}{2\ell} - g.$$

If we take the initial condition to be $\dot{\theta} = 0$ when the bob is at its maximum height, given by $\cos\theta = 1 - h/\ell$ then we find that

$$a = \frac{gh}{\ell} - g,$$

which with the previous result implies that $\frac{1}{2}v^2 = gh$, which is the equation for conservation of energy.

So we write

$$\frac{\ell\dot{\theta}^2}{2} = g(\cos\theta - 1) + \frac{gh}{\ell}$$

and make the half-angle substitution, giving $\cos\theta - 1 = -2\sin^2(\psi)$ and $\dot{\theta} = 2\dot{\psi}$, to obtain

$$\ell\dot{\psi}^2 = -g\sin^2\psi + \frac{gh}{2\ell}$$

so

$$\dot{\psi} = \sqrt{\frac{gh}{2\ell^2}}\sqrt{1 - \frac{2\ell}{h}\sin^2\psi}\,.$$

Set $c^2 = 2\ell/h$ and this becomes

$$\dot{\psi} = \sqrt{\frac{gh}{2\ell^2}}\sqrt{1 - c^2\sin^2\psi},$$

which we write as

$$\frac{d\psi}{dt} = \sqrt{\frac{v^2}{4\ell^2}}\sqrt{1 - c^2\sin^2\psi}.$$

This can be rewritten and integrated to give[10]

$$t = (2\ell/v)\int\frac{d\theta}{\sqrt{(1 - c^2\sin^2\psi)}},$$

where $c^2 = \dfrac{4\ell g}{v^2}$.

This was far from being his only incursion into mechanics. Legendre gave a thorough analysis of three problems in his *Exercises* (1817): the rotation of a solid about a fixed point; the motion of a body attracted to two fixed bodies (whether in plane or in space); and the attraction due to an homogeneous ellipsoid. In his *Traité* (1827) he added two more examples: motion under central forces, and the problem of determining geodesics on an ellipsoid, as well as two geometric problems: the surface area of oblique cones, and the surface area of ellipsoids. So this, his final book and one surely intended to be the definitive treatment of a topic that had engaged him all his life, was a prolonged contribution to the theory of real functions. It began with their definition, moved on to their fundamental properties, tabulated their values, and finally displayed their utility in solving significant problems in applied mathematics.

---

[10]As a check, note that the final result is dimensionally correct.

Computing Values of an Elliptic Integral

Initially the analogy between the trigonometric and elliptic integrals led Legendre in a direction from which he recoiled, but which later investigators were to find highly suggestive. It had been known since the time of Ptolemy that to compute tables of, say, the sine function (or, *mutatis mutandis*, the chord function in Ptolemy's case) one makes repeated use of the addition formulae:

$$\begin{aligned}\sin(u+v) &= \sin u \cos v + \cos u \sin v\\ \cos(u+v) &= \cos u \cos v - \sin u \sin v\end{aligned}$$

their important corollaries:

$$\begin{aligned}\sin 2u &= 2\sin u \cos u\\ \cos 2u &= \cos^2 u - \sin^2 u.\end{aligned}$$

and their inverses:

$$\begin{aligned}\sin\left(\frac{u}{2}\right) &= \left(\frac{1}{2}(1-\cos u)\right)^{1/2}\\ \cos\left(\frac{u}{2}\right) &= \left(\frac{1}{2}(1+\cos u)\right)^{1/2}\end{aligned}$$

together with known values of cosine and sine such as $\cos 0 = 1, \sin 0 = 0$, $\cos\pi/2 = 0$, $\sin\pi/2 = 1$.

Legendre showed how to calculate values of F in Book I, §§17–22 of the *Exercises*, where he went on to give a table of values of elliptic integrals to 14 decimal places.

He defined ϕ_n to be an amplitude such that $F(\phi_n) = nF(\phi)$ and sought equations connecting $\sin(\phi_n)$ and $\cos(\phi_n)$ with $\sin\phi$ and $\cos\phi$, under the condition that $0 \leq c \leq 1$. The equations for $\cos(\phi_2)$, $\sin(\phi_2)$ and $\tan(\phi_2)$ came directly from the addition formula. Then identities between $\sin(\phi_{n+1})$, $\sin(\phi_{n-1})$ and $\sin(\phi_n)$ enabled him to obtain formulae for $\sin(\phi_3)$, $\sin(\phi_4)$, and so on.

The trisection formula, for example, relating $\sin(\phi_3) = a$ to $\sin(\phi) = x$, is

$$\frac{\left(3-4\left(1+c^2\right)x^2+6c^2x^4-c^4x^8\right)x}{1-6c^2x^4+4c^2\left(1+c^2\right)x^6-3c^4x^8} = a,$$

and this struck him as unexpectedly complicated. It is of degree 9 in x, not of degree 3 as a naive analogy with trigonometry suggests. More generally, division into n parts leads to an equation of degree n^2.

This approach to constructing a table of values therefore seemed to be too difficult, and Legendre therefore preferred to go back to a method he had sketched in his (1788a) and his (1792). He had observed in his (1788a) that it was traditional in evaluating arc lengths along ellipses to make a substitution that transformed a given ellipse into another, more circular, one. So Legendre showed how to find accurate approximations when c is nearly 0 or 1, using the fact that the elliptic integral is trivial when $c = 0$ or $c = 1$. Then he showed how to reduce the general case to this one by a transformations that either steadily reduced or steadily increased the value of the modulus.

Legendre considered the transformation $c' = \frac{2\sqrt{c}}{1+c}$. Iterating this transformation gives a series of values $c, c', c'', \ldots$ tending rapidly to the value 1. So iterating it backwards gives a series of values $c, c^0, c^{00}, \ldots$ tending towards the value 0. Explicitly, $c^0 = \frac{1-\sqrt{1-c^2}}{1+\sqrt{1-c^2}}$, or, in terms of the complementary modulus, $c^0 = \frac{1-b}{1+b}$. For any given c successive values $c, c^0, c^{00}, \ldots$ can be found from the sine tables by a quick use of the addition formulae.

For example, the sequences $c, c', c'', \ldots$ and $c, c^0, c^{00}, \ldots$ Show that from $c = 1/2$ the increasing sequence reaches 0.9999999764 in just three steps, and from $c = 1/2$ the decreasing sequence reaches 0.0000004174 in just three steps.

Legendre defined ϕ^0 by the equation $2\sin^2\phi = 1 + c^0\sin^2\phi^0 - \Delta^0\cos\phi^0$, where $\Delta^0 := \sqrt{1 - c^{02}\sin^2\phi^0}$, and found that

$$F_c(\phi) = \frac{1+c^0}{2}F^0\left(\phi^0\right), \text{ where } F^0\left(\phi^0\right) = \int_0^{\phi^0}\frac{d\phi^0}{\Delta^0}. \tag{3.11}$$

He then observed that $\tan\left(\phi^0 - \phi\right) = b\tan\phi$. So evaluating F at ϕ reduces to evaluating F^0 at ϕ^0, which is easier to do because is less than c, and iterating,

$$F_c(\phi) = \frac{(1+c^0)(1+c^{00})\ldots(1+c^{(n)})}{2^n}\cdot F^{(n)}\left(\phi^{(n)}\right). \tag{3.12}$$

The convergence is quite rapid because, as Legendre observed, the value of $c^{(n)}$ becomes negligibly small after a small number of steps, and then $\Delta = 1$ and $F(\phi) = \phi$. Therefore he set $\Phi = \lim\limits_{n\to\infty}\frac{\phi^{(n)}}{2^n}$ and found, on setting the product $\left(1+c^0\right)\left(1+c^{00}\right)\ldots = \alpha$, (it is a constant depending only on the choice of c, not the angle) that $F = \left(1+c^0\right)\left(1+c^{00}\right)\ldots = \alpha\,\Phi$. Therefore $F^1(c) = \frac{\pi}{2}\alpha$, and α was easy to find from logarithm tables. When $\phi = \frac{\pi}{2}$ the limit $\Phi = \frac{\pi}{2}$, and the corresponding value of F is $\alpha\frac{\pi}{2}$.

You can check the rapidity of the convergence with this example given by Legendre. The value

$$c = \frac{\sqrt{2}}{2}\left(\frac{1+\sqrt{3}}{2}\right) = \sin 75°,$$

is "unfavourable to calculation", as Legendre put it, because c is close to 1 (it is in fact 0.9659258262), but Legendre showed that four iterations were enough to show that b^{0000} vanished to seven decimal places, and $\log\alpha = 0.2460561$. He also showed that when c is small

$$c^0 = \frac{1}{4}c^2 + \left(\frac{1.3}{2.4}\right)c^4 + \left(\frac{1.3.5}{2.4.6}\right)c^6 + \cdots$$

from which is followed that it was often enough to use just the first two terms.

3.4 Exercises

1. Choose one or two of the following topics and consult the internet about them: Taylor series, MacLaurin series, Fourier series, differential, fundamental theorem of the calculus, elliptic integral, Lagrange, Fourier, Legendre. How intelligible, and how reliable, was the information you found, and how did you reach this decision?
2. Read Fourier's remarks about the convergence of his series and compare them with some Fourier series expansions. How do you interpret his remarks?
3. How useful would the trigonometric functions have been if we did not have tables for them? What would you have to do to calculate a value of sin $1°$ to four decimal places, and how useful would such information be (think about rounding errors).

Chapter 4
Cauchy and Continuity

Painting of Augustin-Louis Cauchy by J. Roller (≃ 1840).

Augustin-Louis Cauchy (1789–1857)

4.1 Introduction

One of the key figures in the transition from 18th century analysis to modern mathematical analysis is Augustin-Louis Cauchy. Over the next several chapters we shall look at various aspects of his contributions to real and complex analysis, and we shall see that his impact is very much tied up with the complicated, and political, nature of his life.

J. Gray, *The Real and the Complex: A History of Analysis in the 19th Century*, Springer Undergraduate Mathematics Series, DOI 10.1007/978-3-319-23715-2_4

In this chapter we begin to look at his teaching at the École Polytechnique, which is where he introduced many of his fundamental ideas about continuity. This was a novel concept to stress as lying at the foundations of the calculus, and it was part of a wholesale reworking of the subject that he put forward after 1820. But it is also a difficult one, and we shall see that in his attempt to prove the binomial theorem for arbitrary exponents he made a revealing error that he only corrected much later in life, even though Niels Henrik Abel questioned it early on.[1] That he made an error, and that no-one corrected it for some time, show just how difficult the concept of continuity was to be to elucidate, and is one of the principal themes of this book. As we shall see by the time we have taken these developments as far as we can go (Chaps. 22–24) there are many problems connected with the concept of continuity that occupied the minds of many mathematicians, and it is hard to believe that Cauchy can have had a precise idea of what he was ushering in to mathematics.

A particular point to watch for is the decline of verbal definitions and the rise of ones symbolically (using ε, δ, and N) along with the increasing use of quantifiers $\forall$ and $\exists$ and attention to the order in which they occur; all evidence of the need for new levels of precision.

4.2 The Young Cauchy

Augustin-Louis Cauchy began teaching at the École Polytechnique in November 1815, when he replaced Poinsot, who had been in poor health. He was then twenty-six. The next year he was appointed a regular professor and he taught there until 1830.

Although Cauchy was already the author of several important papers, his appointment as a professor was undoubtedly political.[2] Royalists were working across French society to replace supporters of the now-discredited Napoleonic regime, and they correctly saw Cauchy as one of their own. Monge, Lacroix and others among the old guard of the École Polytechnique were forced to resign, and this made it easy to break with tradition and appoint as a professor someone who had not previously taught at the École. A few months later the purge reached the Institut de France when the old Académie des Sciences was reestablished under its old name, Monge and Carnot were removed, and Cauchy and the obscure Louis de Bréguet replaced them.

In these months of 1815 Cauchy won an Académie prize of 3,000 francs for a long paper on the theory of waves, and then resolved an outstanding conjecture of Fermat's that every number is the sum of three triagonal numbers, four square numbers, five pentagonal numbers and generally n n-gonal numbers.[3]

---

[1] Cauchy's error has provoked much discussion among historians of mathematics, see for example Giusti (1984) and Schubring (2005).

[2] See Belhoste (1991, 45–50) upon which the next few paragraphs are based.

[3] The kth n-gonal number is $\frac{1}{2}(n-2)(k^2-k)+k$.

Lagrange had proved this for square numbers in 1770, Legendre for triagonal numbers in 1798, and Gauss had rederived these results in his *Disquisitiones Arithmeticae* of 1801, so to go well beyond what these men had done made Cauchy famous, and would have opened the doors of the Institut to him in any case, but the manner of his appointment turned people against him. Monge in particular was highly respected as a teacher and a founding father of the École Polytechnique, and for Cauchy to join the Académie as part of what many intellectuals and scholars regarded as crude political manipulations without a hesitation struck many as unworthy. "Cauchy", Bertrand wrote later, "found few defenders. He has seen more than one friend who, though naturally tolerant and decent, turned away and refused to call him 'brother'".[4]

Royalists had long disliked the École Polytechnique for its liberal and Bonapartist sympathies, and they seized upon a pretext in April 1815 to send the students home and reform the school. In September 1816 they demilitarised the school structure and completed the purge of the professors, thus incidentally making room to appoint Cauchy as a full professor, and his career prospered.

As was customary, Cauchy was required by the Conseil of the École to provide his students with textbooks. Accordingly, in 1821 he published the *Cours d'analyse*, devoted to algebraic analysis, followed two years later by the *Résumé* (1823), where differential and integral calculus was presented in, if not a rigorous form for the time, then in a form it proved possible to make rigorous.[5]

4.3 Cauchy and Continuity

In 1821, in his lectures at the École Polytechnique Cauchy, defined and discussed a number of concepts with a remarkable, novel, degree of precision. These include: quantities, real functions, infinitely small and infinitely large quantities, continuity of functions, singular values of functions, real convergent and divergent series, rules for convergence, summation of some convergent series, the binomial theorem, imaginary expressions and their moduli, imaginary variables and functions, imaginary convergent and divergent series, rules for convergence, summation of some convergent series, as well as other topics.

Let us look at how some of these terms were defined.

Algebraic analysis, said Cauchy, deals with numbers and quantities. It is not clear if the 'quantities' he had in mind were the abstract algebraic quantities that Lagrange had invoked, or something closer to the familiar measuring numbers associated with, for example, lengths. In any case, they may be variable or constant and are represented by letters. The crucial concept of limit he defined in these words (*Cours*, p. 4)

> When the values successively attributed to the same variable indefinitely approach a fixed value, in such a way as to finish by differing from it by as little as one wishes, this last is called the *limit* of the others.

---

[4]Quoted in Belhoste (1991, 46–47) from Bertrand *Éloges académiques*, 1902, vol. 2, p. 112.

[5]There was only one mathematician more skeptical than Cauchy at this time, and that was Bolzano.

So an irrational number is a limit of a sequence of rational numbers, and a circle is a limit of a suitable sequence of inscribed polygons. An infinitely small number (*Cours*, p. 4) is a variable taking successive numerical values and decreasing indefinitely in such a way that it becomes smaller than any given number. An infinitely large positive number (written $+\infty$, *Cours*, p. 5) is a variable taking successive numerical values and increasing indefinitely in such a way that it becomes larger than any given number. Cauchy defined $-\infty$ similarly. The concept of an *infinitely small* quantity, α, of various orders α^n was important for Cauchy—it is how he expressed many arguments involving limits—and he gave rules for handling them, noting that one must distinguish between decreasing constantly and decreasing indefinitely.

It is perhaps unexpected that Cauchy tangled up the concept of a limit with talk of indefinitely small and infinitesimal quantities. Schubring (2005, 400) has argued that this has to do with a clash between mathematicians and the directors of the École Polytechnique about how to teach the calculus. By 1810 there had developed a general feeling among mathematicians that limits were a better concept than infinitesimals for this task, but the directors preferred the language of infinitesimals; Cauchy's approach was therefore intended a something of a compromise that gave the directors the terminology they wanted and the students access to the limit concept. This does not involve us in accepting that Cauchy, of all people, wrote down some mathematics that he knew to be wrong or inadequate, because it it not simply a matter of terminology: the use of infinitesimals has an intuitive quality that can suggest conclusions and proofs, and can be fairly easily replaced by limiting arguments. This is why the language of infinitesimals persisted well into the 20th century, especially in differential geometry, but it is a seductive language that, as we shall see, began to obscure more than it revealed and had eventually to be replaced in analysis.

Multiply-valued expressions occur in many places in mathematics, most obviously in algebra, and there was a tendency throughout the 19th century to see them as many-valued functions. This became an issue in the early history of functions of a complex variable—not least for Cauchy himself—and we have already met them as integrands in the context of elliptic integrals, where they were to prove even more troubling. Cauchy introduced them in the *Cours* (see p. 7) by saying that, for example, there are two values to $a^{\frac{1}{2}}$, of which the positive one is written $a^{\frac{1}{2}}$ or $\sqrt{a}$. He then gave rules for multi-valued expressions, such as arcsin. Then he turned to give rules for evaluating the limits of certain expressions.

A function is a variable that depends on another variable (*Cours*, p. 19), which is called the independent variable. Cauchy regarded uniform (i.e. single-valued) functions, many-valued functions (such as square or nth roots), and implicit functions as functions.

Cauchy now turned to define the concept of *continuity of a function* (pp. 34–35). This was a radical step to take at a time when almost all mathematicians of the time took differentiability for granted: continuity had never been given a prominent position in the edifice of analysis before. He said that f is continuous between certain limits if, for each x within those limits the value of $f(x)$ is unique and finite and the numerical (i.e., absolute) value of $f(x+\alpha) - f(x)$, where α is infinitely small, decreases indefinitely with α. Equivalently, an infinitely small increase in x produces an infinitely small increase in $f(x)$. This is not continuity at a point, a concept Cauchy never defined, but continuity on an interval. However, he did note that continuity can fail at a point. A little later (p. 41) he defined a function of several variables to be continuous on some domain if it is continuous in each variable separately.[6]

Among the many examples of functions stated without proof to be continuous were linear functions (those of the form $ax + by + cz + \cdots$) A^x, $\sin(x)$, $\arcsin(x)$, and x^α.

Theorems

Cauchy then proved some theorems.[7]

He began with a theorem on the composition of functions (*Cours*, p. 41): If f and g are continuous within certain limits then the composite function $f(g(x))$ is continuous everywhere it is defined.

Then came his intermediate value theorem (*Cours*, p. 43): If $f(x)$ is continuous between x_0 and X, and b lies between $f(x_0)$ and $f(X)$ then there is at least one x between x_0 and X such that $f(x) = b$.

The proof that Cauchy gave in the body of the lectures is no proof at all, only an appeal to the geometric idea that the curve $y = f(x)$ must cross the line $y = b$ because the curve is continuous. But he was aware how to to do much better, as his lengthy Note of some 60 pages added to the lectures (Note 3 of 10) shows. Here he argued that if the interval $[x_0, X]$ of length h is divided into m equal parts then a comparison of successive values in the sequence of values

$$f(x_0), f(x_0 + h/m), f(x_0 + 2h/m), \ldots, f(X - h/m), f(X).$$

shows that there must be an consecutive pair of opposite signs. Let $f(x_1)$ and $f(X_1)$ be such a pair, so

$$x_0 \leq x_1 < X_1 \leq X.$$

Now subdivide again and repeat the argument, to obtaining a new consecutive pair x_2 and X_2 where f takes opposite signs, and note that

$$x_0 \leq x_1 \leq x_2 < X_2 \leq X_1 \leq X.$$

Continue in this fashion and two sequences are obtained:

$$x_0 \leq x_1 \leq x_2 \leq \cdots x_n \leq \ldots$$

and

$$X \geq X_1 \geq X_2 \geq \cdots X_n \geq \ldots.$$

---

[6]Schubring (2005) has insisted on the point that for Cauchy a function is a function of a variable quantity; it will have a value at a point, but it is hard to think of a function of a variable being continuous at a point.

[7]Cauchy's notation has sometimes been changed, but not otherwise the statement of his results.

The first is monotonic non-decreasing, the second monotonic non-increasing and their successive terms differ by no more than

$$h = X - x_0,\ h/m,\ h/m^2, \ldots$$

so their terms, as he put it, "will differ by as little as desired and therefore must converge to a common limit". Let a be that limit. Since the values of f on $x_0, f(x_1), f(x_2), \ldots, f(X_2), f(X_1), f(X)$ always remain of opposite sign "it is clear that the quantity $f(a)$, which must be finite, cannot differ from zero." Therefore there is a number a such that $f(a) = 0$ as required, and the theorem is proved.

As Lützen (2003, 167–168) pointed out, Cauchy's otherwise exemplary account convergence has one fundamental flaw: he had no way of establishing the property of quantities (or real numbers, had he used the term) that modern analysis calls the completeness of the real numbers. He could not explicitly show that a sequence whose terms ultimately differ by arbitrarily small amounts actually has a limit and had to leave this point unexplained and unanalysed.[8]

Cauchy then proceeded to draw out the implication of this result for the numerical solution of equations.

Real Convergent and Divergent Series

A series, said Cauchy (*Cours*, p. 123), (NB: this is not the modern definition) is an indefinite sequence $u_0, u_1, u_2, \ldots$ where the terms are given (one from the other) by a determinate law. It has (partial) sums $s_n = u_0 + u_1 + u_2 + \cdots + u_{n-1}$, n arbitrary. It is convergent, he said, if the sequence of sums has a limit, otherwise divergent, and he gave the example of $1 + x + x^2 + \cdots$, which has a sum only when the absolute value of x is less than 1. For a 'series' $u_0, u_1, u_2, \ldots$ to be convergent—by which Cauchy meant for the *sum* to be convergent—it is necessary that the u_n decrease indefinitely as n increases, but this is not sufficient—for example, if $u_n = \frac{1}{n}$ the series is divergent. Cauchy proved this by bracketing the terms so that it was made of successive runs of n terms

$$\frac{1}{n+1} + \frac{1}{n+2} + \cdots + \frac{1}{2n}$$

whose sum is greater than $\frac{1}{2}$.

A good example of a convergent series that Cauchy gave was $u_n = \dfrac{1}{1.2.3.\ldots.n}$, with the sum $e = 2.7182818\ldots$.

As he had to do if this theory was to be any use, Cauchy gave several tests for convergence. The first applied to series $u_0, u_1, u_2, \ldots$ all of whose terms are positive.

The root test (*Cours*, p. 132): consider $(u_n)^{\frac{1}{n}}$ and let k be the greatest of its limits. Then the series converges if $k < 1$ and diverges if $k > 1$.

---

[8]More precisely, he could and did define what we call a Cauchy sequence, but he could not say in mathematically precise terms that it has a limit.

The sum of a series formed term by term from two convergent series of positive terms is the sum of their sums.

Cauchy also discussed series with positive and negative terms by passing to their absolute values. For them he stated and proved the ratio test (*Cours*, p. 143): the series converges if, in absolute values, $\frac{u_{n+1}}{u_n}$ has a limit $k < 1$ and diverges if $k > 1$. (*Cours*, p. 144) An alternating series of terms that decrease indefinitely in absolute value is convergent.

He then made what has become in the literature in the history of mathematics his most celebrated mistake (pp. 131–132)—but we shall see that he and his contemporaries regarded it differently, even when they noticed it was not correct as stated.

Cauchy's Attempt at the Binomial Theorem

Cauchy now considered the sequence

$$u_0(x), u_1(x), \ldots, u_n(x), \ldots$$

of functions defined on a given interval, and its partial sums $s_n(x) = \sum_{j=0}^{n-1} u_n$. He supposed that the series converged to $s(x)$, and defined the nth remainder $r_n(x)$ by the equation

$$s(x) = s_n(x) + r_n(x).$$

He then offered this theorem with the accompanying proof (*Cours*, pp. 131–132):

When, the terms of series having the same variable,

$$u_0(x), u_1(x), \ldots, u_n(x), \ldots \tag{4.1}$$

the series is convergent, and its different terms are continuous functions of x, in the neighbourhood of a particular value attributed to this variable, then

$$s_n, r_n, \text{ and } s$$

are also three functions of x, of which the first is evidently continuous with respect to x in the neighbourhood of the particular value under consideration. This done, let us consider the increases that the three functions receive when one increases x by an infinitely small α. The increase in s_n will be, for all possible values of n, an infinitely small quantity; and that of r_n becomes insensible at the same time as r_n if one gives n a very large value. Therefore the increase in the function s can only be an infinitely small quantity. From this remark one immediately deduced the following proposition.

Cauchy then proved a theorem on the term by term multiplication of series, which he used along with the above theorem to prove the binomial theorem (pp. 164–165): $(1 + x)^\mu$ can be developed as this power series (when convergent):

$$1 + \frac{\mu}{1}x + \frac{\mu(\mu - 1)}{1.2}x^2 + \cdots.$$

He gave this proof: consider the sum of the series $\Phi(\mu) = 1+\frac{\mu}{1}x+\frac{\mu(\mu-1)}{1.2}x^2+\cdots$ as a function of μ. It is a continuous function of μ when $-1 < x < +1$, and one has $\Phi(\mu).\Phi(\mu') = \Phi(\mu+\mu')$ (by the theorem on the product of series). This functional equation for Φ is solved as follows.

$$\Phi(m) = \Phi(1)^m = (1+x)^m$$

for all integer values of m and therefore for all rational values of m and therefore, because Φ is a continuous function of μ, for all values of μ.

A clear counter-example will show that Cauchy's argument about the sum of a convergent series of continuous functions necessarily converging to a *continuous* function is wrong. Consider the continuous functions $f_n, n = 1, 2, \ldots$ that are defined as follows. For each positive integer n,

$$f_n(x) = \begin{cases} -1 & \text{if } x \leq \frac{-1}{n} \\ nx & \text{if } \frac{-1}{n} \leq x \leq \frac{1}{n} \\ +1 & \text{if } \frac{1}{n} \leq x \end{cases}$$

Define $u_1(x) = f_1(x)$, and $u_n(x) = f_n(x) - f_{n-1}(x)$ when $n > 1$, so

$$s_n(x) = u_1(x) + u_2(x) + \cdots + u_n(x) = f_n(x).$$

The individual terms of the series, $u_n(x)$, are the difference of two continuous functions, and are therefore continuous everywhere, so they meet the requirement of being continuous in the neighborhood of an arbitrary point, in particular the point $x = 0$.

The sum of the series is the limit $s(x) = \lim_{n\to\infty} s_n(x)$, and it is the following function

$$s(x) = \begin{cases} -1 & \text{if } x < 0 \\ 0 & \text{if } x = 0 \\ +1 & \text{if } x > 0 \end{cases}$$

which is continuous except at $x = 0$.

Now consider the remainder $r_n(x) = s(x) - s_n(x)$. Outside the interval $-\frac{1}{n} \leq x \leq \frac{1}{n}$ the remainder is zero. Inside this interval it is defined by

$$r_n(x) = \begin{cases} -1 - nx & \text{if } \frac{-1}{n} \leq x < 0 \\ 0 & \text{if } x = 0 \\ 1 - nx & \text{if } 0 < x \leq \frac{1}{n} \end{cases}$$

It is true that as n becomes indefinitely large the remainder term becomes zero for each value of x, but this does not happen for all x simultaneously as n becomes indefinitely large. On the contrary, for each n there are values of x for which the remainder takes the value 1 (and others for which the remainder takes the value -1).

Evidently, with a modern understanding of the terms Cauchy has made a mistake. Indeed, we have to read the statement of the 'Theorem' charitably, because it invites the interpretation that a series is known to converge in a neighbourhood of a point if it is known to converge at the point, which is trivially wrong (it is enough to take $s_n(x) = nx$ and follow the above counter-example).

It is clear, on looking at Cauchy's argument, that he was unclear about what it is for the terms $r_n(x)$ to become arbitrarily small. Does it mean that

for all $\varepsilon > 0$ there is an N such that for all x in the interval under consideration $n > N$ implies $|r_n(x)| < \varepsilon$?
Or does it mean that
for all $\varepsilon > 0$ and for all x in the interval under consideration there is an N such that $n > N$ implies $|r_n(x)| < \varepsilon$?

For our example the first claim is false, but the second claim is true, which shows that there is a distinction to be made here. However, Cauchy did not do so. Instead, he passed to the limit as n tended to infinity without considering that different terms of the series might behave differently, which our modern argument does. So it is interesting, and perhaps surprising, to see what happened at the time.

The first person to comment was Abel, who took as his guide, he said (Abel 1826, 221) "the excellent work of M. Cauchy "Cours d'analyse de l'école polytechnique", which must be read by every analyst who aims at rigour in mathematical research".[9] Abel observed (1826, 225) that

> it seems to me that the theorem admits exceptions. For example, the series
>
> $$\sin x - \frac{1}{2}\sin 2x + \frac{1}{3}\sin 3x - \cdots$$
>
> is discontinuous for every value $(2m+1)\pi$ of x, m being a whole number. There are, as one knows, many series of this kind.

The series Abel gave had been known since Fourier's time to be copies of the function $y = \frac{x}{2}$ between $-\pi$ and π and discontinuous, as Abel said, at all odd multiples of π, where it takes the value zero, and presumably Abel took this example from Fourier's work. It had indeed been known to Euler, who discussed it in 1783, and it had been picked up and commented on by Lacroix in his book of 1810. For pictures of the graph of this series, see Sect. 4.4.

It is likely that Abel chose his words carefully to indicate that Cauchy's argument was essentially correct but wrong in a way that could be put right, or, in other words, that Cauchy's argument would be true in most cases but fail in some anomalous circumstances that had to be understood. For a discussion of Abel's views in a context of the transition from formal, algebraic, analysis to concept-driven analysis, with an indication of how other mathematicians whom Abel knew saw exceptions to theorems, see the interesting account (Sørensen 2005). We shall look at Abel's attempt at providing a rigorous proof shortly.

---

[9]Page references are to Abel's *Oeuvres complètes*, Vol. 1.

Certainly Cauchy knew this work, but he did not react when Abel published these remarks, and the obvious implication is that he did not consider them to be relevant to his ‘Theorem’? On the other hand, Abel did think this series was an exception.

Let us look at the series in a neighbourhood of an odd multiple of π. Each term is a sine function and therefore continuous everywhere. Each term takes the value 0 at every odd multiple of π, so the series converges to zero at all odd multiples of π.

On any closed interval not containing an odd multiple of π the convergence is uniform. But plainly the sum is continuous only away from the odd multiples if π. Abel read Cauchy’s ‘theorem’ as saying that if the series is convergent at a point x_0 and the individual terms of the series are continuous on a neighbourhood of x_0 then the series is also continuous on that neighbourhood.

One way to focus this conflicting set of views is to stress that Cauchy never expressed a concept of continuity at a point, only on an interval—which gives him a concept more like uniform continuity. If you also take him to mean by convergence on an interval what was later meant by (and what today we mean by) uniform convergence, then his ‘Theorem’ is actually correct. Abel, on the other hand, may well have taken convergence to be what we mean by it today, convergence at a point, in which case Cauchy’s ‘Theorem’ is wrong and his counter-example shows that. However, what Abel wrote when defining continuity and convergence is almost identical to what Cauchy wrote. Thus (Abel 1826, 220):

> An arbitrary series
> $$v_0 + v_1 + v_2 + \cdots + v_m + \cdots$$
> will be said to be convergent, if for continually increasing values of m, the sum $v_0 + v_1 + v_2 + \cdots + v_m$ indefinitely approaches a certain limit; This limit will be called the sum of the series.

and (Abel 1826, 223):

> A function $f(x)$ will be said to be a continuous function of x between the limits $x = a$ and $x = b$ if, for an arbitrary value of x between these limits, the quantity $f(x - \beta)$ indefinitely approaches the limit $f(x)$ for continually decreasing values of β.

It looks as if they wrote the same words and could have meant something different by them.

The picture becomes even more complicated if we look at how Abel himself tackled the binomial theorem.[10] He replaced Cauchy’s faulty theorem with

Theorem IV: If the series

$$f(\alpha) = v_0 + v_1\alpha + v_2\alpha^2 + \cdots + v_m\alpha^m + \cdots$$

is convergent for a certain value δ of α, it will also be convergent for every value less than δ, and if one supposes that α is less than or equal to δ, then the function $f(\alpha - \beta)$ will approach the limit $f(\alpha)$ indefinitely for continually decreasing values of β.

[10] Here we follow Lützen (2003, 178).

He followed this with

Theorem V: Let

$$v_0 + v_1\delta + v_2\delta^2 + \cdots$$

be a convergent series in which $v_0, v_1, v_2, \ldots$ are continuous functions of the same variable quantity x between the limits $x = a$ and $x = b$, then the series

$$f(\alpha) = v_0 + v_1\alpha + v_2\alpha^2 + \cdots$$

where $\alpha < \delta$ will be convergent and a continuous function of x between the same limits.

Theorem V can be used to replace Cauchy's faulty theorem and thereby provide a valid proof of the binomial theorem. However, Abel's proof of Theorem IV does not make clear where the uniform convergence comes into play, and in any case he did not use that property in the crucial Theorem V. As Lützen points out (2003, 178), Kronecker was to observe correctly that Abel's proof is flawed at this point.

It is also worth noting, as Viertel has pointed out (2012, 57–59) that when Enne Heeren Dirksen reviewed the German translation of Cauchy's *Cours d'analyse* in 1829 he explicitly recorded his doubts about this theorem, writing in the third person that "the proof is not sufficient, and it seems to him on a closer consideration that the validity of the theorem itself is in doubt".[11] He came to this opinion, or at least was able to express it, because he handled the equation with more formalism and less reliance on words than Cauchy had used, and he used his notation to show that similar reasoning to Cauchy's could be used to show that $0 = 1$. But he did not produce a counter-example to Cauchy's claim.

Dirksen is described in Schubring (2005, 471–475) as an adherent of the German combinatorial school who took his Ph.D at Göttingen in 1820 and then went to Berlin, where he became a professor in 1824. He worked for a long time on a long, and largely unpublished, manuscript called the *Organon*, and in it he defined continuity of a function this way:

> If x designates an independent variable, $f(x)$ an appropriately defined [function], and x_0 a specific value of x, it is said that $f(x)$ is continuous for the specific value x_0 and remains so insofar as its special value for x_0 of x is completely determined by x and is equal to the value of the limit of $f(x)$ for the limit x_0.

As Schubring observes, this is quite clearly continuity at a point. He goes on to remark that Dirksen, influenced by his mathematical training, which emphasised algebraic formalisms, was the first person to devise a notation that allowed mathematicians to handle two limit processes at once ($n \to \infty,\ x \to 0$), and so to produce the example that Viertel was to discuss later.

Another interesting perspective on this is provided by Dirichlet's investigations, which we shall discuss in later chapters (see Chap. 14).

---

[11]Quoted in Viertel (2012, 57 n. 55).

Very importantly, Cauchy himself returned to this question in 1853, when Charles Briot and Claude Bouquet, two young French mathematicians whom he supported, questioned the 'Theorem', and simply admitted he had been wrong. As he put it,

> As MM. Briot and Bouquet have remarked, this theorem can be proved for power series arranged in ascending powers of a variable, but for other series it cannot be admitted without restriction.

He now argued that

> If one calls n' an integer larger than n the remainder r_n will be nothing other than the limit to which the difference
>
> $$s_{n'} - s_n = u_n + u_{n+1} + \cdots + u_{n'-1} \tag{4.2}$$
>
> will converge for increasing values of n'. Now let us suppose that in attributing to n a sufficiently large value one can make, for all values of x within the given limits, the modulus of (4.2) (whatever n' may be) and therefore the modulus of r_n less than a number ε that is as small as one wishes. As an increase attributed to x can still be supposed to be sufficiently close to zero for the corresponding increase in s_n to have a modulus less than a number as small as one likes, it is clear that it is enough to attribute an infinitely great value to the number n and an infinitely small value to the increase in x, to establish the continuity of the function
>
> $$s = s_n + r_n$$
>
> between the given limits.
>
> But this proof evidently supposes that expression (4.2) satisfies the condition stated above, i.e. that this expression becomes infinitely small for an infinitely large value attributed to the integer n. Moreover, if this condition is met the series (4.1) will evidently be convergent. Consequently, one has the following theorem
>
> **Theorem 1** If the different terms of the series
>
> $$u_0(x), u_1(x), \ldots, u_n(x), \ldots$$
>
> are functions of the real variable x that are continuous, with respect to this variable between given limits, if, moreover the sum
>
> $$u_n + u_{n+1} + \cdots + u_{n'-1}$$
>
> always becomes infinitely small for infinitely great values of integers n and $n' > n$, then the series (4.1) will be convergent and the sum of the series (4.1) will be a continuous function of the variable x between the given limits.

One might still wish for a little more precision, but the sentiment is clear: Cauchy now inserted a clause that makes the appropriate values of n and n' depend on x, which clears up his earlier ambiguity. He then showed that Abel's counter-example does not apply to the new theorem.[12]

The most likely explanation for this confusion is that everyone was confused. The natural idea of continuity on an interval, which might be taken to mean uniform continuity in modern parlance, was not perhaps properly distinguished from continuity

[12] In view of the fact that some mathematicians have attempted to interpret Cauchy's analysis using modern non-standard analysis, which can make Cauchy's 'Theorem' correct, it is worth noting that Cauchy still used the language of infinites in 1853, but now to rectify his previous claim, which makes it unlikely that he had changed his mind about infinites and infinitesimals. See Chap. 22 below.

at a point, which is not a natural intuition to have. Cauchy had perhaps thought, in 1821, that the theorem was true except when it was obviously false, and in 1853 had decided that such slipperiness would not do. But the clearest thing is that even these mathematicians found the matter confusing, and were not, in the end, entirely consistent or clear. The lack of a good notation, and the heavy reliance on verbal statements, obscures delicate questions such as the dependence of $r_n(x)$ on both n and x, and this seems to have allowed mathematicians to misunderstand the problem and to have been misunderstood by their contemporaries.

We shall return to this question when the concepts of continuity and uniform continuity began to be more clearly expressed and disentangled in the 1840s and 1850s (see Chap. 22).

4.4 Abel's 'Exception' and Cauchy's Series

Abel's Series

Abel's exception to Cauchy's theorem is the function

$$Ab(x) := \sum_{1}^{\infty} \frac{(-1)^{k+1} \sin(kx)}{k}.$$

It is the Fourier series for the function $y = x/2$ from $-\pi$ to π. Figure 4.1 is a graph of the first 10 terms of this series.

Figure 4.2 is a graph of the first 100 terms of this series: Notice how nearly straight the diagonal parts are—they are in fact very wiggly smooth curves that oscillate fast but not by very much around the eventual limit curve. Notice too how unhappy Maple gets at the points where the Abel function will be discontinuous.

Figure 4.3 is a graph of the result of differentiating the first 100 terms of this series term by term. It oscillates wildly, and is not at all like the derivative of the limit function (which is the constant function 1/2 everywhere that the Abel function

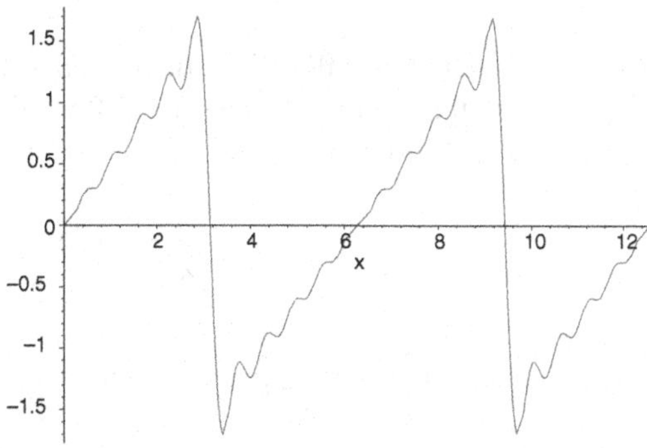

Fig. 4.1 The first 10 terms of the series for $Ab(x)$

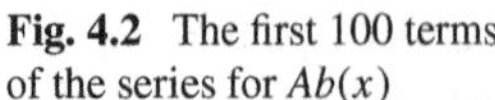

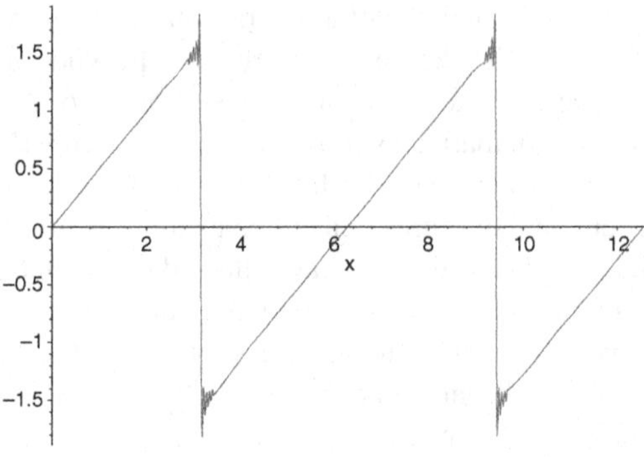

Fig. 4.2 The first 100 terms of the series for $Ab(x)$

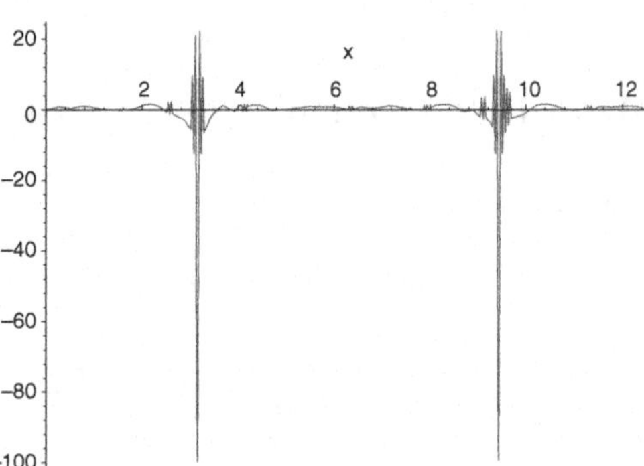

Fig. 4.3 The first 100 terms of the series for $\frac{dAb(x)}{dx}$

is differentiable). This hints at how the theory for the differentiation of series will be much more complicated than that for continuous functions.

Cauchy's Series

Here for comparison are graphs of the first 10 terms (Fig. 4.4) and the first 100 terms (Fig. 4.5) of the very similar series Cauchy used to illustrate his mistake:

$$Ca(x) := \sum_{1}^{\infty} \frac{sin(kx)}{k}.$$

In 1853 Cauchy knew that this series does not converge uniformly. The next graph Fig. 4.6 shows this clearly. It shows the difference between the sum of the first 500 and the sum of the first 30 terms, so it gives an impression of the tail

$$\sum_{30}^{\infty} \frac{\sin(kx)}{k}.$$

It is very clear that the convergence is unlikely to be uniform on the interval $(0, 2\pi)$.

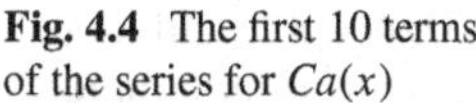

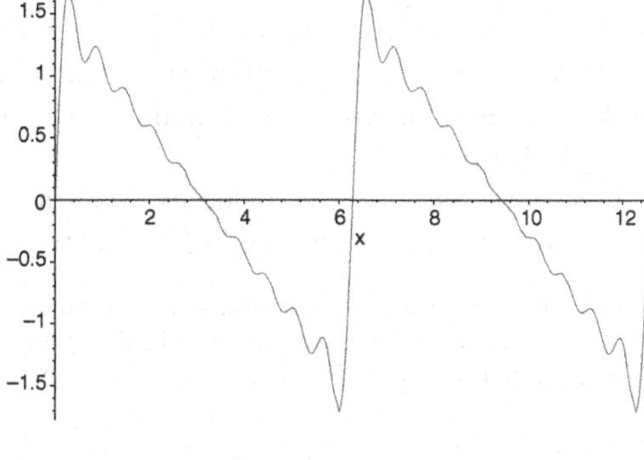

Fig. 4.4 The first 10 terms of the series for $Ca(x)$

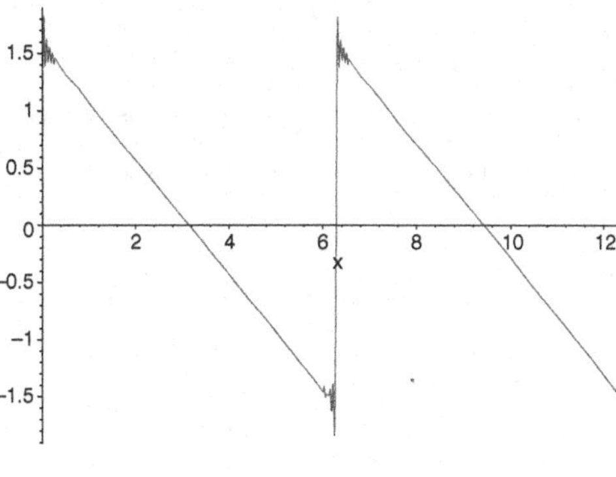

Fig. 4.5 The first 100 terms of the series for $Ca(x)$

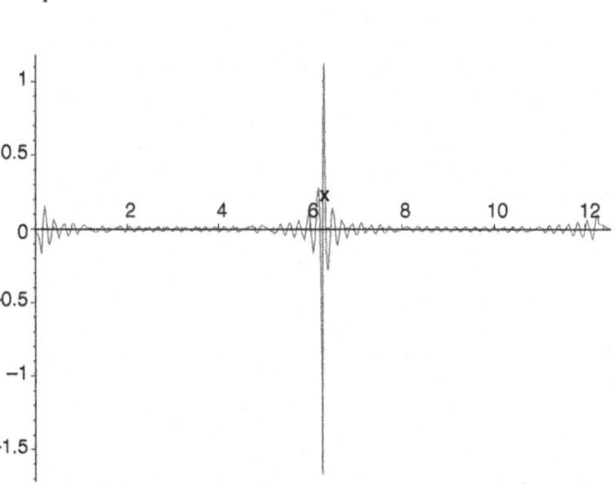

Fig. 4.6 Leading terms of the remainder after the first 30 terms of the series for $\sum_{30}^{500} \frac{sin(kx)}{k}$

A point worth exploring here is this. Consider a small neighbourhood of the origin, one that does not contain an odd multiple of π. To be sure, Abel's series converges uniformly here. A graph of the first N terms, say for $N = 10{,}00{,}000$, will look indistinguishable from a straight line. Does this mean that the slope of the derivative is very nearly constant (say equally indistinguishable to the naked eye)? Term by term differentiation of the truncated series involving just the first N terms is certainly allowed. Consider now the behaviour of the series obtained from Abel's series by forgetting the Nth term. It looks very like the full Abel series, but differs from it by a function that oscillates very fast indeed, going through $N = 10{,}00{,}000$ oscillations as x goes from $-\pi$ to π. This hints at the fact that to require a series of continuous functions to converge to a continuous function is not asking very much of the limit, but to a series of differentiable functions to converge to a differentiable function is asking rather more.

Precisely this point was made by Abel in a letter to his friend Bernt Michael Holmboe in a letter on January 1826.[13] He wrote

> One applies all operations to infinite series as if they were finite, but is this permissible? I think not. – Where is it proved that one gets the differential of an infinite series by differentiating each term? It is easy to give an example for which this is not true, e.g.
>
> $$\frac{1}{2}x = \sin x - \frac{1}{2}\sin 2x + \frac{1}{3}\sin 3x - \cdots.$$
>
> Differentiation gives
>
> $$\frac{1}{2} = \cos x - \cos 2x + \cos 3x - \cdots \text{etc.}$$
>
> a result which is quite false because this series is divergent.

[13] Quoted in Sørensen (2005, 468).

Chapter 5
Cauchy: Differentiation and Integration

5.1 Introduction

Here we look at Cauchy's treatment of differentiation and integration. Cauchy not only reversed a century of tradition that regarded integration as essentially the inverse of differentiation, he brought a raft of new ideas and theorems to the subject. His foundational ideas did not simply underpin existing theory and leave mathematicians free to go about their business as before, they forced a wholesale rewriting of the subject that required new methods and markedly raised standards of rigour.[1] Examples of this that we discuss in this chapter include his mean value theorem and the proof that continuous functions are integrable. Moreover, although Cauchy's emphasis was on what can be done and on what can now be properly established, by defining what a continuous function is, what it is for a function to be differentiable or to have an integral, Cauchy opened the door to the idea that functions might fail to have these properties, an impression that was strengthened by his insistence that functions defined by series may only have these properties on restricted domains.

5.2 Cauchy's Criticisms

Cauchy's Avertissement or preface to his *Résumé* contains the following striking passage:

> The methods that I have followed differ in many respects from those found in works of the same kind. My principal aim has been to reconcile rigor, which I had made a law unto myself in my *Cours d'analyse*, with the simplicity that results from the direct consideration of infinitely small quantities. For this reason, I have thought it necessary to reject the expansions of functions in infinite series whenever the series thus obtained are not convergent; and I

[1] I am indebted to Jesper Lützen for stressing to me in conversation what he calls the 'new architecture' of mathematical analysis that Cauchy introduced, a point he makes in his (2003, 161).

J. Gray, *The Real and the Complex: A History of Analysis in the 19th Century*,
Springer Undergraduate Mathematics Series, DOI 10.1007/978-3-319-23715-2_5

have been forced to return Taylor's formula to the integral calculus. This formula can no longer be admitted as general unless the series that it includes is reduced to a finite number of terms and completed by a definite integral. I am not unaware that the illustrious author of the *Mecanique analytique* [Lagrange] took the formula in question as the basis of his theory of *derived functions*. But, despite all the respect that I have for such a great authority, the majority of geometers now agree in recognizing the uncertainty of the results to which one can be led by the use of divergent series, and I add that in many cases the Taylor theorem seems to furnish the expansion of a function in convergent series, even though the sum of the series essentially differs from the proposed function. For the rest, those who read my book will I hope be convinced that the principles of the differential calculus and its most important applications can easily be set out without the use of series (Cauchy, *Résumé*, pp. 9–10).[2]

Despite the polite, diplomatic language, Cauchy here made it clear that he was offering a profoundly different account of the fundamental processes of the calculus. Compare it with what Lagrange had published only a few years earlier, in a supplement to the lessons he had given on the theory of functions at the École Polytechnique[3]:

Every function of one variable only can always be regarded as an exact derivative; for, if it does not naturally have a primitive function, one can always find one by series ...by resolving the given function in a series of powers of the variable and then taking the primitive function of each term.

In the same volume of the *Journal de l'École Polytechnique* that contained Lagrange's supplement, André-Marie Ampère, after whom the amp is named, published a long note (Ampère 1806) in which he claimed to give "a new demonstration of the Taylor series". He disagreed with Lagrange's assumption, but nonetheless sought to prove that a continuous function is differentiable, indeed infinitely differentiable, and that every function $f(x)$ can be expanded in a series of increasing powers. He began by giving a new definition of the derivative, which he stated in these terms[4]:

The derived function of $f(x)$ is a function of x such that $\frac{f(x+i)-f(x)}{i}$ is always included between two of the values that this derived function takes between x and $x+i$, whatever x and i may be (see footnote 4).

This definition was, he said, 'the most general and the most rigorous possible', and he justified it by basing it on a property of derived functions found by Lagrange and which he stated this way:

Let A and K be the values of $f(x)$ corresponding to $x = a$ and $x = k$. One can always suppose that a and k are chosen in such a way that $f(x)$ never becomes infinite in the interval and that neither does A equal K nor does a equal k, so that $\frac{K-A}{k-a}$ is neither zero nor infinite, and that one can take this quantity for that above or below which [i.e. as an upper or lower bound, JJG] one can always take $(f(x+i) - f(x))/i$, giving to i a value as small as necessary.

---

[2]Quoted in Bottazzini, *The* Higher Calculus, 118.

[3]See Lagrange (1806, 364), quoted in Bottazzini, *The* Higher Calculus, 118.

[4]See Ampère (1806a, 151), quoted in Bottazzini, *The* Higher Calculus, 119.

Cauchy respected Ampère's work and his lectures at the École Polytechnique, he even taught courses with him—they alternated in alternate years—and he cited Ampère among those to whom he is indebted in both the *Cours d'analyse* and in the ***Résumé***. But, unpersuaded as he was by the whole idea of Lagrange's approach, Cauchy reversed the whole formulation of the problem. He defined the derivative of a function (if it exists) as a suitable limit, with the result that Ampère's definition became a property of the derivative, expressible as a theorem.

5.3 Cauchy's Definitions

Differentiation

Cauchy began by repeating the definitions of limit, infinitesimal, and continuity from the *Cours d'analyse*. Then he wrote

> When the function $y = f(x)$ remains continuous between two given limits of the variable x and one assigns to this variable a value included between the two limits in question, an infinitely small increase in the variable produces an infinitely small increase in the function itself. As a consequence, if one then sets $\Delta x = i$, the two terms of the *ratio of differences* $\frac{\Delta y}{\Delta x} = \frac{f(x+i)-f(x)}{i}$ will be infinitely small quantities. But, while these two terms will indefinitely and simultaneously approach the limit zero, the ratio itself can converge towards another limit, either positive or negative. This limit, *when it exists* [my emphasis, JJG], has a determinate value for every particular value of x, but it varies with x. ...The form of the new function that will serve as the limit of the ratio $\frac{f(x+i)-f(x)}{i}$ will depend on the form of the proposed function $y = f(x)$. To indicate this dependence, we give the new function the name of derived function, and designate it, with the aid of an accent, by the notation y' or $f'(x)$.[5]

In his next lecture Cauchy connected this with the familiar $\frac{dy}{dx}$. He defined the differential of the function $f(x)$ as "the limit towards which the first member of the equation

$$\frac{f(x+\alpha h) - f(x)}{\alpha} = \frac{f(x+i) - f(x)}{i}h,$$

where $i = \alpha h$ converges when the variable α approaches indefinitely close to zero while the quantity h remains constant" (***Résumé***, p. 27). When $f(x) = x$, the equation $df(x) = hf'(x)$ reduces to $dx = h$, whence $df(x) = f'(x)dx$, or, equivalently, $dy = y'dx$. Therefore, said Cauchy, the first derivative can be written as dy/dx, that is, as the ratio between the differential of the function and that of the variable.

Cauchy's thorough-going re-write of the foundations of the calculus is based on the next result, which is his mean value theorem.[6]

> If, the function $f(x)$ being continuous between the limits $x = x_0, x = X$, one denotes by A the smallest and by B the largest of the values that the derived function $f'(x)$ assumes in this interval, the ratio of finite differences

[5]Cauchy, ***Résumé***, pp. 22–23, in Bottazzini *The* Higher Calculus, 120.

[6]See the ***Résumé***, pp. 44–46.

$$\frac{f(X) - f(x_0)}{X - x_0} \tag{4}$$

will necessarily be contained between A and B.

He gave this proof.

Let us denote by δ, ε two very small numbers, the first being chosen in such a way that for numerical values of k less than δ and for an arbitrary value of x between x_0 and X the ratio

$$\frac{f(x+k) - f(x)}{k}$$

is always greater than $f'(x) - \varepsilon$ and always less than $f'(x) + \varepsilon$. If between these limits one interposes $n - 1$ new values of the variable x, let us say

$$x_1, x_2, \ldots, x_{n-1}$$

in such a way as to divide the difference $X - x_0$ into elements

$$x_1 - x_0, x_2 - x_1, \ldots, X - x_{n-1}$$

which, all being of the same sign, have numerical values less than δ, the fractions

$$\frac{f(x_1) - f(x_0)}{x_1 - x_0}, \frac{f(x_2) - f(x_1)}{x_2 - x_1}, \ldots, \frac{f(X) - f(x_{n-1})}{X - x_{n-1}} \tag{5}$$

will be found, the first between the limits $f'(x_0) - \varepsilon$, $f'(x_0) + \varepsilon$, the second between the limits $f'(x_1) - \varepsilon$, $f'(x_1) + \varepsilon$, ...all being greater than the quantity $A - \varepsilon$ and all less than the quantity $B + \varepsilon$. (Here, Cauchy used the result that if $\frac{a}{b} < \frac{c}{d}$ then $\frac{a}{b} < \frac{a+c}{b+d} < \frac{c}{d}$ which was better known then than now.) Moreover, the fractions (5) having denominators of the same sign, if one divides the sum of their numerators by the sum of their denominators one obtains a mean fractions, that is to say one lying between the smallest and the greatest of the ones considered (see *Analyse Algébrique*, Note II, Theorem XII). The expression (4), with which the mean coincides, will therefore itself be contained within the limits $A - \varepsilon$ and $B + \varepsilon$, and as this conclusions persists however small the number ε shall be, one can say that expression (4) will be contained between A and B.

Cauchy immediately deduced that

Corollary If the derived function $f(x)$ is itself continuous between the limits $x = x_0, x = X$, in going from one limit to the other the function varies in such a manner as to always remain between the two values A and B and successively takes all the intermediate values. Therefore any mean quantity between A and B will be a value of $f'(x)$ corresponding to a value of x between the limits x_0 and $X = x_0 + h$, or, which comes to the same thing, a value of x of the form

$$x_0 + \theta h = x_0 + \theta(X - x_0),$$

θ denoting a number less than unity. In applying this remark to expression (4), one concludes that between the limits 0 and 1 there is a value of θ that satisfies the equation

$$\frac{f(X) - f(x_0)}{X - x_0} = f'[x_0 + \theta(X - x_0)]$$

or, which comes to the same thing, the following:

$$\frac{f(x_0 + h) - f(x_0)}{h} = f'(x_0 + \theta h) \tag{6}$$

> This last formula being substituted whatever the value of x represented by x_0 provided that the function $f(x)$ and its derivative $f'(x_0)$ remain continuous between the extreme values $x = x_0, x = x_0 + h$, one will generally have, under this condition,
>
> $$\frac{f(x+h) - f(x)}{h} = f'(x + \theta h) \tag{7}$$
>
> whence, on writing Δx in place of h, one deduces
>
> $$f(x + \Delta x) - f(x) = f'(x + \theta \Delta x)\Delta x. \tag{8}$$
>
> It is essential to observe in equations (7) and (8) that θ always denotes an unknown number less than unity.

Grabiner (1981, 115) has observed that in the course of proving this theorem, an interesting shift occurred away from the language of infinitesimals or limits and towards the arithmetic of inequalities when Cauchy began his proof by writing

> Designate by δ and ε two very small numbers: the first being chosen in such a way that, for numerical values of i less than δ, and for any value of x between x_0 and X, the ratio of $(f(x+i) - f(x))/i$ always remains greater than $f'(x) - \varepsilon$ and $f'(x) + \varepsilon$.

This is the first appearance of the delta–epsilon notation, although it is modelled on more verbal expressions of the same kind in the *Cours d'analyse*, and it became a growing presence in Cauchy's work.[7]

This proof merits some careful discussion.[8] Cauchy assumed that a single choice of δ and ε could be made which works simultaneously for all of $x_1, x_2, \ldots, x_{n-1}$. This will not always be the case. He also assumed that the derived function $f'(x)$ is bounded in the interval. This too is not always the case. These assumptions can be made to hold if the derivative is assumed to be continuous in the interval $[x_0, X]$. Contrariwise, when Cauchy does assume the continuity of the derivative, in the corollary, he need not have done so. The intermediate value property holds for the derivative even if it is not continuous (a result later proved by Darboux). Cauchy's genuine insights nonetheless left quite a delicate muddle to sort out!

These two examples, taken from Bressoud (1994), illuminate these subtleties.

1. Bressoud (1994, 89): The function $f(x) = x^2 \sin(x^{-2})$ is differentiable in $[0, 1]$ but there is no ε for which there is a δ such that $|x - a| < \delta$ implies $\left|\frac{f(x)-f(a)}{x-a} - f'(x)\right| < \epsilon$ for all $a \in [0, 1]$.
2. Bressoud (1994, 81): The function $f(x) = x^{4/3} \sin(1/x)$ is differentiable in $[-1, 1]$ but the derivative is not bounded there. Note, by the way, that for this function $f'(0) = 0$, contrary to what any attempt at graphing it might suggest and consistent with the fact that the formal expression for the derivative does not makes sense on substituting the value $x = 0$. This underlines the fact that the derivative is defined not as an expression but as a limit. The expression does have a limiting value as $x \to 0$—but you must prove that. Check that the derivative is not continuous at $x = 0$.

---

[7]See also Bottazzini, *The* Higher Calculus, 120.

[8]See Bressoud (1994, 81).

Cauchy was also aware that there are problems with quantities defined as limiting values, and commented that singular values, those obtained only as a limit, may present real difficulties, as for example with x^x at $x = 0$. He therefore gave a number of tests for convergence appropriate to that delicate case.

Lagrange had started his account with a (flawed) argument to show that every function admits a Taylor series expansion. Cauchy's much more careful defence of this idea was restricted to functions that, with their first n derivatives, are continuous within the z-interval $[0, h]$, and it comes at the conclusion of his lectures on the infinitesimal calculus (*Résumé*, pp. 211–219). There he gave what is nowadays known as Cauchy's form of the Taylor series with remainder:

$$f(x+h) = f(x) + hf'(x) + \frac{h^2}{2!}f''(x) + \cdots + \frac{h^{(n-1)}}{(n-1)!}f^{(n-1)}(x) + \frac{h^n}{n!}f^{(n)}f(x+\theta h),$$

$0 \leq \theta \leq 1$.

The remainder can also be written as an integral:

$$\int_x^{x+h} \frac{(x+h-z)^n}{n!} f(z)dz,$$

and he observed that when the integral tends to zero with increasing values of n it serves to expand the function $f(x+h)$ in series ordered according to the ascending integral powers of the quantities x and h.

Cauchy concluded with a remarkable observation that is the first hint that the world of continuous, and even differentiable, functions may be much larger than the world of functions defined by power series. If the Taylor series $f(0)+xf'(0)+\frac{x^2}{2!}f''(0)+\cdots$ is convergent, then one might think that its sum is the function $f(x)$, and in particular if all the terms of the Taylor series vanish than the function itself vanishes—but this is not necessarily the case. In fact, he said,[9]

> to be certain of the contrary it is sufficient to observe that the second condition will be fulfilled if we suppose $f(x) = e^{-1/x}$, and the first if we suppose $f(x) = e^{-x^2} + e^{-(1/x)^2}$. However, the function $e^{-(1/x)^2}$ is not identical to zero, and the series derived from the last supposition does not have the binomial $e^{-x^2} + e^{-(1/x)^2}$ as its sum, but its first term $e^{-(1/x)^2}$.

The observation is profound. Hitherto, every mathematician had believed that every function can be expanded in a Taylor series, trivial cases aside, and that their task was to prove it—if indeed they felt that a proof was called for. Cauchy had now shown that it was possible to define a function that did not agree with its Taylor series. This not only destroyed the foundations of the Lagrangian calculus, it opened up the much more disturbing question of the relationship between a representation of a function and the function itself. By then Cauchy may well have known that much graver problems were to present themselves with the representation of a function by its Fourier series. Finally, this example suggested that functions may not, after

[9] Cauchy *Résumé*, 229–230.

all, submit naturally to the operations of the calculus. Cauchy's new definitions and theorems opened up a domain that only they were precise enough to allow mathematicians to explore.

The Definite Integral

In the century since Newton and Leibniz the fundamental theorem of the calculus had become regarded as allowing the integral of a function to be regarded as the opposite of its derivative; the integral of a function $f(x)$ is a function $F(x)$ with the property that $\frac{d}{dx}F(x) = f(x)$. As such, it is determined up to an additive constant.

Cauchy firmly reversed this trend, and restored an independent existence on the integral. In the second part of the *Résumé* of 1823 he defined the integral as a limit of sums of areas.

The Cauchy integral of a function of a real variable $y = f(x)$ that is continuous in a given interval was defined as follows. He divided the interval into n equal subintervals $[x_{i-1}, x_i]$, and considered the sum $S = \sum_i (x_i - x_{i-1}) f(x_{i-1})$. The integral will be the limit of this sum as n tends to infinity.

To prove that this limit exists, Cauchy considered a different sum:

$$S' = \sum_i (x_i - x_{i-1}) f(x_{i-1} + \theta_{i-1}(x_i - x_{i-1})).$$

In this expression, the function f is evaluated at the point where it takes its average value (by the mean value theorem). Now,

$$f(x_{i-1} + \theta_{i-1}(x_i - x_{i-1})) = f(x_{i-1} + \varepsilon_i),$$

for some ε_i. So

$$S' = \sum_i (x_i - x_{i-1}) f(x_{i-1}) + \sum \varepsilon_i (x_i - x_{i-1}).$$

The second part of the sum can be made arbitrarily small because f is continuous, so every ε will 'differ very little' from zero. Here, not for the first time, as we have seen, Cauchy failed to see that choosing a δ that works simultaneously for every subinterval (and depends only on the ε) cannot be guaranteed but must be proved to be possible. In fact, as later mathematicians were to show, this can be done because a continuous function on a closed interval is uniformly continuous.

By repeating the reasoning and implicitly reverting to the 'Cauchy criterion' that he had demonstrated in his *Cours d'analyse* Cauchy concluded[10]

> if we let the numerical values of these elements decrease indefinitely by increasing their number, the value of S will end by being sensibly constant or, in other words, it will end by attaining a certain limit that will depend uniquely on the form of the function $f(x)$ and the extreme values x_0, X attributed to the variable x. This limit is what we call a definite integral.

[10]Cauchy, *Résumé*, p. 125, Bottazzini, *The* Higher Calculus, 144.

By later standards, Cauchy should have considered the effect of dividing the interval into unequal pieces, which he did not. But he did consider certain kinds of discontinuous functions, those in which the function becomes infinite at one or more points of the interval (his argument allowed finitely many such points). So it is not correct to say, as the literature sometimes does, that Cauchy defined the integral only for continuous functions.

Cauchy's method was to consider the obvious limit, and to say that the function has an integral if this limit exists (Cauchy *Resumé*, p. 146). More precisely, let us suppose that the function $f(x)$ is defined on the interval $[x_0, X]$ and continuous there except at the interior point $x = a$. Then it makes sense to consider the integrals $\int_{x_0}^{a-\varepsilon} f(x)dx$ and $\int_{a+\varepsilon}^{X} f(x)dx$ and their sum

$$\int_{x_0}^{a-\varepsilon} f(x)dx + \int_{a+\varepsilon}^{X} f(x)dx.$$

Cauchy defined the limit of this sum when it exists as $\varepsilon \to 0$ as the principal value of the integral $\int_{x_0}^{X}$, and said it was a singular integral.

There are two ways to think of this. One is that mathematics is about functions. Something one does with functions is integrate them, and here is a way of integrating some functions that do not immediately seem to be amenable to integration. The second way is that mathematics is about operators, such as integration, that operate on functions, and here is a way of making that operator work on a wider class of functions. The difference in points of view may not be very apparent in this example, but we shall see that analysts were at least as concerned with getting a theory of the integral that worked well as with understanding functions.

The Indefinite Integral

This work led Cauchy to the concept of the indefinite integral. In the Avertissement to the *Résumé* Cauchy had written[11]

> it seemed to me necessary to demonstrate the existence of the integrals or primitive functions in general before making their various properties known. To this end, it was first necessary to establish the idea of integrals taken between given limits or definite integrals.

Now that Cauchy had defined the integral without reference to differentiation, he had to reconsider the formula that is sometimes called the fundamental theorem of the calculus:

$$\int_a^b f(x)dx = F(b) - F(a),$$

where $F'(x) = f(x)$.

Cauchy argued as follows. Set $T(x) = \int_{x_0}^{X} f(x)dx$, so

$$T(x) = (X - x_0)f(x_0 + \theta(X - x_0)), |\theta| < 1.$$

[11] *Résumé*, 10.

Then

$$\int_{x_0}^{X+\alpha} f(x)dx - \int_{x_0}^{X} f(x)dx = \alpha f(x + \alpha\theta),$$

whence

$$T(x + \alpha) - T(x) = \alpha f(x + \theta\alpha).$$

If f is finite and continuous in the neighbourhood of a point x, the same will be true for T, and by passing to the limit as $\alpha \to 0$, Cauchy deduced that $T'(x) = f(x)$.

Cauchy then showed that, if $w'(x) = 0$ then $w = constant$ on an interval and hence (*Resumé* p. 154) that "the general value of y, the solution of the equation $dy = f(x)dx$, is given by $y = \int_{x_0}^{X} f(x)dx + constant$". This is what he called the indefinite integral.

Cauchy concluded the *Résumé* with a discussion of differentiation under the integral sign.

Chapter 6
Cauchy and Complex Functions to 1830

6.1 Introduction

Mathematicians in the 1810s did not feel that there was an identifiable single subject, albeit perhaps in an imperfect state, that could be called complex function theory. Rather, there were the topics involving multiple integrals, functions of several variables, mappings from $\mathbb{R}^2$ to $\mathbb{R}^2$, and similar topics, in all of which complex variables could be used, if only as a convenient notation for pairs of variables.

Cauchy's first involvement in these matters is his *Mémoire* of 1814 on the evaluation of double integrals. By the 1820s he had begun to consider that there might be a coherent topic involving complex functions of a complex variable (a topic he had not discussed in his *Cours d'analyse* or the *Resumé*). In the great *Mémoire* (1825) he produced the integral and residue theorems that are today named after him, but even then he does not seem to have appreciated the depth of his discovery, perhaps because he was still not thinking geometrically about complex variables. A deeper appreciation had to wait until the 1840s, by which time other, younger, French mathematicians were taking up the subject, so the history of this part of the subject follows a much more disorganised course than does the impact of Cauchy's work on real analysis.

Yet, as we shall see (especially in Chaps. 11, 16, 17, 19, and 20) by the 1870s complex function theory was on its way to becoming one of the dominant fields of mathematics, and its origins are the other principal theme of this book.

6.2 'The Passage from the Real to the Imaginary'

A curious feature of mathematical analysis in the years around 1800 was the use of complex variables to evaluate real definite integrals. The practice had begun with Euler and continued with Laplace, Poisson, and Legendre, in his *Exercises de calcul intégral* (1811). In the *Mémoire* on this topic that he presented in 1814 Cauchy commented that many of the integrals had been evaluated for the first time "by

J. Gray, *The Real and the Complex: A History of Analysis in the 19th Century*,
Springer Undergraduate Mathematics Series, DOI 10.1007/978-3-319-23715-2_6

means of a kind of induction" based on "the passage from the real to the imaginary" (Cauchy 1814, 329), and that no less a figure than Laplace had remarked that the method "however carefully employed, leaves something to be desired in the proofs of the results".[1] Cauchy accordingly set himself the task of finding a "direct and rigorous analysis" of this dubious passage.

Cauchy 'Misses' the Cauchy–Riemann Equations

Cauchy's *Mémoire* came in two parts. The first, entitled 'The equations that authorise the passage from the real to the imaginary', offered a defence of the method, based on apparent validity of the equation

$$\frac{\partial[f(y)\frac{\partial y}{\partial x}]}{\partial z} = \frac{\partial[f(y)\frac{\partial y}{\partial z}]}{\partial x}. \tag{6.1}$$

According to Cauchy, this equation could be verified "directly by a single differentiation" (Cauchy 1814, 337), and the whole first part of the memoir proceeds in a formal way that makes what later eyes would see as a series of tacit assumptions about properties of the functions in question. He then substituted the equations $y = M(x, z) + \sqrt{-1}N(x, z)$ and $f(y) = P' + \sqrt{-1}P''$ into Eq. (6.1) and obtained

$$\frac{\partial S}{\partial z} = \frac{\partial U}{\partial x}, \quad \frac{\partial T}{\partial z} = \frac{\partial V}{\partial x} \tag{6.2}$$

where

$$S = P'\frac{\partial M}{\partial x} - P''\frac{\partial N}{\partial x}, \quad U = P'\frac{\partial M}{\partial z} - P''\frac{\partial N}{\partial x},$$
$$T = P'\frac{\partial N}{\partial x} + P''\frac{\partial M}{\partial x}, \quad V = P'\frac{\partial N}{\partial z} + P''\frac{\partial M}{\partial z}.$$

According to Cauchy, nothing more needed to be done except to show how Eq. (6.2) can be used, and he filled the rest of this part of the memoir with examples. At no stage, even when he set $M(x, z) = x$ and $N(x, z) = z$, did Cauchy observe that from (6.2) one could obtain the equations

$$\frac{\partial P'}{\partial x} = \frac{\partial P''}{\partial z}, \quad \frac{\partial P'}{\partial z} = -\frac{\partial P''}{\partial x}. \tag{6.3}$$

Since these are the Cauchy–Riemann equations, some historical comments are in order. The Cauchy–Riemann equations were first written down by d'Alembert in his celebrated essay on hydrodynamics, *Essai d'une nouvelle théorie de la résistance des fluides* (1752, Chap. 4, §45). They were created as part of an argument about the solutions of flows in the plane (vector fields, as we would say today), and even in Euler's hands they remained part of a body of ideas that allowed the temporary

[1]See Cauchy (1814, 329–330).

appearance of imaginary quantities on condition that they vanish at the end. They had no foundational significance, and Cauchy gave them none either.

Cauchy's main aim was the evaluation of improper integrals. When the functions S, T, U, V are regular within the rectangle of integration, Cauchy integrated the equations to obtain

$$\int (S'' - S')dx = \int (U'' - U')dz, \int (T'' - T')dx = \int (V'' - V')dz,$$

where the primes indicate the values assumed by the functions at the limits of integration. He then applied this to a number of particular cases.

Memoirs presented to the Institut were reviewed by members of the Institut, and in reviewing Cauchy's Legendre remarked that it "merely conforms to the ordinary rules of analysis, which is not a subject of any difficulty". Siméon Denis Poisson was similarly unimpressed. In the long interval that intervened between the submission of the *Mémoire*—1814—and its eventual resubmission—1825—Cauchy added numerous footnotes protesting at this judgement, some of which reflect his own deepening understanding of what was involved.[2] But in any case his chief interest as far as this paper was concerned was in double integrals.

6.3 Cauchy, Double Integrals, and Integrals Round a Rectangle

Mathematicians of the period, with their somewhat formal attitude to the processes of the calculus, not only tended to regard a double integral as a nested pair of single integrals but to be indifferent to the order of integration. This formal approach precluded interpreting double integrals in terms of line integrals along the boundaries of a rectangle in the plane of complex numbers; they were strictly about integrals on a rectangle in the real plane. However, it was known that in some cases this attitude was too naive and led to conflicting results, and it was this problem that Cauchy addressed in the second part of the *Mémoire*. Legendre was more impressed by what he read here, and indeed Cauchy was much more original.

Cauchy showed varying the order of integration of a double integral can yield two distinct definite values when the function becomes infinite or indeterminate for certain values of the variables within the domain of integration. To investigate the matter further, he introduced what he called singular integrals ("intégrales singulières"), which he defined (Cauchy 1814, 334, see Sect. 5.3) as limits as the end points of the integral tend to the same fixed value "without the integrals being zero." Singular integrals are a profound idea. They were to crop up repeatedly in real analysis and, in an intriguing different way, in Cauchy's later work on complex integrals.

[2]The *Mémoire* was published in 1827.

Cauchy first supposed that the singular value of the integrand was in a corner of the rectangle. So, to evaluate the integral as $\int_a^{a'}\int_b^{b'}\frac{\partial K}{\partial z}dxdz$, where $K = \phi(x,z)$ becomes indeterminate for $x = a$, $y = b$, he split the integral into two parts:

$$\int_a^{a'}\int_b^{b'}\frac{\partial K}{\partial z}dxdz = \int_b^{b'}\int_a^{a'}\frac{\partial K}{\partial z}dzdx + A,$$

where A is the singular integral $-\int_0^{\varepsilon}\phi(a+\xi, b+\zeta)d\xi$, where ε is a 'very small' quantity and ξ equal to zero after the integration.

Cauchy took this example

$$K = \frac{z}{x^2 + z^2}$$

and the integral $\int_{\varepsilon}^{1}\int_0^1\frac{\partial K}{\partial z}dzdx$. He integrated this first with respect to z and then with respect to x to obtain the value $\frac{\pi}{4}$. But, on integrating first with respect to x and then with respect to z, he found it was necessary to add the singular integral A given by $-\arctan(\varepsilon/\zeta)$, which for $\zeta = 0$ equals $\frac{-\pi}{2}$. Therefore the value of the integral was $\frac{\pi}{4} + A = -\frac{\pi}{4}$. Cauchy then considered the general case, when the singular point (x,z) is inside the rectangle of integration, before applying his result to the evaluation of many improper integrals.

Cauchy and Contemporary Definitions of the Complex Numbers

Discussions about the meaning, nature, possibility and even existence of complex numbers form an oft-told tale in the history of mathematics. From 1813 Gergonne's *Annales* carried numerous articles for and against Argand's geometrical interpretation of complex numbers, and Argand himself published definitive papers in 1806 and 1814, where his ideas were explained in detail along with a new proof of the fundamental theorem of algebra.[3] Oddly, in this debate Cauchy's name does not usually appear. What in fact did he know, or believe, about complex numbers?

In his late fifties (see Cauchy 1849, 175) he recalled that when he was working as a young engineer at Cherbourg between 1810 and 1812 he was told that one Henri–Dominique Truel had found a way of representing complex numbers in a plane in 1786. Some historians, for example Petrova (1974), claim that Cauchy knew nothing of the debates around Argand's work, even though he was back in Paris by then, others that he found it insufficiently rigorous, and both interpretations are supported by the fact that Cauchy did not mention Argand's work when he gave a proof of the fundamental theorem of algebra (Cauchy 1817a, b), nor did he do so in his *Cours d'analyse*, as we shall shortly discuss.

It is worth observing that mathematicians in the 1810s regarded mathematical operations of all kinds rather formally, and considered square, cube and even nth roots as functions, even though they could be many-valued. Cauchy would have seen that what is manageable in the real case, say an expression such as $\sqrt{1-k^2x^2}$,

[3]See Gilain (1997).

becomes much less so in the complex case. It is usually easy to treat two real square roots, one positive one negative, separately, but two complex ones are harder to control once the variable x is allowed to be complex. With such functions in mind it is reasonable to believe that the difficulty is not in defining a complex number but in defining functions of a complex variable. Controversies had arisen, for example between Lagrange and Poisson, over the value of

$$(2\cos x)^m = \sum_{k=0}^{m} \frac{m!}{(m-k)!k!}\cos(m-2k)x\,, \tag{6.4}$$

when $m = 1/3$ and $x = \pi$. An application of de Moivre's formula provides three different complex values:

$$2^{1/3}\frac{(1+i\sqrt{3})}{2},\ -2^{1/3},\ 2^{1/3}\frac{(1-i\sqrt{3})}{2},$$

while the power series development of $2^m \cos^m x$ according to (6.4) gives yet another different value: $\frac{1}{2}\left(2^{1/3}\right)$, which is the arithmetic mean of the first and third of the previous values.

Complex functions in Cauchy's *Cours d'analyse*

Cauchy defined imaginary numbers in his *Cours d'analyse* (1821a, 173–176) much as we would define an ordered pair of real numbers, that is as "symbolical expressions" of the form $\alpha + \sqrt{-1}\beta$, where α and β are real quantities. But then, arguing that a symbolic expression signifies nothing, he gave over 55 pages to defining the arithmetical operations upon them and establishing their properties. This is one of the passages in the book where his concern for rigour is most apparent, particularly so in his treatment of the multi-valuedness nature of rational powers of imaginary numbers. Here his approach greatly surpassed anything else in the contemporary mathematical literature. It forms another step in his slow journey from formalism to a geometrical understanding of complex function theory, one he may not have fully appreciated as he took it.

Existing controversies surely showed Cauchy the pitfalls in any naive and formal approach. It was not enough simply to replace real by complex numbers, and real by complex variables, in formulae, even when that seemed possible. A proper understanding was needed of how to pass in general from a function $f(x)$ to a function $f(x+iy)$. Evidently there is no unique answer if the problem is posed at that general level: one simply writes $f(x+iy) = f(x, y) = f(x) + ig(y)$ where the function $g(y)$ is arbitrary. Yet, in some tantalising way, there is a single 'correct' answer for most known functions, such as the exponential and trigonometric functions. Prevailing conventions, he said, 'are insufficient for fixing in a precise manner the meaning of the expressions A^x, $\log x$, $\sin x$, $\cos x$, $\arcsin x$, $\arccos x$ when the variable x becomes imaginary' (Cauchy 1821a, 240).

Cauchy proceeded carefully. He defined the concepts of imaginary variable and imaginary limit, and introduced a special symbol $(())$ for denoting any value of a multi-valued function (Cauchy 1821a, 7–12). For certain functions defined by a power series he concluded that it is enough to replace the real variable x by the complex variable $x + iy$, but not without a great deal of careful work. He first defined imaginary valued functions of a real variable, and extended the theorems already proved in the *Cours d'analyse* to them. He then proved that a geometric series with imaginary terms is convergent when $|z| < 1$, by considering pairs of real series, and remarked that its sum is a continuous function of z in the given interval, by appealing to his flawed theorem (see Sect. 4.3). Then came the extension of his convergence tests and the proof of the binomial theorem to the complex case, at which point the multi-valuedness question for functions had to be confronted. He answered it by resorting to their expansions in convergent series (1821a, 308–309). The definition of the corresponding inverse functions $\log x$, $\arcsin x$, $\arccos x$ was discussed in the final part of the chapter.

But Cauchy did not extend his procedure 'by analogy' to *any* real function $f(x)$ that was represented by a convergent power series by considering the imaginary function $f(x + \sqrt{-1}y)$ as defined by the corresponding imaginary series (provided that it was convergent). It remained unclear in his work, and one must suppose to Cauchy himself, how to define and study the general imaginary function of an imaginary variable—and it was remain so for some twenty more years. So, for example, in his (1823b, 330) Cauchy observed that one seemingly could define an imaginary expression by saying that $f(x + t\sqrt{-1})$ represents a function $\phi(x, t)$, continuous with respect to x and t and satisfying both the condition $\phi(x, 0) = f(x)$ and the equation $\frac{\partial \phi}{\partial t} = \frac{\partial \phi}{\partial x}\sqrt{-1}$. Evidently, such a function ϕ satisfies Laplace's equation, but, said Cauchy, in view of what one knows about Laplace's equation this merely leads to a vicious circle.

Cauchy repeated the proof of the fundamental theorem of algebra that he had given in his (1817b) in Chap. X of his *Cours d'analyse*, and cited Legendre (1809) as a source of his proof. He denoted the square of the modulus of the polynomial

$$f(x) = \phi(u, v) + \chi(u, v)\sqrt{-1}$$

by $F(u, v)$, and regarded it self-evidently as a continuous function of u and v. Therefore, he said, its minimum value is non-negative and is attained a certain number of times. Then he showed, by a reductio ad absurdum argument, that this minimum must be zero, which proves the theorem. He then deduced that a polynomial of degree n could be s written as a product of n complex linear factors.

Cauchy's 'Masterwork'

By the time he came to write his memoir (Cauchy 1825) that some have hailed as his masterwork, Cauchy was prepared to take greater risks, or, perhaps, had acquired greater confidence. In his still unpublished 1814 Mémoire he had explained what was meant by the notation $\int_{x_0}^{X} f(x)dx$, where x_0 and X are *real* bounds and $f(x)$ "a real or imaginary function of the variable x". Now he was willing to let the variable x be complex, writing (1825, 43) that "it is necessary to represent by the notation

$$\int_{x_0+y_0\sqrt{-1}}^{X+Y\sqrt{-1}} f(x)dx$$

the limit or one of the limits to which the sum of the products of the form

$$\sum \left(x_{i-1} + y_{i-1}\sqrt{-1}\right) f\left(x_{i-1} + y_{i-1}\sqrt{-1}\right)$$

converges". This adapts his definition of a real integral to the complex setting. When x and y are monotonic continuous functions on the interval $[t_0, T]$, say $x = \phi(t)$, $y = \psi(t)$, the integral becomes

$$I = A + B\sqrt{-1} = \int_{t_0}^{T} \left(\phi'(t) + \sqrt{-1}\psi'(t)\right) f\left(\phi(t) + \sqrt{-1}\psi(t)\right) dt \qquad (6.5)$$

This is an integral with real bounds, and Cauchy proved, by considering what happens when the functions $x = \phi(t)$ and $y = \psi(t)$ are varied slightly, that the integral (6.5) is independent of the nature of these functions, provided $f(x + y\sqrt{-1})$ remains "finite and continuous" within the limits of integration.

We recognise this result as the ancestor, and perhaps a close ancestor, of the Cauchy integral theorem.

Here we encounter several features of what can be called Cauchy's mature writing style. The concept of a continuous function of an imaginary variable is not to be found explicitly defined in this, or any of his previous papers. Moreover, despite the way the theorem was stated, Cauchy tacitly assumed that the function f has a continuous derivative. This unfortunate habit would have been easy to correct when Cauchy came to write this (and other) papers; the fact that he did not suggests that the confusion went deeper than poor writing. Finally, the phrase of the "finite and continuous" is not adequate to the problem at hand. Here is seems to mean single-valued and infinite, but it may be better to interpret as 'well-behaved' in a way that Cauchy could feel but not yet capture precisely.

Some historians have suggested that Cauchy's lack of sensitivity to the meaning of the terms involved was simply continuing to follow the old tradition of regarding continuous functions as differentiable, because they were de facto given by analytical expressions.[4] And while it may be unwise to attribute to Cauchy, who was always intellectually on the move, a steady conviction about anything mathematical, this view seems implausible given that Cauchy had been at pains in the *Resumé* to distinguish a function from its Taylor expansion, and to define continuity and

[4]See Kline (1972, 637) and Remmert (1991, 196).

differentiability in distinct ways. A more plausible alternative is that Cauchy took it for granted that a continuous function of a complex variable satisfied the Cauchy–Riemann equations. This implies that he was willing to suppose that functions were differentiable in the course of a proof without saying so explicitly in the statement of a theorem, but such, confusingly, does seem to be the case on several occasions.

Be that as it may, the deeper question is whether, in 1825, Cauchy was in possession of a theorem that says a complex integral is path independent in any domain where the function does not become infinite. A fair summary of what Cauchy wrote (1825, 20–24) would be that if one considers two different paths within the rectangle (x_0, y_0), (X, Y) and the function $f(x + y\sqrt{-1})$ does not become infinite for values x, y lying within the domain enclosed by the paths, then the corresponding values of the integral I are equal, i.e. the integral I is independent of the path of integration. But this could simply mean that the integral is independent of paths within the rectangle in the familiar real plane. Even the formalism of path integrals involving complex variables can be read as an explanation of how to escape from the troubling world of complex variable and complex functions to the safer world of real paths and mappings from $\mathbb{R}^2$ to $\mathbb{R}^2$. His theorem need not be about complex integrals at all.

Cauchy next investigated the case when the function $f(x + y\sqrt{-1})$ becomes infinite at the point where $x = a, y = b$, corresponding to the value $t = \tau$. He introduced the limit of

$$f = \lim(x - a + (y - b)\sqrt{-1})f(x + y\sqrt{-1})$$

as x tended to a and y tended to b, which he denoted $f^* = \varepsilon f(x + y\sqrt{-1})$, ε being an infinitely small number. In his *Résumé* he had given a theory of the principal value of an integral, which he now used to show that the principal value of the integral I was $2\pi\sqrt{-1}f^*$. Using results from his *Mémoire* of 1814, he deduced that

$$f(z) = \frac{f^*}{z - a - ib} + \omega(z),$$

where $\omega(z)$ remains finite in a neighbourhood of the origin and so contributes nothing to the value of the integral. So, treating the imaginary function $f(x+y\sqrt{-1})$ as he had treated function of two real variables having a point of infinity within the rectangle of integration, he obtained the expression $f^* = \varepsilon f(x + y\sqrt{-1})$ and the value $2\pi i f^*$ for the integral. The extension to a finite number of simple infinities was immediate, and for higher order infinities, say one of order m, Cauchy used a limiting argument to deduce that

$$f^* = \frac{f^{(m-1)}(a + b\sqrt{-1})}{1.2.\ldots(m - 1)}.$$

Here we see, at the same ancestral level, the Cauchy residue theorem.

Cauchy's first use for his theorems, in his (1825), was to real (improper) integrals. But within a year he began publishing his *Exercises de mathématiques*, which can be thought of as a personal journal whose issues appeared monthly until 1830. From the very first issue onwards, Cauchy developed his new calculus in a more systematic way. He called it the "calculus of residues", he introduced a special symbol for indicating "the extraction of the residue" of a function, and he found numerous applications for it. What he did not do was to appreciate the fundamental character of his integral theorem, which was to become one of the basic theorems of complex function theory. He neither used nor quoted from his 1825 paper and his integral theorem for two decades—and Cauchy was not at all averse to quoting from his own papers.

Hans Freudenthal found this "utterly strange and hard to explain" (1971, 139), and asked:

> Did Cauchy not trust the variational method of proof? Was he bothered by the (unnecessary) condition he had imposed on the paths, staying within a rectangle? Did he not notice that the statement could be transformed into the one about closed paths that he most needed? Or had he simply forgotten about that mémoire détaché? In any case, for more than twenty-five years he restricted himself to rectangular paths or circular-annular ones (deriving from the rectangular kind by mapping), thus relying on the outdated 1814 mémoire rather than on that of 1825.

One can go further: nor did Cauchy mention the integral theorem in a paper he presented to the Academy on August 3, 1846 (Cauchy 1846a) which is commonly credited by historians as containing his first general formulation of the integral theorem.

In this case it seems that historians may have been to eager to read a modern interpretation into an old result, for there is no reference in Cauchy's paper (1846a) to either complex functions or complex integration. It concerns the question of when the integrals of some differentials are independent of the path of integration, and Green's work on potential theory is clearly in the background.[5] It is likely that Cauchy had picked up the news of Green's work (1828) from the young William Thomson (later Lord Kelvin) who visited Paris in spring 1845 and showed a copy to Joseph Liouville, and, possibly, to Cauchy, whom he visited several times.[6]

Of course, as for instance Remmert (1991, 198–199) has observed, one can immediately derive the Cauchy integral theorem from his paper (1846a)—but Cauchy did not. He wrote only of real variables and real functions, giving no reason to let the historian infer that he was writing about complex functions at all.

On the other hand, Kline (1972, 638) is surely right when he wrote that "Cauchy must have thought long and hard to realize that some relations between pairs of real functions achieve their simplest form when complex quantities are introduced". In fact, in a steady stream of papers in the autumn of 1846 Cauchy turned to the complex interpretation of his theorems. For example, in Cauchy (1846b, 136) he extended his theorem of residues to a plane domain bounded by a closed curve. In a

[5]As Kline suggested (1972, 640), and Freudenthal agreed (1971, 140).

[6]For Thompson's stay in Paris, and his contacts with Liouville and Cauchy, see Lützen (1990, 135–146).

paper published a week later he explained the result in a more geometrical way that allowed him to speak of a curvilinear integral along a line *OP*, and he observed that

> when one arrives at the point *P* the value of this latter integral is generally independent of the straight line or curve that one has followed (1846c, 146–147).

Cauchy also observed that his results assumed that the functions he was dealing with were all single-valued, and he made some vague remarks about how many-valued functions could be dealt with, but this was to remain a weakness in his work.

Politics Intervenes

Not only was Cauchy himself unsure for many years where the deepest significance of his results lay, a confused publication regime further complicated the reception of his ideas. In July 1830 riots in Paris led to the overthrow of the Bourbon monarchy, to which Cauchy felt he had taken an oath of allegiance.

He left for Turin, where the Bourbon still retained influence and had the support of the King of Sardinia. The Jesuits were powerful at court, Cauchy's interpretation of Catholicism was close to theirs, and in January 1832 the King re-opened the chair of physics at the University of Turin and appointed Cauchy to it. Cauchy was also welcome at the Academy of Sciences in Turin, and presented some important papers there that will be discussed below in Chap. 10. Then in July 1833 he left to serve as a tutor of the son of the exiled king Charles X in Prague before returning in Paris in 1838. He lost his teaching position at the École Polytechnique when he left, and he was unable to obtain a position for several years after his return because of his refusal to swear an oath to the new regime. His absence from Paris undoubtedly diminished the impact of his work.

Bruno Belhoste, whose biography of Cauchy (Belhoste 1991) is well worth reading, notes that the early 1830s were momentous for French mathematics in other ways: Fourier died that year, Galois in 1832, Legendre in 1833. As Belhoste writes: "all foretold the end of an era. Cauchy's absence from Paris was now painfully felt."[7]

6.4 Exercises

1. Grabiner has argued that many of the techniques of $\epsilon - \delta - N$ analysis can be found in the work of Lagrange and others of Cauchy's predecessors. In the light of this information, what exactly was the originality of Cauchy's contributions to the foundations of analysis in the *Cours* and the *Resumé*?
2. What do you understand by Lützen's idea of an architecture of mathematical analysis?
3. Write a review of Cauchy's mémoire (1825) on integration in the complex domain, indicating what he has achieved, what is satisfactory and what is unsatisfactory in it, and what could be done next, from the standpoint of one of his contemporaries.

---

[7]See Belhoste (1991, 157). Belhoste documents a few occasions when Cauchy returned to Paris, and it does not seem that Cauchy's departure for Italy was intended to be an absolute break.

Chapter 7
Abel

Niels Henrik Abel (1802–1829)

7.1 Introduction

One of the great excitements in the mathematical world of the 1830s and 1840s was the discovery by Abel and Jacobi independently of elliptic functions. These are complex-valued functions of a complex variable, and as such they were new and proved an important stimulus for the growth of a theory of complex functions. Equally importantly, they had properties akin to the trigonometric functions—they were, however, not singly but doubly periodic, as will be explained below—and this made them attractive to study. And, as Legendre had already shown, they were likely to have many applications in mechanics.

In this chapter we look at how their discovery burst onto the French and German communities, and then look at the short life of Abel and indicate some of the key properties of the functions he discovered.

J. Gray, *The Real and the Complex: A History of Analysis in the 19th Century*,
Springer Undergraduate Mathematics Series, DOI 10.1007/978-3-319-23715-2_7

7.2 The Prize of 1830

Niels Henrik Abel of Christiania in Norway and Carl Gustav Jacob Jacobi, who was a Professor at the University of Königsberg, came independently to the subject of Legendre's elliptic integrals in 1827. They very quickly entered into a rivalry, the drama of which was heightened by the rapid stream of their publications. Legendre was evidently delighted with the attention his favourite subject was receiving. He could see that with Abel and Jacobi taking it up, its fame was assured. In the first supplement to his *Traité*, dated 12 August 1828, Legendre wrote that

> Until then geometers had taken almost no part in this kind of research, but scarcely had my work seen the light of day, scarcely could its title have become known to scientists abroad, when I learned with as much astonishment as satisfaction that two young geometers, MM. Jacobi of Königsberg and Abel of Christiania had succeeded in their own individual work in considerably improving the theory of elliptic functions at its highest points.

So in 1829, the year he turned 77, Legendre, by then quite ill, worked hard to ensure that the Grand Prize of the Academy of Sciences in Paris went to the two men who had recently transformed his theory. He made sure that the Paris Academy asked him and Poisson to report on it, and Poisson presented their report on Jacobi's work on 21 December 1829.[1] The report, which was based largely on Jacobi's book, the *Fundamenta nova*, which had been published in April 1829, and it and the excitement surrounding the work of Jacobi and Abel, led to the German being elected a corresponding member of the Academy on the 8th of February 1830. Then on 28 June 1830 the Academy announced that its Grand Prize in mathematics, worth 3,000 francs, would be divided equally between Jacobi and the family of the late Niels Henrik Abel of Christiania, for Abel had died on April 6 1829 at the age of 26.[2]

The theory of elliptic functions, as the new complex functions created by Abel and Jacobi are called, grew to be one of the central domains of 19th century mathematics. Both men had transformed Legendre's subject of elliptic integrals in the same ways, which added to the sense of competition: they had inverted the integrals by treating their upper end points as the independent variable, as will be described below, and let the endpoints be complex. The results of these changes were elegant and profound, and were to be a major on the later development of complex function theory. Indeed, Jacobi was to remark later that the introduction of imaginaries alone was enough to solve all the riddles of the early theory (see Dirichlet (1852, 10)).

Part of the charm of elliptic functions was that they were a rich generalisation of the familiar trigonometric functions; they were interconnected by a rich collection of identities. And it rapidly transpired that they had many uses: in geometry, in number theory, in mechanics, and naturally in astronomy because planets and asteroids traverse ellipses. There was therefore was a working theory of approximations to

[1] *Procès-verbaux des séances de l'Académie* 9 (1921), 373; Poisson *Rapport* (1831).

[2] Among those passed over for the prize on this occasion was Évariste Galois.

elliptic integrals in theoretical astronomy, which indeed was the occasion of Gauss's only publication on the subject (1818).

The idea of inversion was suggested so strongly by the familiar example of the trigonometric functions that one can almost wonder how Legendre missed it—the answer being, of course, that the integral is sufficiently fascinating. Moreover, it arises naturally in the mechanical problems Legendre studied. And no-one would ignore the logarithm multi-function just because its inverse, the exponential function, has simpler mathematical properties. That said, as is well-known, the integral

$$u(x) = \int_0^x \frac{ds}{\sqrt{1-s^2}}$$

reduces, on substituting $s = \sin\tau$ and $t = \arcsin(x)$, to

$$u(x) = \int_0^t d\tau = t = \arcsin(x).$$

In this context, it is natural to invert the answer, and write $x = \sin(u)$. This gives a well-understood single-valued function instead of a multi-function. We will see below why inversion of an elliptic integral suggests that the variables should also become complex. That step was taken by Gauss in work he did not publish, Jacobi, and Abel. Legendre knew very well that his tables for his real-valued elliptic integrals allowed him to infer the values of their inverse functions, but never studied the inverse functions directly. In Jacobi's opinion that[3]

> The fear that the mathematician most chiefly concerned with the determination of numerical values had of the imaginary, was the reason that Legendre was prevented from taking the most important step in modern analysis, the introduction of doubly periodic functions.

Abel had also picked up an interesting clue. In 1801 Gauss had published his *Disquisitiones Arithmeticae*, the book that transformed number theory and helped make his name. In its seventh chapter he wrote about the algebra of the problem of dividing a circle into n equal parts—the topic is called cyclotomy—and in §335 he observed of the principles of this theory that "Not only can they be applied to the theory of circular functions, but also to many other transcendental functions, e.g. those which depend on the integral $\int \frac{dt}{\sqrt{1-t^4}}$." Gauss never kept his promise to return to this topic and give a full explanation, but Abel took the remark as a reason to look at that integral in the same way as the integral defining the circular or trigonometric functions.

[3]Quoted in Koenigsberger (1904b, 54) and Krazer (1909, 55n).

7.3 Abel and his Travels

Abel was lucky enough to have an inspiring mathematical teacher in his last year at school, Bernt Michael Holmboe.[4] Holmboe lent him his own copies of the classic works of Euler on the differential and integral calculus, and when Abel was only 16 began to give him private lessons in the works of Lagrange. His school report on Abel said that the youth was "an excellent mathematical genius". When he went to university, Abel's talents were recognised by the small group of scholars intent on raising the level of science in Norway, among the Christopher Hansteen, whose interests were in terrestrial magnetism and who encouraged him to think of a career in mathematics. Among Abel's first ventures into original research was an erroneous 'proof' that a polynomial equation of degree 5 is solvable by radicals. Hansteen knew that this was an outstanding problem, and that neither he nor anyone in Christiania was competent to pronounce upon Abel's argument, so he sent him to Copenhagen and the mathematician there, Ferdinand Degen. Nor could Degen find the flaw, but his sensible advice to test the method on examples soon revealed that there was one.[5]

Abel wrote to Holmboe about the trip, and here he also wrote for the first time about his work on elliptic integrals. Here too had found something unexpected, and this time Degen could not help him. Abel wrote[6]:

> You remember the little paper which treated the inverse functions of the elliptic transcendentals, where I proved something impossible; I asked him to read it from one end to the other; but he could not discover any false conclusion nor understand where the fault might be; God knows how I will get out of it.

The most likely interpretation of the 'mistake' is that it was no mistake at all, but a novel property.[7] If we briefly call the new function $f(x)$ then, by analogy with the trigonometric functions, there ought to be an equation connecting $f(x)$ and $f(x/n)$, and there is. However, as an equation for $f(x/n)$ for a given $f(x)$ it is of degree n^2, and not n, as the analogy would lead one to suspect.[8] Whatever the explanation, the letter is explicit that Abel had already taken one crucial step that Legendre had not, and inverted the integrals.

It was obvious to all who knew him that Abel's tremendous talent for mathematics could not be adequately supported in his native Norway, and with some effort funds were found for him to study abroad. He left with three friends on September 7th 1825 to go to Berlin, partly because Abel's poor German was better than his French. There Abel met the mathematical entrepreneur and Privy Councillor Leopold Crelle.

---

[4]On Abel's short life see Stubhaug (2000). The earlier Ore (1957) is more reliable on the mathematics, which is described in greater detail and depth in Houzel (2004).

[5]In due course Abel's proof that the quintic equation is not solvable by radicals was one of the great breakthroughs opening the door to Galois's work.

[6]Quoted in Ore (1957, 65).

[7]See e.g. Krazer (1909, 56).

[8]The actual number is n^2 if n is odd and $2n^2$ if n is even. A sign enters the expression for the general solution that must be positive when n is odd but may be positive or negative when n is even.

As a young man, Crelle had hoped to be a great German mathematician and build up German mathematics, but he realised that he lacked the ability and turned to other careers. He became an engineer, and constructed the railway from Berlin to nearby Potsdam, but he retained his love for mathematics and his belief that the subject should not be shackled to applications, which in his view had stunted its growth in Germany hitherto. In 1825 he was busy creating a new journal, which would be one of the very few anywhere to be exclusively devoted to mathematics, and his meeting with Abel could not have been more fortunate for each man. Abel gained access to Crelle's extensive library, and he found in Crelle a supporter who would translate his many papers—no less than seven were to appear in the first volume of Crelle's new journal alone. Crelle gained the allegiance of a mathematician who would immediately place his new journal, the *Journal für die reine und angewandte Mathematik* (or, *Journal for pure and applied Mathematics*), at the forefront of the subject.[9]

Abel wrote on a variety of subjects, but often with an eye to the algebraic implications. Even before leaving Norway he had recovered from his early mistake, and had a paper ready showing that the general polynomial equation of degree five is not solvable by radicals.[10] In Germany he routinely scrutinised the polynomial equations of higher degree that occurred in his work to see if they might be solvable by radicals. As noted in Sect. 4.3 above he became aware of Cauchy's work on real analysis, and wrote his own paper (Abel 1826) on the convergence of the binomial series, where commented on the lack of continuity of certain Fourier series.

He did not go to visit Gauss in Göttingen, perhaps because Crelle advised against it. Crelle's opinion of Gauss's genius was, of course, high, but he tempered it with a realistic appraisal of the man's style, saying that everything Gauss man wrote was "an abomination—gruel—so obscure that it is almost impossible to understand anything" and Hansteen added that Gauss "is like the fox who covers his tracks in the snow with this tail".[11] Instead, Abel went to Paris, where he arrived on July 10, 1826. This should have been the highlight of his travels, his visit to the centre of the mathematical world, but it was not. He found the French mathematicians rather unapproachable, and almost comically his best contact turned out to be the young Lejeune Dirichlet, who called on Abel under the mistaken impression that he was a fellow German. When Abel met Cauchy he found him mad and impossible to deal with; Cauchy generally had a habit of talking only about his own work. In a letter to Holmboe (24 Oct) he remarked that "He is extremely Catholic and bigoted, which is strange for a mathematician.".[12] Perhaps the biggest set back for Abel was his

---

[9]As commentators noted, despite its title it concentrated on pure mathematics, to the point that wits called it the *Journal für die reine unangewandte Mathematik*, the *Journal for purely unapplied mathematics*. It became, and remains, one of the leading journals for mathematics in the world.

[10]He seems not to have known at this stage of the comparable work of Ruffini from 1799.

[11]Both remarks are in Bjerknes (1885, 92).

[12]See Abel (1902, 45–46).

realisation that Cauchy was the only person actively engaged in pure mathematics, everyone else having been swept up in an enthusiasm for magnetism and other topics in physics. He did meet Legendre, who was then 74, but his dismissive comment in a letter was that he found him "as old as stone".

7.4 Abel on Elliptic Functions

In December 1826 Abel took up the hint he had found in Gauss's *Disquisitiones arithmeticae*. As we saw above, the arc length of the lemniscate is given by the simplest of elliptic integrals:

$$s = \int_0^x \frac{dt}{\sqrt{(1-t^4)}}. \tag{7.1}$$

He was able to show, as he wrote to Crelle, that indeed the lemniscatic arc can be divided by ruler and compass into n equal arcs for precisely the same values of n as the circular arc, exactly as Gauss had hinted, and to Holmboe he said that he now saw as clear as day how Gauss had come to discover his results.[13] But Abel was not happy in Paris, and in January 1827 he was back in Berlin.

1827 was the year his great work on elliptic functions began, and his first major paper on elliptic functions, the 'Recherches sur les Fonctions elliptiques', appeared in Crelle's *Journal*. It is the first published account of the theory of elliptic functions. They are presented, albeit formally, as complex functions of a complex variable, but also in close analogy with the trigonometric functions. He introduced three functions (the analogues of the sine and cosine functions), and derived addition and division laws for them. He ended the paper with expansions of the new functions in infinite series and as infinite products.

Abel wrote the general elliptic integral of the first kind not quite as Legendre had done but in the slightly different form:

$$\alpha = \int_0^x \frac{dt}{\sqrt{(1-c^2t^2)(1+e^2t^2)}} \tag{7.2}$$

Consequently, the inverse function $x = \phi(\alpha)$ therefore satisfies

$$\frac{d\phi}{d\alpha} = \sqrt{(1-c^2x^2)(1+e^2x^2)}.$$

[13]See Abel (1881, 2, 261). The circumference of the lemniscate is divisible into m equal parts *by ruler and compass alone* (italics Abel's) if m is of the form 2^n or a prime of the form $2^n + 1$, or if m is a product of numbers of these two kinds. See Houzel (2004).

Abel then defined the functions

$$f(\alpha) := \sqrt{1 - c^2\phi^2(\alpha)}, \quad \text{and} \quad F(\alpha) := \sqrt{1 + e^2\phi^2(\alpha)}. \tag{7.3}$$

He noted that each of the equations $\phi(\alpha) = 0$, $f(\alpha) = 0$, $F(\alpha) = 0$ have infinitely many roots which can be determined, and for any integer m, $\phi(m\alpha)$ can easily be found in terms of $\phi(\alpha)$, etc.[14]

The division problem is, however, harder, and he showed that the equation for $\phi(\alpha)$, $f(\alpha)$, or $F(\alpha)$ in terms of $\phi(n\alpha)$ is of degree n^2. The analogy with the trigonometric functions is evident, but the degree of the equation may be unexpected, and the memoir is focussed on understanding this equation.

Abel defined

$$\frac{\omega}{2} = \int_0^{1/c} \frac{dt}{\sqrt{(1 - c^2t^2)(1 + e^2t^2)}} \tag{7.4}$$

The function ϕ is positive in the range $0 < \alpha < \frac{\omega}{2}$, $\phi(0) = 0$ and $\phi(\frac{\omega}{2}) = \frac{1}{c}$. Moreover, ϕ is an odd function, $\phi(-\alpha) = -\phi(\alpha)$. To explain the degree being n^2, Abel introduced complex variables. He replaced α formally by $i\beta$, so $ix = \phi(i\beta)$, and noted that $\beta = \int_0^x \frac{dt}{\sqrt{(1+c^2t^2)(1-e^2t^2)}}$ is real and positive between $x = 0$ and $x = 1/e$. He inverted this integral and defined

$$\frac{\tilde{\omega}}{2} = \int_0^{1/ie} \frac{dt}{\sqrt{(1 + c^2t^2)(1 - e^2t^2)}}, \tag{7.5}$$

where x is positive in the range $0 < \beta < \frac{\tilde{\omega}}{2}$. Therefore $\phi(i\alpha) = i\phi(\alpha)$, $F(i\alpha) = f(\alpha)$, and $\phi(\frac{i\varpi}{2}) = \frac{i}{e}$.

The functions ϕ, f, F are now known along the real and imaginary axes. To find their general values Abel established these addition formulae:

$$\phi(\alpha + \beta) = \frac{\phi(\alpha)f(\beta)F(\beta) + \phi(\beta)f(\alpha)F(\alpha)}{R},$$

$$f(\alpha + \beta) = \frac{f(\alpha)f(\beta) - c^2\phi(\alpha)\phi(\beta)F(\alpha)F(\beta)}{R},$$

$$F(\alpha + \beta) = \frac{F(\alpha)F(\beta) + e^2\phi(\alpha)\phi(\beta)f(\alpha)f(\beta)}{R},$$

where $R = 1 + c^2e^2\phi^2(\alpha)\phi^2(\beta)$.

---

[14] Compare the equation for $\sin(nx)$ given $\sin(x)$.

He argued that if r denote the right-hand side of the first equation then the expression for $\frac{dr}{d\alpha}$ is symmetric in α and β, and so $\frac{dr}{d\alpha} = \frac{dr}{d\beta}$, from which it follows that r is a function of $\alpha + \beta$ that can be found by setting $\beta = 0$. This implies that $r(\alpha + \beta)$ = $\phi(\alpha + \beta)$.[15]

Abel now deduced a slew of other formulae then follow from these addition formulae, formulae for $\phi(\alpha + \beta) + \phi(\alpha - \beta)$, for $f(\alpha + \beta) + f(\alpha - \beta)$ for $\phi(\alpha \pm \frac{\tilde{\omega}}{2})$, and so forth. From

$$\phi(\alpha + \tfrac{\tilde{\omega}}{2}) = \phi(-\alpha + \tfrac{\tilde{\omega}}{2}) \text{ and } \phi(\alpha + \tfrac{i\tilde{\omega}}{2}) = \phi(-\alpha + \tfrac{i\tilde{\omega}}{2})$$

he deduced that

$$\phi(\frac{\omega}{2} + \frac{i\tilde{\omega}}{2}) = f(\frac{\omega}{2} + \frac{i\tilde{\omega}}{2}) = F(\frac{\omega}{2} + \frac{i\tilde{\omega}}{2}) = \frac{1}{0}. \tag{7.6}$$

which in due course became the recognition that these functions have poles at $\frac{\omega}{2} + \frac{i\tilde{\omega}}{2}$. The formulae at once imply the functions are periodic:

$$\phi(\alpha + \omega) = -\phi(\alpha), \phi(\alpha + 2\omega) = \phi(\alpha), \tag{7.7}$$

$$\phi(\alpha + i\tilde{\omega}) = -\phi(\alpha), \phi(\alpha + 2i\tilde{\omega}) = \phi(\alpha), \tag{7.8}$$

(the formulae for f and F are similar), and Abel observed (§5)

> The formulae that have been established make clear that one will have all the values of the functions $\phi(\alpha)$, $f(\alpha)$, $F(\alpha)$ for all real or imaginary values of the variable if one knows them for the real values of this quantity lying between $\frac{\omega}{2}$ and $\frac{-\omega}{2}$ and for the imaginary values of the form $i\beta$, where β lies between $\frac{\tilde{\omega}}{2}$ and $-\frac{\tilde{\omega}}{2}$.

The quantities 2ω and $2i\tilde{\omega}$ are periods for the new functions, which are therefore said to be doubly periodic. The zeros of $\phi(\alpha + i\beta)$, for example, can be found by looking in the rectangle $\frac{-\omega}{2} < \alpha < \frac{\omega}{2}$, $\frac{-\tilde{\omega}}{2} < \beta < \frac{\tilde{\omega}}{2}$ and appealing to the double periodicity. By the addition formula, the equation $\phi(\alpha + i\beta) = 0$ reduces to two equations on separating out real and imaginary parts:

$$\phi(\alpha) f(i\beta) F(i\beta) = 0\,, \quad \phi(i\alpha) f(\beta) F(\beta) = 0.$$

These equations in turn yield $\alpha = m\omega$, $\beta = n\tilde{\omega}$. The similar equations for f and F imply $f(i\beta)F(i\beta) = 0$ $\alpha = (m + \frac{1}{2})\omega$, $b = (n + \frac{1}{2})\tilde{\omega}$, and yield the values for which ϕ is infinite. So the zeros of $\phi(x)$ are $x = m\omega + ni\tilde{\omega}i$. Abel likewise solved the equation $\phi(x) = \frac{1}{0}$ and found $x = (m + \frac{1}{2}\omega) + (n + \frac{1}{2}i\tilde{\omega})$, and the equation $\phi(x) = \phi(a)$ in terms of a and found $x = (-1)^{m+n}a + m\omega + ni\tilde{\omega}$.

Thus far, the analogy with the trigonometric functions had guided Abel's work rather smoothly. He was also able quite easily to express $\phi(m\alpha)$ as a rational function of $\phi(\alpha)$, $f(\alpha)$, and $F(\alpha)$. He set $\phi(n\alpha) = \frac{P_n}{Q_n}$ and found recurrence relations for P_n and Q_n, and dealt similarly with $f(n\alpha)$, and $F(n\alpha)$. However, the converse question—express, say, $\phi(\alpha/m)$ as a function of $\phi(\alpha)$, $f(\alpha)$, and $F(\alpha)$—proved to be harder;

[15] Abel also observed that these results followed from those in Legendre's *Exercises*.

what Abel did here was important in the later work on the theory of equations, and influenced the future development of Galois theory (see Houzel 2004).

As is well known, the equation for $\sin(u/3)$ given $\sin u$ is the cubic equation

$$4x^3 - 3x + \sin u = 0,$$

because $\sin u = 3\sin(u/3) - 4\sin^3(u/3)$, and clearly the angles $(u/3)$, $(u+2\pi)/3$ and $(u+4\pi)/3$ are such that their sines satisfy this equation. However, in the elliptic case, the division equation that gives $\phi(u/3)$ in terms of $\phi(u)$ is of degree 9. This is because some of the roots in the elliptic case are complex, whereas as in the trigonometric case they are real. Once the elliptic function is treated as a function of a complex variable, which displays its double periodicity, the values of u that satisfy the equation for $\phi(u/3)$ will all clearly be of the form $u_0 + (m\omega + \tilde{m}\tilde{\omega})/3$, and so there are 9 of them. This makes it clear why the number of solutions is unexpectedly large.

Abel's ideas about complex variables were entirely formal, but that said he presented a fully developed theory of elliptic functions as complex functions, analogous to the trigonometric functions. We shall now see that Jacobi's way in to the subject was significantly different, and closer to Legendre's.

Chapter 8
Jacobi

Carl Gustav Jacob Jacobi (1804–1851)

8.1 Introduction

In this chapter we look at the life of Jacobi, and indicate some of the key properties of elliptic functions that he discovered by taking a different route to Abel. Then we look at how he published his work, the correspondence with Legendre, and his major book the *Fundamenta nova* of 1829. As we shall see, Jacobi also found important and entirely new applications of elliptic function theory to number theory and to geometry.

Carl Gustav Jacob Jacobi was born into a banker's family in Potsdam in 1804. He went to the University of Berlin at the age of 16, by which time he had already

J. Gray, *The Real and the Complex: A History of Analysis in the 19th Century*,
Springer Undergraduate Mathematics Series, DOI 10.1007/978-3-319-23715-2_8

been reading the works of Euler, Lagrange, and Laplace, and in 1825, at the age of 20, he submitted his Ph.D. and received permission to begin his Habilitation, the then-necessary and sufficient qualification for teaching in any German university (it was not a sufficient condition for being paid). In May 1826 Jacobi transferred to the University at Königsberg, where he became an associate professor at the end of 1827. The University had a small but significant science faculty: Franz Neumann and Heinrich Dove taught physics, Gauss's friend the astronomer Bessel was also there, and these high-powered colleagues sparked a lifelong interest in applied mathematics. Ironically, the topic that made his name internationally and which interests us most, elliptic functions, was the only one where success did not come easily to him. Dirichlet, his close friend in later life, told this story in his obituary of Jacobi:

> One of his friends who noticed him in a bad mood one day, received this answer when he asked why he was out of sorts: You see me on the point of returning this book (Legendre's *Exercises*) to the library, with which I've been exceedingly unlucky. On other occasions when I have studied an important work this has stimulated me to some thoughts of my own and there has always been something in it for me. This time I have come away quite empty handed and have not been inspired in the least (Jacobi, *Werke*, I, 7).

Later, but richer, was Dirichlet's comment.

8.2 Jacobi and the Transformations of Elliptic Integrals

Jacobi took a different route from Abel's into the subject: the transformation of one elliptic integral into another by rational changes of the variable.[1] In early 1827 he had not seen Legendre's *Traité* and he did not know that Legendre had found a transformation of order 3, but by June of that year he had found become the first to discover the existence of transformations of every degree. His two notes on his discoveries appeared in the September issue of the *Astronomische Nachrichten*, edited by another of Gauss's friends, Heinrich Christian Schumacher.

In the first note Jacobi considered elliptic integral

$$\int_0^{\phi} \frac{dt}{\sqrt{(1-c^2\sin^2 t)}} = m\int_0^{\psi} \frac{dt}{\sqrt{(1-k^2\sin^2 t)}}$$

as a function of ϕ, and sought to transform it into another involving a new variable ψ. The transformation will produce values of the new modulus, k, and the number m. Jacobi claimed there are nth order transformations for every integer n that can be found by writing $\sin\phi$ as a rational function of $\sin\psi$: say, $\sin\phi = \frac{U}{V}$, where U is a polynomial in odd powers of $\sin\psi$ up to the nth, and V is a polynomial in even powers of $\sin\psi$ up to the $(n-1)$th. However, at this stage he could only prove his claim when $n = 3$ or 5; the rest of his claim was strictly speaking only a conjecture.

[1]See Cogliati (2014).

He also pointed out that the composition of two transformations yields solutions to the multiplication problem of order n relating two elliptic integrals

$$\int_0^{\phi} \frac{dt}{\sqrt{(1 - c^2 \sin^2 t)}} = n \int_0^{\psi} \frac{dt}{\sqrt{(1 - c^2 \sin^2 t)}}.$$

In his second note Jacobi he showed how his new transformations, like Legendre's, could assist in calculating tables of values for the integrals.

There has always been a question about how Jacobi came to the idea of inverting the elliptic integrals. He took this step in the December edition of Schumacher's journal, when he proved his general transformation formula.[2] He argued that the effect of the substitution $y = \frac{U}{V}$ is to transform

$$\frac{dy}{\sqrt{((1 - a_0 y)(1 - a_1 y)(1 - a_2 y)(1 - a_3 y))}} = \frac{dy}{\sqrt{(f_4(y))}}$$

into

$$\frac{dy}{M\sqrt{((1 - b_0 y)(1 - b_1 y)(1 - b_2 y)(1 - b_3 y))}} = \frac{dy}{\sqrt{(g_4(y))}}$$

where, if

$$T^2 = V^4 \frac{f_4(\frac{U}{V})}{g_4(x)},$$

then $\frac{T}{M} = V\frac{dU}{dx} - U\frac{dV}{dx}$ and M is to be a constant.

To carry out the transformation of

$$\frac{dy}{\sqrt{((1 - y^2)(1 - \lambda^2 y^2))}}$$

into

$$\frac{dx}{M\sqrt{((1 - x^2)(1 - k^2 x^2))}},$$

he denoted the integral $\int_0^{\phi} \frac{dt}{\sqrt{(1-k^2 sin^2 t)}}$ by $F(\phi)$ and defined

$$K := \int_0^{\pi/2} \frac{dt}{\sqrt{(1 - k^2 \sin^2 t)}}.$$

[2]See Jacobi (1827).

He now remarked that if $F(\phi) + F(\psi) = F(\sigma)$ then σ is the expression given by Legendre (see above, (3.6)) and writing $F(\phi) = \xi$, Jacobi said

$$F(\phi) = \int_0^x \frac{dt}{\sqrt{((1-t^2)(1-k^2t^2))}}$$

and, inverting the elliptic integral,

$$x = \sin(am(\xi)).$$

He now reinterpreted formulae such as the one for σ in terms of ϕ and ψ so that they apply to $x = \sin(am(\xi))$, and found formulae first for V in terms of x, then for U in terms of x, and finally for M. The coefficients of U and V and the expression for M are full of terms like $\sin am \frac{2K}{2n+1}$ and $\sin am \frac{2nK}{2n+1}$, so it is quite evident that Jacobi could never have proved his general claim with inverting the elliptic integral. When, then, did he discover it?

The Legendre–Jacobi Correspondence

For this we can turn to the famous correspondence between Legendre and Jacobi, which began when Jacobi wrote to Legendre on 5th August 1827 to inform him of the discoveries that Schumacher was shortly to publish.[3] He described his work on the transformation of elliptic integrals, mentioned in passing that Gauss had discovered a transformation of order 7, and ended with some remarks about number theory that were to prove significant later.

Legendre replied on 30 November 1827. He observed that he had already found the transformation of order 3, and welcomed the one of order 5, but confessed that he could not find polynomials U and V for the 7th order transformation and doubted that they existed. He therefore urged Jacobi to find and publish a proof. But unwittingly Jacobi's favourable reference to Gauss upset the older man. Legendre launched a string of complaints: Gauss had claimed the law of quadratic reciprocity as his when he, Legendre, had published it in 1785 (in fact, Legendre's statement was correct but Gauss's proof was the first correct proof). Gauss had tried to claim priority over Legendre for the method of least squares (again, this was correct but Gauss had not published his method, only put it to use). Now, apparently, Gauss was claiming to have discovered one of Jacobi's ideas in 1808. "Such excessive impudence", Legendre wrote, "was unbelievable in a man of Gauss's abilities", and Jacobi should not worry about it.

Legendre then took the step of communicating Jacobi's results to the Académie des Sciences in the warmest terms on 5 November 1827. His report was published in the *Globe* on the 29th of November, which delighted Jacobi. He next wrote to

[3] See Legendre and Jacobi (1875).

Legendre on 12 January to inform him of the content of Abel's 'Recherches', the first part of which, with the inversion of the elliptic integrals, was now in print, and drew attention to the striking division of the lemniscate. He then informed Legendre that that the next edition of Schumacher's journal would carry his proof of the existence of the general transformation. Legendre replied on 9 February 1828 to say that he had already seen Abel's paper, although he was happy to have had it analysed in Jacobi's language, which was closer to his own, and he went on:

> It is a great satisfaction to me to see two young mathematicians, like you and him, cultivate successfully a branch of analysis which has for so long been my favourite object of study and which has not attracted the attention in my country that it deserves.

In his reply to this letter (12 April 1828) Jacobi described his route to his original discovery, making it clear that he had had the idea of inverting the integrals before he wrote his second letter to Schumacher, in August 1827.

Does this establish that Jacobi came to the idea of inverting the elliptic integral a month before he could have seen Abel's paper? Krazer was willing to accept Jacobi's testimony, but Carl Anton Bjerknes was the first to raise the idea that Abel's work might have provided the necessary inspiration for Jacobi, which he later covered up. This in turn might be a Norwegian reaction to the way, at the time, that history seemed to have favoured Jacobi's approach over Abel's, a matter naturally influenced by the fact that Abel had died so young. But it was not the view of Sophus Lie, who was working on the second edition of Abel's collected works at the time, and was the cause of some friction between him and Bjerknes. The difficulty is that Jacobi himself admitted that his published proofs were different from his discovery method, and based his first study of the transformations on the inverted integrals. Moreover, Ore has argued convincingly (1957, 180–190) that Jacobi could well have seen Abel's 'Recherches' a little earlier when, under pressure from Legendre, to provide a proof, and with his application for a Professorship at Königsberg in the air, he suddenly found that a rival had transformed the whole subject. But even if the conclusion were to be that human weakness guided Jacobi's hand, and the idea of inversion "in all probability was revealed to him through reading Abel's 'Recherches'" (Ore 1957, 184), what he did with the idea was still to be remarkable.

8.3 Abel and Jacobi as Rivals

Abel was certainly alarmed when he first heard of Jacobi. As Hansteen described to Schumacher in a letter that Schumacher quoted to Gauss (quoted in Ore 1957, 189):

> Abel sends herewith an article about elliptic transcendents, which he asked to have printed as soon as possible, since Jacobi is on his heels. The other day, when I handed him the last number of the *Astronomische Nachrichten*, he became quite pale and was compelled to run to the confectioner's shop and take a schnapps of bitter to counteract his alteration. For several years he has been in possession of a general method which he communicates in this paper, and which includes more than Jacobi's theorems.

His countermove was to send to Schumacher's journal an article on the transformation of elliptic functions but one written from a vastly more novel standpoint: "My Knockout of Jacobi" he called it in a letter to Holmboe. Off stage one participant was not impressed: Gauss replied to Schumacher (23 May 1828), who told Crelle, who in turn told Abel, that "he (Abel) has come about one third of the way that I have gone in my researches, with the same aim and even with the same choice of notation." True to form, Gauss did nothing to encourage either man.

Jacobi's response to the 'knockout' was generous. He wrote to Legendre on 9 September 1828 to say that. "It is as far above my praise as it is above my own work", and that he valued it particularly for supplying the rigorous proofs his own work lacked. But he too switched journals to acknowledge the presence of the other man, and pointed out in a brief note in Crelle's *Journal* 3, 1828, that he knew there were $p+1$ transformations of order a prime p and gave them explicitly as the result of substituting q^p or any of the p distinct values of $q^{1/p}$ into

$$\sqrt{k} = \frac{2q^{1/4} + 2q^{9/4} + 2q^{25/4} + \cdots}{1 + 2q + 2q^4 + 2q^9 + \cdots}$$

The Jacobi–Legendre correspondence also shows that Jacobi immediately began to use Abel's ideas in his own study of transformations. He even admitted to Legendre (12 April 1828) that he could not do without Abel's analysis. It is in his letters that several of the striking infinite series that Jacobi was so adept at handling appear for the first time, and which will entertain us below. In Jacobi's opinion the most remarkable of these was

$$\sqrt{\left(\frac{2K}{\pi}\right)} = 1 + 2q + 2q^4 + 2q^9 + 2q^{16} + \cdots$$

where K is a complete integral and $q = \exp\left(\frac{-\pi K'}{K}\right)$, but he also found this striking identity concerning eighth powers:

$$\left((1-q)(1-q^3)(1-q^5)\ldots\right)^8 + 16q\left((1+q^2)(1+q^4)(1+q^6))\ldots\right)^8$$
$$= \left((1+q)(1+q^3)(1+q^5)\ldots\right)^8.$$

Jacobi's first proofs of these identities depended on the expression for sin *amu* as a quotient of infinite products, which at that time had only been proved by Abel, and Legendre expressed himself pained by this admission:

> which testifies to your candour, a quality that accords so well with real talent (…but) having done justice to the beautiful work of M. Abel I place it far below your own discoveries and would like the glory of them, that is to say of their proofs, to belong entirely to you.

Jacobi later found independent proofs of the crucial theorems, as he told Legendre on 16 June 1828.

By now Crelle was helping to get Abel a job somewhere in Europe, and Jacobi wrote to Crelle full of praise for Abel's work. It was known that Abel had contracted tuberculosis, and in September 1828 Legendre, Poisson, Lacroix, and Baron Maurice (a Swiss mathematician and member of the Paris Academy of Sciences) wrote from the Institut de France to the Swedish King Karl Johan urging him to find a place for Abel in Stockholm. In October Legendre finally wrote directly to Abel, who replied (25 November 1828) that

> Jacobi will certainly perfect to an undreamed-of degree the theory of elliptic functions but even mathematics in general. No-one can esteem him more highly than I do.

Abel was by then back in Norway, where his health collapsed towards the end of February, and he died on the 6th of April, 1829, at the age of only 26.

It was inevitable that Abel's reputation would decline for a while after his death as Jacobi continued to develop the subject of elliptic functions his way. The paper that eventually restored Abel's reputation was a short note in Crelle's *Journal* in 1828, in which he discussed integrals of the form $\int \frac{rdx}{\sqrt{R}}$, where $r(x)$ is an arbitrary rational function and $R(x)$ is an arbitrary polynomial. This was but a fragment of a much longer paper he had submitted to the Académie des Sciences in Paris in 1826 dealing magnificently with the general case of arbitrary integrands, but at the time of his death the paper was sitting as good as forgotten in a pile of papers in the care of Cauchy.

8.4 Jacobi's *Fundamenta Nova*

In 1829 Jacobi published the book that was to become the definitive account of the theory of elliptic functions for at least a generation, his *Fundamenta nova theoriae functionum ellipticarum* (or, *New foundations of a theory of elliptic functions*). It opens with the transformation problem for elliptic integrals. He wrote: "The problem we shall propose is generally this: to find a rational function y of x such that"

$$\frac{dy}{\sqrt{A + By + Cy^2 + Dy^3 + Ey^4}} = \frac{dx}{\sqrt{A + Bx + Cx^2 + Dx^3 + Ex^4}}.$$

He then went on to the related multiplication problem, which is to find solutions to

$$\frac{dy}{\sqrt{A + By + Cy^2 + Dy^3 + Ey^4}} = \frac{Mdx}{\sqrt{A + Bx + Cx^2 + Dx^3 + Ex^4}}$$

where the quantity M, which may depend on the parameters $A, B, \ldots, E$, does not depend on x.

He gave the example of the transformation of order 3:

$$U = x(a + a'x^2), \quad V = 1 + b'x^2.$$

In this case the moduli λ and k are related by this equation, where $u^4 = k$ and $v^4 = \lambda$:

$$u^4 - v^4 + 2uv(1 - u^2v^2) = 0$$

and the multiplier is $M = \frac{v}{v+2u^3}$. He also showed that the transformation of order 5 produces this relation between u and v:

$$u^6 - v^6 + 5u^2v^2(u^2 - v^2) = 0,$$

where $M = \frac{v(1-uv^3)}{(v-u^5)}$. Equations like these between two moduli he called modular equations (§24); they later caught the attention of Galois, because the cases of low degree can be remarkably simplified.

Jacobi sketched an argument to show that if one substitutes $y = \frac{U}{V}$, where U and V are polynomials in x of degrees p and m respectively, and $p \geq m$, then the integrand in y reduces to another of the same form, possibly multiplied by a constant, provided that the coefficients of U and V can be suitably chosen. When this is the case, the transformation is said to be of order p, and Jacobi showed by a counting argument that this can be done for any value of p. More specifically, he showed that there are more coefficients in the polynomials U and V than the number of equations they have to satisfy—a typically 19th Century approach. In the book, however, Jacobi solved these problems by inverting the elliptic integral

$$u = \int_0^{\varphi} \frac{dt}{\sqrt{1 - k^2 \sin^2 t}} = \int_0^{x} \frac{ds}{\sqrt{(1 - s^2)(1 - k^2s^2)}}$$

(where $\sin \varphi = x$) to obtain elliptic functions, solving the problems there, and then transferring them back to the elliptic integral s. Jacobi showed that transformations of every order exist, and investigated the number of each order. He studied the effect of a transformation on the moduli, and showed how the composition of two transformations can be made to yield a solution to the multiplication problem.

To study the elliptic functions, he followed Legendre in calling ϕ an angle and regarding it as the amplitude of the function u. He wrote $\phi = am(u), x = \sin(am(u))$, and called x an elliptic function. He defined the complete elliptic integrals

$$K = \int_0^{\pi/2} \frac{dt}{\sqrt{1 - k^2 \sin^2 t}}$$

and

$$K' = \int_0^{\pi/2} \frac{dt}{\sqrt{1 - k'^2 \sin^2 t}},$$

where k and ℓ are complementary moduli, that is, $k^2 + k'^2 = 1$. Elliptic functions with a purely imaginary argument were defined by introducing a new variable defined by the transformation $\sin\phi = i.\tan\psi$, $\cos\phi = \sec\psi$. He then used an addition theorem to define elliptic functions for arbitrary complex variables, and finally deduced that the new functions were doubly periodic, with periods $4K$, which is real, and $4iK'$, which is purely imaginary, at least when the modulus k is real. Both are imaginary when the modulus itself is imaginary.

Jacobi showed, as Legendre had earlier, that the complete elliptic integrals satisfy the differential equation

$$k(1-k^2)\frac{d^2Q}{dk^2} + (1-3k^2)\frac{dQ}{dk} - kQ = 0$$

and he used this differential equation to the study properties of the multiplier M.

Jacobi devoted the second half of the book to the study of the new elliptic functions. He showed how to write them as quotients of one infinite product by another, and how investigated how the complete elliptic integrals depend on the modulus. This led him to such striking formulae as (§36, nr 7):

$$k = 4\sqrt{q}\left(\frac{(1+q^2)(1+q^4)(1+q^6)\ldots}{(1+q)(1+q^3)(1+q^5)}\right)^4$$

where $q = \exp(\frac{-\pi K'}{K})$.

In §52 Jacobi introduced a quantity that was to play a more important, foundational role in later versions of the theory: the theta function, here defined by the formula (p. 198)

$$\theta(u) = \theta(0)\exp\left(\int_0^u Z(t)dt\right)$$

where Z was defined in terms of the complete and incomplete elliptic integrals of the first and second kinds by the formula

$$Z(u) = \frac{F^1E(\phi) - E^1F(\phi)}{F^1}.$$

In what follows we stick to his later definition, as he did, which is related to this one but so obscurely that I omit it.[4]

[4]See Whittaker and Watson (1927, 479).

He investigated his theta function as a function of a complex variable and showed that it was periodic but not doubly periodic. It satisfies

$$\theta(u+2K)=\theta(u), \text{ and } \theta(u+2iK')=-\exp\left(\frac{\pi(K'-iu)}{K}\right)\cdot\theta(u).$$

Jacobi also noted that the theta function satisfies the heat equation.

The book is full of power series in terms of the variable q and identities involving the theta function and the elliptic functions. These included some that he had already shown to Legendre, such as

$$\sqrt{\left(\frac{2K}{\pi}\right)}=1+2q+2q^4+2q^9+2q^{16}+\cdots$$

where K is a complete integral, and the striking identity concerning eighth powers:

$$\left((1-q)(1-q^3)(1-q^5)\ldots\right)^8+16\left(q(1+q^2)(1+q^4)(1+q^6)\ldots\right)^8$$
$$=\left((1+q)(1+q^3)(1+q^5)\ldots\right)^8.$$

He also found series for

$$\sin am\left(\frac{2Kx}{\pi}\right) \text{ and } \theta(0)\theta\left(\frac{2Kx}{\pi}\right),$$

and came close to writing the function $\sin am$ as a quotient of two theta functions.

8.5 Elliptic Functions and Complex Functions

Can the first theories of elliptic functions be regarded as theories of functions of a complex variable? There is no question that they promoted the development of that theory, as this book will show, but that cannot obscure the fact that the theories were entirely formal. Abel and Jacobi did not discuss what they took a complex variable to be—or even a complex number—and surely thought of a complex variable as an expression of the form $x+iy$, to which the usual rules of arithmetic apply and $i^2=-1$. The difficulty comes with defining a complex function, and so justifying such arguments as this: $\int_0^u \frac{dt}{\sqrt{1-t^4}}=v$ because the substitution $s=it$ implies that $\int_0^{iu} \frac{-ids}{\sqrt{1-s^4}}=v$, and so $u(v)=iu(iv)$. If the addition theorem is proved only for real variables, or perhaps also for purely imaginary ones if one accepts the validity of the substitution just described, is it then proved for all imaginary variables? What, if so, is the corresponding theory of complex integrals?

Abel and Jacobi would very likely have appealed to the common-sense meaning of the formulae that resulted. They might have argued that the fact that the integral is not well defined once the variables have become complex but is an infinitely many-valued expression is somewhat beside the point. They could have pointed to $\log x$, which loses its unique meaning once x can be complex, and argued that $\log x$ always arises in a context that will make clear what it means. Nonetheless, they might have said, just as mathematical common-sense enables one to interpret $\int_1^x \frac{dt}{t}$, and to know that $\frac{d \log x}{dx} = \frac{1}{x}$. so too does simple formal algebra enable you to work with (many-valued) elliptic integrals and complex-valued elliptic functions.

Moreover, the early developments of complex function theory, then almost exclusively in the hands of Cauchy, were confused. Doubly confused, in fact: mathematically confused, in that Cauchy's own ideas were in a turbulent state of development, and confused as regards their dissemination. This may be why, in the first flood of invention, neither Abel nor Jacobi mentioned Cauchy's work, and nor did Legendre suggest that they might consult it. In fact, it is clear that in the first phase of the creation of a theory of elliptic functions there was no theory of complex integrals. Furthermore, the fact that Cauchy's theory of complex integrals made no mention of multi-valued integrands was to drive people, including Jacobi in the 1830s, to seek other foundations for elliptic functions altogether, and to abandon the starting point of elliptic integrals.

Other features of what was to become complex function theory are also lacking in the elliptic function theory of the 1820s. Legendre's classification of elliptic integrals into three kinds, and Abel's later study of arbitrary algebraic integrals, remained purely algebraic, lacking the idea that the integrands either have no poles, or simple poles, or poles of higher orders. Similarly, the elliptic functions become infinite, but there is no understanding of poles. We can conclude that Abel's and Jacobi's theories of elliptic functions were purely formal.

But that does not mean the theories were not novel and exciting; plainly they were. The new functions were an extension of the trigonometric functions and an addition to the short list of functions of a complex variable that were known at the time. Learning to use these functions was a task mathematicians found intrinsically attractive as well as useful in diverse fields. However formal, they were a theory of complex integrals and intrinsically complex valued functions of a complex variable, and they were to suggest to later generations of mathematicians that they way to handle them would be to incorporate them into a general theory of complex functions.

8.6 Two Applications

An Application to Number Theory

Fermat had claimed, without proof, that every positive integer is the sum of at most four squares. In a letter to Goldbach in 1749 Euler suggested that the most natural way to prove Fermat's theorem would be to use power series; an argument Lagrange was able to conclude in 1770.[5] Jacobi, without knowing of Euler's letter, first stated his result about the number of distinct ways a number is a sum of four squares in a letter to Legendre on 9 September 1828. He announced it publicly in Crelle's *Journal* (vol. 3, 1828) and gave his proof on the last page of his *Fundamenta Nova* in 1829.

Jacobi found that the coefficients r_n in one of his power series were essentially the number of ways n could be expressed as a sum of four squares. From his knowledge of the function that this power series represented he deduced this identity for r_n:

$$r_n = 8\sum\nolimits_{d|n,4\nmid d} d$$

In words, r_n is eight times the sum of the divisors of n that are not multiples of 4.[6]

Jacobi's observation to Legendre concerned the identity

$$\begin{aligned}\left(\frac{2K}{\pi}\right)^2 &= 1 + \frac{8q}{1-q} + \frac{16q^2}{1-q^2} + \frac{24q^3}{1-q^3} + \dots \\ &= 1 + \frac{8q}{(1-q)^2} + \frac{8q^2}{(1-q^2)^2} + \frac{8q^3}{(1-q^3)^2} + \dots \\ &= 1 + 8\sum \phi(p)\left(q^p + 3q^{2p} + 3q^{4p} + 3q^{8p} + \dots\right)\end{aligned}$$

where the sum is taken over all odd numbers p and $\phi(p)$ is the sum of the factors of p. The series on the right hand side contains every power of q, and by comparison with the fourth power of the series for $\sqrt{\frac{2K}{\pi}}$ one finds that every number is the sum of four squares. In the *Fundamenta Nova* Jacobi went further to deduce that the number of ways a number is the sum of four squares is 8 times the sum of its divisors that do not divide 4.

To explain Jacobi's results, we need to look a little more closely at how he obtained them.[7] Jacobi worked with a power series (later) denoted

$$\theta_3(0,q) := 1 + \sum_{n=1}^{\infty} 2q^{n^2}.$$

---

[5]See Fuss (1845, vol. 1, 522–524), letter OO 861, also Weil (1984, 179).

[6]Note that on this count d^2 and $(-d)^2$ are different.

[7]See the very helpful exposition by Eric Conrad: Jacobi's Four Square Theorem, at http://www.math.ohio-state.edu/~econrad/Jacobi/sumofsq/sumofsq.html.

If you regard this as a series $1+\sum_j a_j q^j$ you see that the coefficient a_j is the number of ways the integer j can be expressed as a sum of one square (i.e., a single square), where we count n^2 and $(-n)^2$ as distinct. Therefore the coefficient of j in the power series expansion of $\theta_3(0,q)^m = 1+\sum_j a_j q^j$ represents the number of ways j can be written as a sum of m squares, where we count the order of the summands.[8]

If you write down the first few terms of $\theta_3(0,q)^4$ you get $1+8q+24q^2+32q^3+24q^4+48q^5+96q^6+64q^7+24q^8+104q^9+144q^{10}+\cdots$. This already makes it likely that every number can be expressed as a sum of four squares (Lagrange's theorem), but it doesn't prove it—it is possible that the coefficient of some high power of q might vanish—nor does it enable us to write down the number of ways any given number j can be expressed as a sum of squares: we need a formula for the coefficient of q^j. But Jacobi knew that the theta function was (singly) periodic, so he did exactly what Fourier had shown was the natural thing to do: express a periodic function as a Fourier series. He scaled the variable so that the elliptic function went from having period of $4K$ to a period of 2π, by using the variable $\frac{2Kx}{\pi}$, and obtained these Fourier series:

$$\frac{2K}{\pi}\mathrm{sn}\left(\frac{2Kx}{\pi},k\right) = \frac{4}{k}\sum_{j=0}^{\infty}\frac{q^{(2j+1)/2}}{(1-q^{2j+1})}\sin(2j+1)x.$$

$$\frac{2K}{\pi}\mathrm{cn}\left(\frac{2Kx}{\pi},k\right) = \frac{4}{k}\sum_{j=0}^{\infty}\frac{q^{(2j+1)/2}}{(1+q^{2j+1})}\cos(2j+1)x.$$

$$\frac{2K}{\pi}\mathrm{dn}\left(\frac{2Kx}{\pi},k\right) = 1+4\sum_{j=0}^{\infty}\frac{q^j}{(1+q^{2j})}\cos 2jx.$$

Knowledge about elliptic functions therefore provides knowledge about these power series. For example, the series for the function dn, with $z=0$ is

$$\frac{2K}{\pi} = 1+4\sum\frac{q^j}{1+q^{2j}}$$

and the series for its square is

$$\left(\frac{2K}{\pi}\right)^2 = 1+8\sum\frac{q^j}{(1+(-q)^j)^2}.$$

And, splendid to say,

$$\theta_3(0,q) = \sqrt{\left(\frac{2K}{\pi}\right)}.$$

[8]You can check that in this way the number $5 = 1^2+2^2$ can be represented in 8 different ways: switch the order of 1 and 2 , replace 1 with -1, or 2 with -2.

Jacobi expanded the series for $\frac{2K}{\pi}$ as a power series and found that it came out as

$$1 + 4\sum_{j=1}^{\infty}(d_1(j) - d_3(j))q^j,$$

where $d_1(j)$ is the number of divisors of j that are congruent to 1 mod 4 and $d_3(j)$ is the number of divisors of j that are congruent to 3 mod 4. This gave him his Two Squares theorem: The number of ways a positive integer can be a sum of two squares is equal to four times the difference of the numbers of divisors congruent to 1 and 3 modulo 4.

When Jacobi expanded the series for $\left(\frac{2K}{\pi}\right)^2$ he found that it came out as

$$1 + 8\sum_{k|j, 4\nmid k|}^{\infty}\left(\sum k\right)q^j,$$

where the sum is taken over all divisors of j that are not divisible by 4. This gives Jacobi's four squares theorem: The number of ways an integer can be written as the sum of four squares is equal to eight times the sum of all its divisors which are not divisible by 4. Lagrange's theorem follows by observing that 1 divides every integer and 1 is not divisible by 4. So every number has at least 8 representations as a sum of four squares.

An Application to Geometry

Jacobi also boosted the new theory of elliptic functions with an attractive proof of a striking geometrical theorem called the Poncelet closure theorem. Jean Victor Poncelet had given his own difficult and delicate proof in his *Traité des propriétés projectives des figures* (1822); Jacobi's was much more direct.[9]

Take any two conics C and D with C inside D, say, and suppose for simplicity that both of them are circles, centres C_0 and D_0 respectively. An arbitrary point P_0 on D is chosen, and from it a tangent to C is drawn that meets D again at a point P_1. The construction is repeated starting with P_1 to obtain P_2, from P_2 obtain P_3, and so on, and in this way a sequence of points P_n is obtained on D. Poncelet's theorem asserts one of two things happens: either the sequence of points is infinite, or for some n, $P_n = P_0$, and the sequence closes, and therefore there is an n-gon inscribed in D and circumscribing C. This is merely a logical dichotomy; but Poncelet also showed that if there is an interscribed n-gon for some choice of initial point P_0 then there is an an interscribed n-gon for any choice of initial point P_0. In other words, the behaviour of the sequence is determined by the two conics and their relative positions, but not by the initial point.

[9]The theorem has a complicated history that cannot be discussed here, see Bos et al. (1987) and, in greater detail, A. del Centina's article 'Poncelet's porism, a long story of renewed discoveries' to appear in *Archive for history of exact sciences*. The underlying reason for the appearance of elliptic functions is explained in Griffiths and Harris (1978).

The German geometer Jakob Steiner had already investigated the special cases $n = 5$ and $n = 6$, and Jacobi followed his his lead and investigated the relationship between ϕ_n and ϕ_{n-1} where ϕ_n is the angle $P_n C_0$ makes with $C_0 D_0$. He found that

$$\tan\left(\frac{\phi_{n+2} + \phi_n}{2}\right) = \frac{R-a}{R+a} \tan \phi_{n+1}$$

where R is the radius of D and $a = |C_0 D_0|$.

"In this form of the equations" said Jacobi (1828c, 285), "it springs to the eyes at once that they coincide with those for the multiplication of elliptic functions".[10] He set $u = \int_0^\phi \frac{dt}{\sqrt{1-k^2 sin^2 t}}$, where $\phi = am(u)$, and k is an explicit function of R and a, and deduced that $\phi_n = am(u + nt)$. The necessary and sufficient condition for the n-gon to close is that $am(u + nt) = am(u)$, in which case nt is a period, and this is independent of u.

This proof was so unexpected and simple that later writers have referred to it as Jacobi's geometric proof of the multiplication formula for elliptic functions—a complete reversal of his original intention.

[10]For example, those given earlier by Legendre, see Sect. 3.3 above.

Chapter 9
Gauss

Carl Friedrich Gauss(1777–1855)

9.1 Introduction

After Gauss died in 1855 mathematicians who looked at his unpublished papers were astonished to discover how much he had known and never revealed. His sympathy for the work of Bolyai and Lobachevskii, for example, was decisive in awakening the first positive readings of their work on the new, non-Euclidean, geometry. As surprising was the extent of his investigations into the theory of elliptic functions, an effect enhanced by the fact that he had made considerable progress with them even before Abel and Jacobi were born. He began by inverting the simplest elliptic integral, the so-called lemniscatic integral, thereby creating a new family of complex functions, and he rapidly became convinced that it would be necessary to build up a theory of complex functions in general, as we shall see.

J. Gray, *The Real and the Complex: A History of Analysis in the 19th Century*, Springer Undergraduate Mathematics Series, DOI 10.1007/978-3-319-23715-2_9

9.2 Lemniscatic Functions

Gauss worked on the subject of elliptic functions intermittently for 35 years, and because he scarcely published on the topic the historian is in the fortunate position of having an extensive array of working notes and drafts that document how he came to his discoveries. We also have a diary[1] that Gauss kept in his early years as a record of his discoveries. These documents were published after his death in various of the volumes that make up Gauss's *Werke*, and have been frequently and expertly analysed over the years by Schlesinger (1912), Klein (1926–1927), Geppert (1927), and, more recently, Cox (1984). His progress is strikingly logical, and the questions that he could not readily answer became the focuses of particular interest for him and guided his research.

From Gauss's diary we learn that Gauss first considered the lemniscatic integral

$$x = \int_0^z \frac{dt}{\sqrt{(1-t^4)}}$$

in September 1796. He treated the integrand as a function of t, expanded it as a power series and integrated it term by term, thus obtaining x as a power series in z. He then inverted this series to obtain z as a power series in x, thus inverting the lemniscatic integral. In January 1797 he read Euler's posthumous paper (Euler 1786) on elliptic integrals and Stirling's book *Methodus differentialis* of 1730, and learned the remarkable result that if

$$A = \int_0^1 \frac{dx}{\sqrt{(1-x^4)}} \text{ and } B = \int_0^1 \frac{x^2dx}{\sqrt{(1-x^4)}}$$

then $AB = \frac{\pi}{4}$. He recorded in his diary that he now began "to examine thoroughly the lemniscate" (Diary entry nr. 51). He decided to call the function z of x defined above sl for sinus lemniscaticus and introduced the corresponding cosine, $cl(x) = sl(\omega/2 - x)$. Euler's addition formula for the lemniscatic integral could therefore be written as

$$sl^2 + cl^2 + cl^2 sl^2 = 1.$$

From a draft Gauss wrote on or before 1801 we see that he derived formulae for $sl(2x)$, $sl(x_1 + x_2)$, and, for example, for $sl(3x)$, which, setting s for $sl(x)$, is:

$$sl(3x) = \frac{s(3 - 6s^4 - s^8)}{1 + 6s^4 - 3s^8}.$$

[1]Reprinted, with an English translation, a commentary and corrections in Dunnington (2004).

This shows that division of the lemniscatic sine by 3 leads to an equation of degree 9, whereas division of the ordinary sine leads only to a cubic equation.[2] Gauss also wrote down the corresponding expression for $sl(nx)$, and on March 19 1797 he noted in his diary (entry nr. 60): "Why dividing the lemniscate into n parts leads to an equation of degree n^2". The (unstated) reason is that all but n of the roots are complex. This observation prompted Gauss to regard sl and cl as functions of a complex variable, as the draft we are following shows (see Gauss *Werke* 3, 407). The change of variable from x to ix in the lemniscatic integral establishes that $sl(ix) = isl(x)$, and so Gauss wrote down formulae for $sl(x_1 + ix_2)$, from which it follows that $cl(ix) = \frac{1}{cl(x)}$.

Gauss defined $\frac{\varpi}{2}$ to be the value of the complete lemniscatic integral $\int_0^1 \frac{dx}{\sqrt{1-x^4}}$ and observed that $sl(\varpi) = 0$, so from the addition formula it follows that as a real function, sl is a periodic function with period 2ϖ. Therefore the complex function sl had two distinct periods, 2ϖ and $2i\varpi$, and so, when m and n are integers,

$$sl(x + (m + in)2\varpi) = sl(x).$$

This is the first occurrence of the Gaussian integers, $m + in$ (see *Werke* 3, 411–412).

Gauss now knew all the points where sl and cl were zero or infinite, and so he proceeded to investigate $sl(x)$ and $cl(x)$ as quotients of two infinite series. He set

$$sl(x) = \frac{M(x)}{N(x)}$$

and considered the functions M and N as functions of a complex variable in their own right. Much as one might with the trigonometric functions on first acquaintance, he computed special values of them from the known values of $sl(x)$ and $cl(x)$ and formulae such as the addition formula. One result struck his as particularly intriguing. He calculated $N(\varpi)$ to 5 decimal places and its log to 4, and noted that this seemed to be $\pi/2$. He connected this to Euler's remarkable result that $AB = \pi/4$. Because $A = \varpi/2$, $2B$ must be π/ϖ, giving another connection between ϖ and π. Gauss now wrote "Log hyp this number = 1.5708 = $\frac{1}{2}$ of the circle" (*Werke*, X.1, 158). In his diary (March 29, 1797, nr. 63) he checked this coincidence to 6 decimal places, and commented that this "is most remarkable and a proof of this property promises the most serious increase in analysis".

He returned to these calculations for ϖ a year later, in July 1798, when he had a better grasp of $M(x)$ and $N(x)$, and he commented in his Diary (July, nr. 92): "we have found out the most elegant things exceeding all expectations and that by methods which open up to us a whole new field ahead". Not for the first time the diary reflected some understandable optimism. It was to take Gauss a year to enter this new field, and the key was an old love of his, the arithmetico-geometric mean.

---

[2]These and other formulae come from Gauss *Werke* 3, 403–412.

9.3 The Arithmetico-Geometric Mean

The arithmetico-geometric mean is defined for two real numbers a and b as follows: set $a_0 = a$ and $b_0 = b$, and recursively

$$a_{n+1} = \frac{1}{2}(a_n + b_n) \quad \text{and} \quad b_{n+1} = \sqrt{(a_n b_n)}.$$

The two sequences $\{a_n\}$ and $\{b_n\}$ converge to the same value, called the arithmetico-geometric mean (or agm) of a and b, written $M(a, b)$. It is clear that

$$M(\lambda a, \lambda b) = \lambda M(a, b), \text{ so } M(a, b) = aM(1, b/a),$$

so it is enough to study $M(1, x)$ as a function of a single variable.

On May 30 1799 Gauss wrote in his Diary (nr. 98): "We have found that the arithmetico-geometric mean between 1 and 2 is π/ϖ to 11 places, which thing being proved a new field in analysis will certainly be opened up." The proof, however, was elusive, as a letter from Pfaff to Gauss in November 1799 (*Werke*, X.1, 232) makes it clear.

One of Gauss's earliest results about the agm was a power series for $M(1, 1+x)$. He derived it in 1800, by setting $x = 2t + t^2$, so

$$\begin{aligned} M(1, 1+x) &= M(1, 1+2t+t^2) = M(1+t, 1+t+t^2/2) \\ &= (1+t)M\left(1, 1+\frac{t^2}{2(1+t)}\right) \end{aligned} \tag{9.1}$$

so

$$M(1, 1+2t+t^2) = (1+t)M\left(1, 1+\frac{t^2}{2(1+t)}\right).$$

Gauss then wrote $M(1, 1 + x) = 1 + ax + bx^2 + \cdots$ and used the method of undetermined coefficients to obtain two expressions for M in the above equation. By expanding both sides and collecting like terms he obtained the power series expansion of $M(1, 1+x)$, and found that it begins

$$M(1, 1+x) = 1 + \frac{x}{2} - \frac{x^2}{16} + \frac{x^3}{32} - \frac{x^4}{1024} + \cdots. \tag{9.2}$$

This allowed Gauss to find other, related power series, such as the one for $M(1 + x, 1 - x)$, and its reciprocal, which turned out to be most important in all the future developments.[3] It has this strikingly regular expansion:

---

[3]Schlesinger, in Gauss's *Werke*, vol. X.2, 63, suggested that it was the fact that the reciprocal of $M(1, \sqrt{2})$ occurs which led Gauss to consider not the agm in general but its reciprocal.

$$K(x) := \frac{1}{M(1+x, 1-x)} = 1 + \frac{x^2}{4} + \frac{9x^4}{64} + \frac{25x^6}{256} + \cdots ; \tag{9.3}$$

the coefficient of x^{2n} is $\left(\frac{1.3.5...(2n-1)}{2.4.6...2n}\right)^2$. Gauss showed formally that the function $K(x)$ satisfies the linear ordinary differential equation

$$(x^3 - x)K''(x) - (3x^2 - 1)K'(x) + xK(x) = 0 \tag{9.4}$$

and noted that another independent integral of this differential equation is $M^{-1}(1, x)$.

Gauss connected this result with an earlier calculation of his. As soon as he had considered the functions sl and cl as functions of a complex variable, he had written down the formal Fourier series expansion of $\frac{1}{\sqrt{1+\sin^2 V}}$:

$$\frac{1}{\sqrt{1+\sin^2 V}} = a + b\cos 2V + c\cos 4V + d\cos 6V + \cdots \tag{9.5}$$

and observed that

$$a = 1 - \frac{1^2}{2^2} + \frac{1^2.3^2}{2^2.4^2} - \cdots = \frac{2}{\pi}\int_0^1 \frac{dx}{\sqrt{1-x^4}}$$

which is, of course ϖ/π. Now he expanded $\frac{1}{\sqrt{1-y\sin^2 V}}$, and wrote down that the reciprocal of $M(1, \sqrt{1+y})$ is the V—free part (as he called it) of the expansion of $\frac{1}{\sqrt{1-y\sin^2 V}}$. This gave him that

$$M(1, \sqrt{1+y}) \cdot \int_0^1 \frac{dV}{\sqrt{1+y^2\sin^2 V}} = \pi/2.$$

Writing $y = -k^2$, this becomes

$$M(1, \sqrt{(1-k^2)}) \cdot \int_0^1 \frac{dV}{\sqrt{1+k^4\sin^2 V}} = \pi/2.$$

So Gauss had shown that the complete elliptic integral, regarded as a function of the modulus k, was given by the reciprocal of the agm at the complementary modulus, $k' = \sqrt{1-k^2}$.

The significance of the clue given by the value of π/ϖ is now revealed: when $k = i$ the above expression for $M(1, x)$ reduces to

$$M(1,\sqrt{2})\cdot\int_0^{\pi/2}\frac{dV}{\sqrt{1+\sin^2 V}}=\pi/2.$$

With the known value of $\varpi/2$ for the complete lemniscatic integral

$$\int_0^{\pi/2}\frac{dV}{\sqrt{1+\sin^2 V}}=\int_0^1\frac{dx}{\sqrt{1-x^4}},$$

the above equation becomes $M(1,\sqrt{2})=\pi/\varpi$. This gave Gauss the conceptual explanation that he sought for the numerical coincidence that had intrigued him for so long, and it had been obtained by generalising the original question about complete lemniscatic integrals to the setting of complete elliptic integrals. The enigmatic value of $N(\varpi)$ was simultaneously explained from the power series expansion of N, for the value of $\int_0^{\pi}\frac{d\phi}{\sqrt{1-k^2\cos^2\phi}}$, a complete elliptic integral of the first kind, is $\pi N(k)$.

9.4 Elliptic Integrals

By May 6, 1800 Gauss believed that he had led the theory of elliptic integrals "to the summit of universality" (Diary entry nr. 105). However, he soon noted that the theory was "greatly increased and unified", becoming "most beautifully bound together and increased infinitely" (May 22, diary entry nr. 106). These entries, and two more written on or before June 3, show that Gauss had now proceeded to invert the general elliptic integral, subject only to the restriction that the real modulus is k.

Gauss remarked in May 1800 (Diary entry nr. 108) that the new elliptic functions are doubly periodic and have zeros and poles and can be expanded as quotients of power series. His approach was two-fold: in one direction he inverted an elliptic integral depending on a parameter k; in the other he developed doubly periodic functions directly as quotients of entire functions. This raised the question of whether the two approaches define exactly the same objects.

It was easy enough for Gauss to mimic the arguments for the lemniscatic integral in the general case, and conclude that the general elliptic function can be represented as a quotient. It proved much harder to show that every doubly periodic function defined as a quotient arises as the inverse of a suitable elliptic integral. The problem is to find the real modulus k from the periods, and Gauss answered this question affirmatively with a long argument involving the agm that will not be repeated here. We can, however, note, that the formula connecting the complete integral and the agm had already given him an expression for a period in terms of the modulus. Gauss now found an expression for the modulus as a function of the quotient of the periods (called the modular equation) and argued that this function could take any value.

Gauss now proceeded to push beyond the case where the modulus is real and to allow it to be complex. This restriction had arisen because of the way elliptic integral s had arisen in geometrical and mechanical problems, but once the modulus is complex the ratio of the periods can no longer be purely imaginary, and so the agm must itself become a complex-valued function of a complex variable.

But this forces the investigator to decide at every stage which square root $\sqrt{a_n b_n}$ to choose. This raises a very difficult problem indeed, but one Gauss quickly solved. He wrote in his diary on 3 June 1800 (diary entry nr. 109) that connection between the infinitely many means has been completely cleared up, and two days (diary entry nr. 110) that "We have now immediately applied our theory to elliptic transcendents", confirming the interpretation that the way Gauss took to generalising his theory of elliptic functions was to make the agm into a complex function. But we cannot be sure that Gauss had worked out the full solution at this date—the entry may be another of Gauss's optimistic ones—and unfortunately, nothing survives from 1800 to indicate what Gauss's solution was. The best indication we have is from a note from as late as 1825, although Schlesinger and Cox argue that the Diary entry nr. 109 of June 3 1800 enables one to date the discovery to 1800.

9.5 Gauss and Complex Functions

I argued earlier that the work of Abel and Jacobi was purely formal in its handling of complex numbers and complex functions. The case of Gauss is rather different. It is clear from the way he presented complex numbers in his *Disquisitiones Arithmeticae* of 1801 that he was comfortable with their representation as points in the plane. This is already one step beyond the formal, algebraic approach.

Gauss also left elliptic integrals behind when he came to consider elliptic functions with a complex modulus. He did not start with the integrals and invert them, but worked exclusively on generalising the power series side of the real theory to the complex case. This suggests that he was not willing to deal with complex integrals, but that he was willing to deal with complex power series, and indeed with linear differential equations in the complex domain, as other work of his confirms.

Schlesinger suggested (1912, 184) that Gauss's mostly unpublished theory looked like this by 1828:

> The first third was the general theory of functions arising from the (hypergeometric series), the second the theory of the agm and the modular function, and finally the third, which Abel published before Gauss, was the theory of elliptic functions in the strict sense.

The hypergeometric series

$$F(\alpha, \beta, \gamma, x) = 1 + \left(\frac{\alpha.\beta}{1.\gamma}\right) x + \left(\frac{\alpha(\alpha+1)\beta(\beta+1)}{(1.2)\gamma(\gamma+1)}\right) x^2 + \cdots \tag{9.6}$$

(where the variable x is a complex variable of modulus less than 1 so that the series converges) was the subject of a paper Gauss published in 1812. There he showed that the series is intimately connected to the trigonometric functions, and that it arises in the Fourier series expansion of $(a^2 + b^2 - 2ab\cos\theta)^{-n}$, so it is of practical use in astronomy. He also used it to give a quick proof of formulae that, he said, Euler had worked very hard to obtain, such as the formula that had intrigued him for so long, $A \cdot B = \pi/4$.

More importantly, Gauss had gone further in the second part of this paper, which for some reason he did not publish. He showed that the hypergeometric series satisfies the hypergeometric differential equation

$$\left(x - x^2\right)\frac{d^2F}{dx^2} + (\gamma - (\alpha + \beta + 1)x)\frac{dF}{dx} - \alpha\beta F = 0. \tag{9.7}$$

From this he deduced that the hypergeometric equation defines a function of a complex variable x everywhere except at the points $x = 0, x = 1$, and $x = \infty$. This function can be represented inside the unit circle by the hypergeometric series, but unlike the series it makes sense outside it.

Gauss's discussion of the following paradoxical result (his term) shows how good his insight into the nature of complex functions already was. He first set $x = 4y - y^2$, and $\gamma = \alpha + \beta + 1/2$, when F becomes equal to $F(2\alpha, 2\beta, \alpha + \beta + 1/2, y)$, and then changed y to $1 - y$ and found that his earlier results implied that

$$F\left(2\alpha, 2\beta, \alpha + \beta + \frac{1}{2}, y\right) = F\left(2\alpha, 2\beta, \alpha + \beta + \frac{1}{2}, 1 - y\right) \tag{9.8}$$

"which equation is certainly false", as he said. He then resolved the paradox by distinguishing between the function F and the infinite series that represents the function only when the variable is less than one in absolute value. The series takes a unique value for each value of the variable for which it is defined and convergent, whereas the function does not. The false equation is meaningless as a result about series because they have distinct domains of convergence, and misleading as a result about functions because the functions are many-valued. One could no more deduce a false result here, he remarked, than one could validly infer from $\arcsin(1/2) = 30°$ and $\arcsin(1/2) = 150°$ that $30° = 150°$.

Even in this unpublished paper Gauss did not explain how the hypergeometric series arose in the complex theory of elliptic functions. As we saw above (see Eq. (9.4)) he knew in 1800 that the functions $y_1 = \frac{1}{M(1+x,1-x)}$ (denoted K, above) and $y_2 = \frac{1}{M(1,x)}$ satisfy the differential equation:

$$\left(x^3 - x\right)\frac{d^2y}{dx^2} + \left(3x^2 - 1\right)\frac{dy}{dx} + xy = 0 \tag{9.9}$$

This is not a hypergeometric equation, but on making the transformation $x^2 = z$ it becomes the equation

$$z(1-z)\frac{d^2y}{dz^2} + (1-2z)\frac{dy}{dz} - \frac{y}{4} = 0 \tag{9.10}$$

which is a hypergeometric equation with $\alpha = \beta = 1/2, \gamma = 1$—indeed, it is Legendre's equation. And, as Schlesinger pointed out, Gauss's derivation of the hypergeometric equation for the hypergeometric series follows exactly his derivation (*Werke* vol. X.1, pp. 177–180) of the power series for $\frac{1}{M(1+x,1-x)}$ described above. Schlesinger therefore conjectured that it was Gauss's recognition of the great generality of the hypergeometric equation that caused him to change his direction of research. Gauss's theory of the modular function $k = k(\frac{K}{K'})$ and the connection with differential equations (such as Legendre's) mark a profound insight into the connections between elliptic functions and a burgeoning theory of complex functions of any kind.

Was Gauss therefore right to say in 1828 that Abel had come only one-third of the way? After all, as Gauss wrote to Bessel (Gauss 1880, 477), Abel "has followed the same path that I embarked upon in 1798, so that the great coincidence of the results is not to be wondered at ...". This ignores Abel's consideration of the extent to which the division equations are solvable algebraically, but Gauss would have noticed that Abel did not seem to appreciate the infinite series and product representations he had found, and discussed elliptic functions with a real modulus only.

On the other hand, very soon Jacobi's *Fundamenta nova* and Abel's 'Précis' displayed a general theory of the transformations of elliptic functions that surpassed anything Gauss had written down. Jacobi's book, and Abel's paper (1828a), did develop a theory of theta functions, albeit only for elliptic functions with real modulus, or (in Abel's case) a purely imaginary modulus. Even so, the theories of Abel and Jacobi remained entirely formal in their treatment of complex functions.

Gauss's theory of elliptic functions was not only of greater scope that Abel's and Jacobi's, it was deeper mathematically, because Gauss was clear about the importance of the complex domain. As he put it in a letter to Bessel on December 18, 1811 (*Werke*, X.1, 366), he asked of anyone who wished to introduce a new function into analysis to explain:

> if he would apply it only to real quantities, and imaginary values of the argument appear so to speak only as an offshoot, or if he agrees with my principle that the imaginaries must enjoy equal rights in the domain of quantities with the reals. Practical utility is not at issue here, but for me analysis is an independent science that through the neglect of any fictitious quantity loses exceptionally in beauty and roundness and in a moment all truths, that otherwise would be true in general, have necessarily to be encumbered with the most burdensome restrictions.

In this regard, the most difficult question is to understand what is meant by an integral in the complex domain. In the same letter to Bessel he wrote that the value of a complex integral depends on the path between its end-points, and

> The integral $\int_a^b \phi x.dx$ along two different paths will always have the same value if it is never the case that $\varphi x = \infty$ in the space between the curves representing the paths. This is a beautiful theorem whose not-too-difficult proof I will give at a suitable opportunity .... In any case this makes it immediately clear why a function arising from an integral $\int_a^b \phi x.dx$ can have many values for a single value of x, for one can go round a point where $\varphi x = \infty$ either not at all, or once, or several times. For example, if one defines $\log x$ by $\int_a^b dx/x$, starting from $x = 1$, one comes to $\log x$ either without enclosing the point $x = 0$ or by going around it once or several times; each time the constant $+2\pi i$ or $-2\pi i$ enters; so the multiplicity of logarithms of any number are quite clear.

Gauss was never to find the suitable opportunity, but this letter makes it clear that he had obtained the first crucial insight into the integration of functions of a complex variable. The theory is necessarily geometrical, and cannot be simply formal.

9.6 Exercises

1. Write down a short summary of the properties of Abel's elliptic functions, explaining what is meant by double periodicity, and the zeros and infinities of these functions.
2. Jacobi's theorems about the representation of numbers fit into a context of similar theorems of this kind. What can you find out about Fermat's conjecture that every number is the sum of n n-gonal numbers (three triangular numbers, four square numbers, and so on)?
3. What was the importance of the hypergeometric equation by the 1820s?

Chapter 10
Cauchy and Complex Function Theory, 1830–1857

10.1 Introduction

Cauchy's publications in the 1830s were scattered over several journals, one of which he created for the purpose, while he lived first in Turin and then Prague, where he attended the Bourbon Court as a tutor to the Duke of Bordeaux, the grandson of the former Bourbon monarch Charles X. As Belhoste describes (1991, Chap. 10) this was an unrewarding experience dealing with a petulant boy that diminished even Cauchy's energies for research. Cauchy concentrated on mathematical optics and the theory of light, but, as ever, he worked on many subjects, including the theory of functions defined by power series in a complex variable. When Dauphin turned 18 in September 1838 Cauchy's services were no longer required, and he returned to Paris with the title of Baron and a number of mathematical notebooks that would inspire him for several years.

In this chapter we look at Cauchy's erratic progress towards recognising the appropriate defining features of a complex function and his final appreciation of the "Cauchy–Riemann equations" and their implications.

10.2 Complex Functions and Power Series Expansions

By the time Cauchy took up his academic position in Turin he had already published a long paper in three parts (Cauchy 1830–1831) in the Italian journal *Biblioteca Italiana*, in which he had set out his approach to rigorising the calculus. This provocative paper generated a hostile reaction among Italian mathematicians still attracted to the old, Lagrangian, approach to the calculus, and Cauchy next attempted, rather disingenuously, to deflect these attacks in a memoir he gave to the Turin Academy of Sciences. In it he claimed that his methods were necessary to help reduce the labours of astronomers. That is unlikely, but by setting out the principles underlying his

J. Gray, *The Real and the Complex: A History of Analysis in the 19th Century*,
Springer Undergraduate Mathematics Series, DOI 10.1007/978-3-319-23715-2_10

theory of the expansion of functions in power series his work now began to change the opinions of his Italian colleagues.

Cauchy was aware of a fundamental connection between complex functions and power series expansions, but he struggled for some twenty years to formulate it precisely. In his *Cours* (1821, 151–152, 280–282) he had shown how to determine the radius of convergence of a given (real or imaginary) power series. His first attempt at the deeper result, that a complex function can be represented on some domain by a convergent power series, was presented in a Mémoire he presented and published in Turin (Cauchy 1831). There he said that:

> A function $f(x)$ can be developed (by Stirling's theorem) in a convergent series of ascending powers of x when, the modulus of x being less than or equal to ρ, the function $f(x)$ remains finite and continuous for the modulus ρ or any modulus smaller than the real or imaginary variable x.

This is inadequate.[1] The theorem is formulated only for a neighbourhood of the origin, and, more damagingly, the function was required only to be "finite and continuous". This vague phrase, often modified in the course of a proof, was used by Cauchy to mean that the function was never infinite and was not multi-valued. But even this is not enough because no condition was imposed on the derivative—indeed it does not even require that the function *have* a derivative. Moreover, there is a considerable difference between real and complex differentiability, as perhaps Cauchy might have suspected: his famous example of a function that is zero when x is negative or zero and equal to e^{-1/x^2} when x is positive shows that a real function may be infinitely differentiable and still not equal its power series expansion. But if Cauchy's loose way with words can annoy us today, it is nonetheless the case that his attention to rigour, even here, greatly exceeded that of his contemporaries—as its mixed reception indicates.[2]

It is helpful to look ahead briefly, to 1839 (when Cauchy was back in Paris), because Cauchy restated the Turin theorem on several occasions in the 1830s. In particular, he applied it to the solution of differential equations when he proved the existence of solutions to what we nowadays call the 'Cauchy problem' by expanding them as convergent series. Then in 1839 (Cauchy 1839, 486) he announced a new version of the theorem. Now he claimed that a complex function had a power series expansion whenever the modulus of the variable x remains less than the smallest value at which the function or its derivative cease to be finite and continuous. As Cauchy explained in a note he added to the paper, this hypothesis on the derivative $f'(x)$ was needed to make sure that one can change the order of the integrations used in the proof.

He argued that when the variable x takes the imaginary value $z = Ze^{p\sqrt{-1}}$ one has the identity (which we would call the Cauchy–Riemann equations!)

[1] See Peiffer (1978, 11–12).

[2] Cauchy found a more sympathetic audience in Piola, who translated Cauchy's Turin Mémoire into Italian and published it in the *Opuscoli* that he edited (Cauchy 1834).

$$\frac{\partial f(z)}{\partial Z} = \frac{1}{Z\sqrt{-1}}\frac{\partial f(z)}{\partial p}.$$

Integrating twice with respect to Z and p he obtained

$$\int_{-\pi}^{\pi} f(z)dp = 2\pi f(0)$$

and, under the further hypothesis that $f(0) = 0$, that $\int_{-\pi}^{\pi} f(z)dp = 0$. The theorem then follows on substituting $z\frac{f(z)-f(x)}{z-x}$ for $f(z)$ (assuming $x \neq z$, $|x| < Z$), developing $\frac{z}{z-x}$ in a power series, and finally by integrating the result term by term.

Cauchy's examples of functions, continuous together with their first derivatives, were $\cos z$, $\sin z$, and e^z, which therefore could be expanded in power series for any z. Whereas, according to Cauchy, functions like $\sqrt{1+z}$, $\frac{1}{1-z}$, $\log(1+z)$, and $\arctan z$ cease, with their first order derivatives, to be continuous functions of z at the moment when the modulus becomes equal to 1. Therefore their power series are divergent when the modulus is greater than 1. Finally, functions like $e^{1/z}$, e^{1/z^2}, and $\cos(1/z)$ were discontinuous at $z = 0$ together with their derivatives and consequently not capable of expansion in power series.

So still in 1839 Cauchy was still insisting on continuity, rather than differentiability, as the crucial property for a function to admit a power series expansion; differentiability appeared as an artefact of the proof. In a second proof he gave of the theorem, he even claimed that his theorem (1840, 331) "immediately gives the rules for the convergence of series provided by the development of explicit functions and reduces the law of convergence to the law of continuity."

In 1833 the Bourbon Court had moved to its domains in the Czech lands, and Cauchy was called to Prague to act as tutor for the son of Charles X. There, in 1834, he was joined by his wife and family. His royal duties were to leave him little time for research or publication until 1839 when his tutee, the Duke of Bordeaux, turned 18, and they came to an end. Cauchy, who turned 50 that year, was now free to move back to Paris permanently and to pick up the career he had walked away from in 1830.

He returned to an uncongenial place. Many of his older colleagues, among them Ampère, Fourier, and Legendre, had died; Poisson soon followed them in 1840. A new generation had grown up who had not been taught by Cauchy and did not know much, if anything, about the papers Cauchy had published while abroad. So he decided to republish the results that he had obtained in Turin and Prague, as well as his more recent results, by resuming his *Exercises* under the new title *Exercises d'analyse et de physique mathématique*. He also used the *Comptes rendus des séances* of the Paris Academy, recently established in 1836, to publish a note or a memoir almost weekly, a number of which dealt with topics in complex analysis to which we now turn.[3]

It is these papers in which Cauchy set out to rewrite his theory of functions of a complex variable and establish it securely for the first time.

10.3 Laurent's Theorem

On 28 November 1841 Cauchy considered the uniqueness of a power series expansion that contains negative powers of the variable. He noted (Cauchy 1841, 359) that he had proved uniqueness when no negative powers are involved in his *Cours d'analyse* (Cauchy 1821, 144–145), and that that proof ceases to be valid when negative powers of the variable occur. Now he stated and proved that

1. if the sum of a convergent series of positive whole powers of the variable is 0 then all its terms are equal to 0;
2. if the sum of a convergent series of positive and negative whole powers of the variable is 0 for a given modulus of the variable, then all its terms are equal to 0.

From (1) and (2) the following uniqueness theorems could be easily derived:

3. a continuous function can be expanded in a convergent series of positive, whole powers in a unique manner [one can only wonder what Cauchy meant by the word 'continuous' in this setting, and what degree of confusion there might have been in his mind];
4. a continuous function can be expanded a unique manner in a convergent series of positive, null and negative whole powers of the variable for a given value of its modulus and any value of its argument.

Cauchy proved theorems (1) and (3) using the arguments he had used two decades before in his (1821). To prove Theorem (2) he multiplied both sides of the equation

$$a_0 + a_1 x + a_2 x^2 + \cdots + a_{-1} x^{-1} + a_{-2} x^{-2} + \cdots = 0$$

by $e^{inp}dp$ and then integrated with respect to p between 0 and 2π, thus obtaining a_n for any positive, null or negative n. From this theorems (2) and (4) followed immediately.

Other matters then occupied him, and he only returned to the topic in a paper (Cauchy 1843a) that he presented to the Academy on 31 July, 1843 when he gave an "elementary" proof of his uniqueness theorem (4) (one, that is, that did not resort to integration).

[3]In this period, and for the rest of his life, Cauchy published something almost every week on whatever subject had caught his attention. An indication of their quality is provided by a letter Jacobi wrote to Dirichlet that is quoted in Schubring (2005, 430): "Cauchy has become unbearable. Every Monday, broadcasting the known facts he has learned over the week as a discovery. I believe there is no historical precedent for such a talent writing so much awful rubbish."

This paper led Pierre Alphonse Laurent to submit to the Academy a "fragment" of a more extended work. Laurent was former pupil of the École Polytechnique, who had graduated in 1832 as one of the best students of his class and eventually worked as a military engineer in Algeria before returning to France around 1840. He now became interested in mathematics, and submitted a paper to the Academy. Cauchy and Liouville were charged with reviewing it, and although they recommended that the paper be published the Academy did not accept the recommendation and Laurent's original memoir was never published. Its principal results were included in the (posthumous) article (Laurent 1843). When the same misfortune befell a second paper Laurent turned to applied mathematics, until his untimely death in 1854.[4]

However, at the next meeting of the Academy Cauchy (1843b) claimed that his papers (Cauchy 1837a, b) gave him priority over Laurent in the matter. Only on 30 October did he present his (and Liouville's) report on Laurent's memoir to the Academy. Cauchy then went on to argue that the main theorem in Laurent's paper could be deduced from a theorem he (Cauchy) had stated in Cauchy (1840, 186–187) and elsewhere. To support his claims Cauchy added to the report a note of his own (Cauchy 1843c) to show that "the easiest way" to obtain Laurent's theorem was by reformulating the results in Cauchy's (1840). If Laurent's ideas can be disentangled from Cauchy's presentation, it would seem that he had realised that a function $f(x)$ that is 'finite and continuous' whenever $|x|$ lies between two limits (i.e. in an annulus) can be expanded as a sum of two convergent series, one in ascending and one in descending powers of x. From this, Laurent had deduced results about the separation of the roots of an algebraic equation. Sadly, Cauchy's less than generous response, coupled with his extraordinary rapidity of thought, was to prove typical of the way Cauchy would take the work of others as an opportunity to promote his own.

10.4 Cauchy's Path to the Cauchy–Riemann Equations

In the crucial year 1846, albeit in a vague and incoherent manner, Cauchy was acquainted with the basic facts about complex functions. He had restated the integral theorem in a more general setting, he was adept at using the concept of a residue and the integral formula (see below). Yet a comprehensive theory of complex functions was still lacking, and an understanding of multivalued functions and their behaviour near branch points continued to elude him. More surprisingly, he was still far from clear what was the best way to explain what a complex variable might be.

His use of paths in the (complex) plane should surely have rendered inadequate even in his own eyes the conception of complex quantities that he had given in his *Cours* (1821a)—but which he had apparently not yet abandoned—in which he had regarded imaginaries as symbolic expressions. However, in 1847 he proposed to make this theory more palatable by reducing the letter i to no more than a real quantity. In his opinion

[4]See his obituary in Bertrand (1890).

> This done, there is no need to torture the mind to discover what the symbolic sign $\sqrt{-1}$ could represent, for which German geometers substitute the letter i.

The symbol was merely a tool, he continued, that enables the mathematician to quickly reach the real solution of questions. But, he continued, the theory would become even clearer if everything, including the letter i, was reduced to real quantities. (Cauchy 1847a, 313). So, inspired by Gauss's and Kummer's work in algebraic number theory, Cauchy called two real polynomials equivalent if they give the same remainder when divided by $i^2 + 1$. Thus, to use much more modern terminology, calculations with complex numbers could be reduced to calculations (mod $i^2 + 1$) in the ring $\mathbb{R}[i]$ where i was 'a real but indeterminate quantity'. The meaning of the term 'real' here is distinctly ambiguous.

In an expanded version of this paper (1847), published as usual in his reborn *Nouveaux exercises de Mathématiques*, Cauchy extended the theory of equivalences to functions defined by means of convergent series. By considering, for example, the exponential series $e^x = \sum_n \frac{x^n}{n!}$ and writing ix in place of x, he rapidly obtained Euler's relation:

$$e^{ix} = \cos x + i \sin x \; mod(i^2 + 1).$$

It took two more years until, "after new and arduous reflections", as he put it, (Cauchy 1849, 152) he changed his mind again and came out in favour of the theory of geometric quantities, that is to say, the interpretation of complex numbers (and variables) in the complex plane adhered to by most of his contemporaries. In support of this view he cited several of its advocates, including Buée (1806) and Argand, but he mentioned neither Hamilton (1837) nor Gauss (1831). He said instead that he had been led to his 'new' conception by using the ideas that Saint Venant had presented in a note on geometric sums published in 1845. But now, in this geometrical setting, Cauchy began re-establishing the results he had been obtaining since the 1820s.

Cauchy began to introduce the general concept of a function $Z(z) = X + iY$ of a geometric quantity $z = x + iy$ rigorously in a way that he had failed to do in (Cauchy 1821a) when writing about imaginary functions of imaginary expressions. As he now put it (Cauchy 1853, 359–360):

> When one substitutes geometric quantities for imaginary expressions, the imaginary variables are nothing other than variable geometric quantities. It remains to know how functions of an imaginary variable can be defined. This last question has often embarrassed geometers, but every difficulty disappears when, guided by analogy one extends to functions of geometric quantities the definitions generally adopted for functions of algebraic quantities. In this way one initially comes to some singular conclusions which are nonetheless legitimate, as I shall indicate in a few words. Two real variables, or, in other terms, two algebraic variable quantities, are said to be functions, the one of the other, when they vary simultaneously in such a way that the value of the one determines the value of the other. [In this way] Z will be counted as a function of z when the value of z determines the value of Z. Now, it suffices for this that X and Y will be functions determined by x and y.

He extended the concept of continuity to these functions. He began by recognizing that

> in truth, the definition of a continuous function given in this work (Cauchy 1821) and generally adopted today by geometers, is particularly applied to real or imaginary functions of real variables, but nothing prevents it being extended to the case where the variables are geometric quantities (Cauchy 1853, 367).

Before doing so, in order to preserve single-valuedness in the case of multi-valued functions he made it clear that "it was easier" to consider each branch of the function as a separate, well-determined function of the variable. Then he repeated word for word the definitions he had given in his (1821, 34–35), merely translating them into geometrical terms (Cauchy 1853, 376).

In spite of Cauchy's claim, as Felice Casorati later remarked (1868, 71) a generalization of this type was not suggested by a "truly profound analogy". Analogy alone could only propose a theory of functions of two real variables; the crucial identification of functions of a single *complex* variable involved a new idea.

> It is thus that Cauchy himself, while conserving the given definition of a function of a complex variable $x + iy$, found it necessary to introduce an epithet to designate, among all the functions comprised in the definition, only those which, like all the functions resulting from ordinary systems of operations performed on the combination $x + iy$, enjoy the property of having for every value of $x + iy$ a derivative independent of the value $\frac{dy}{dx}$.

The epithet Cauchy introduced was 'monogenic', meaning that the real and imaginary parts of a complex function satisfied the Cauchy–Riemann equations. One almost feels like giving a sigh of relief at this point.

10.5 Final Stages

All this time, multi-valued functions, such as the logarithm function and the square root function, continued to trouble Cauchy, and he tended to deal with them by cutting up the domain and treating each branch separately Freudenthal (1971a, 141) has aptly remarked that "from 1821 he [Cauchy] treated multi-valued functions with a kill-or-cure remedy: if branched at the origin, they would be admitted in the upper half-plane only [although] according to Freudenthal, "fortunately he more often than not forgot this gross prescription, which if followed would lead him into great trouble,...". Indeed, Cauchy's flood of publications contains a large share of reformulations and repetitions, and, as Belhoste documented in his biography of Cauchy, his failure to present a coherent version of his theory of complex functions in one place caused difficulties for his contemporaries. Joseph Bertrand, in his review of Valson's biography of Cauchy, which was filled with unstinting praise for both the man and his work, remarked that

> Cauchy, throughout his career, preserved both his rapidity of thought and his powers of invention and penetration. His genius always made him the master of the most difficult problems in a few moments. But every research involves trial and error and fruitless attempts, which Lagrange, Jacobi, and Gauss doubtless knew as well as him. What distinguished Cauchy, whose genius was without doubt equal to theirs, was that he informed the public of them minutely and at length.

Only as late as 1851 did Cauchy begin to realize fully that the crucial concept involved in his theorems on complex functions was not continuity but analyticity related to Cauchy–Riemann equations. The explicit definition of monogenic functions permitted Cauchy to specify the classes of functions to which his theory could be applied and to completely relieve his uncertainties about the correct formulation of his theorem on the expansion of functions in series. Referring to "the necessity of mentioning the derivative of a function of z" (Cauchy (1851a, 304)) he began to approach the correct hypothesis one has to require i.e. the existence of a unique derivative $f'(z)$ (independent from the path). This condition is enough to ensure the analyticity of the function.[5]

For monogenic and monodromic functions, Cauchy then reformulated his previous results including his integral theorem and the calculus of residues.

10.6 A Summary of Cauchy's Ideas on Complex Analysis

In his *Mémoire* of 1825 on complex functions, Cauchy had defined the integral of a complex function along a path in the complex plane, and showed that it was the same on two paths that did not enclose an infinity of the function.

He obtained his integral theorem and his integral formulae for the case when a function is infinite at a point p inside a

$$f^* := \int_{rect} f(x + y\sqrt{-1}) = \varepsilon f(x + y\sqrt{-1}) = 2\pi\sqrt{-1}Res(f).$$

Recall from §6.3 that when Cauchy had first considered the case when the function $f(x+y\sqrt{-1})$ becomes infinite for values $x = a$, $y = b$ (in 1825) he had considered the limit, denoted f^*, of $f = \lim(x-a+(y-b)\sqrt{-1})f(x+y\sqrt{-1})$ as x tended to a and y tended to b or, as he put it, without noticeable error $f^* = \varepsilon f(x+y\sqrt{-1})$, ε being an infinitely small number. The formula extends by simple addition to finitely many points in the rectangle where the function becomes infinite.

In the case of higher order infinities (or, in modern terms, of a pole of order m) his integral theorem became

$$f^* = \frac{f^{(m-1)}(a + b\sqrt{-1})}{1.2.\dots(m-1)}.$$

The integral formula is the result of applying the integral theorem to the integral $\int\limits_{rect} \frac{f(z)dz}{z-t}$ where f is an analytic function inside and on the rectangle. The formula says

[5]It was a considerable step forward when in 1900 Goursat proved the Cauchy integral theorem, where he stated that "by assuming Cauchy's point of view it is sufficient to build the theory of analytic functions, to suppose the continuity of $f(z)$ and the existence of the derivative"—note, not that the derivative is continuous (Goursat 1900, 16).

$$f(t) = \frac{1}{2\pi i} \int\limits_{rect} \frac{f(z)dz}{z-t}.$$

It yields formulae for all the derivatives of the function f at the point t and opens the way to a Taylor series expansion of an analytic function, so once Cauchy was clear what the definition of a complex analytic function was (and that it requires that the function have a continuous derivative) he could show (in 1851) that a complex analytic function has a Taylor series expansion.

Chapter 11
Complex Functions and Elliptic Integrals

Joseph Liouville (1809–1882)

11.1 Introduction

This chapter considers how elliptic functions and complex functions were first brought together. This was an important step for both subjects, which, as Jacobi noted in his lectures, seemed to be kept apart by the complications resulting from the two-valued nature of the integrand in the elliptic integrals.

J. Gray, *The Real and the Complex: A History of Analysis in the 19th Century*, Springer Undergraduate Mathematics Series, DOI 10.1007/978-3-319-23715-2_11

11.2 Liouville and His Student Hermite

The prime mover was Joseph Liouville, who had already shown himself to be a versatile mathematician, and was the editor of a successful journal. In 1843 he was reading the first edition of Abel's *Oeuvres*, and discussing the theory of elliptic functions with Charles Hermite, who was then a student of his. Liouville recorded in one of his working notebooks that Hermite had shown him that there cannot be a function with two distinct real periods. For, if there is such a function it has a Fourier expansion with respect to one of the periods. The existence of the other period determines properties of the coefficients that quickly show that the function must be a constant. Hermite would have found a similar argument in some papers by Jacobi that he was studying intensely at the time.

In his notebook, Liouville argued by contradiction.[1]. He supposed that the function $f(x)$ has two real periods α and β. With respect to α the function can be represented this way:

$$f(x) = A_0 + \sum_i A_i \cos\left(\frac{2i\pi x}{\alpha}\right) + \sum_i B_i \sin\left(\frac{2i\pi x}{\alpha}\right).$$

Periodicity with respect to β means that

$$f(x) = f(x+\beta),$$

and so

$$f(x) = A_0 + \sum_i A_i \cos\left(\frac{2i\pi\,(x+\beta)}{\alpha}\right) + \sum_i B_i \sin\left(\frac{2i\pi\,(x+\beta)}{\alpha}\right).$$

By equating the ith coefficients and adding the squares of the resulting equations, Liouville found that

$$\left(A_i^2 + B_i^2\right)\left(\sin^2\frac{2i\pi\beta}{\alpha} + \left(1 - \cos\frac{2i\pi\beta}{\alpha}\right)^2\right) = 0.$$

The ratio α/β must be irrational, or α and β would both be integer multiples of a smaller period and would not therefore be distinct. So, when $i \neq 0$ it must be the case that $A_i = B_i = 0$, and the function reduces to a constant.

Liouville soon saw that this proof establish a much more interesting result: no doubly periodic function is everywhere finite. For, let such a function $f(x+iy)$ with a period α be expanded as a series in x of the form

[1] See Lützen (1990, Chap. 13)

$$f(x) = \sum_i A_i \cos\left(\frac{2i\pi x}{\alpha} + \varepsilon_i\right)$$

If β is also a period, then

$$A_i \cos\left(\frac{2i\pi x}{\alpha} + \varepsilon_i\right) = A_i \cos\left(\frac{2i\pi x}{\alpha} + \varepsilon_i + \frac{2i\pi\beta}{\alpha}\right)$$

So if α/β is irrational and $i \neq 0$ it must be the case that $A_i = 0$, and again the function is a constant.[2] This result is the special case of Liouville's theorem for doubly periodic functions.

Liouville found the generalisation of his principle from doubly periodic functions to all complex functions as early as 1844—but he did not publish it.[3]

Charles Hermite (1822–1901)

Hermite's contribution was thought to have been lost for a long time, and was known about only through of a memoir he presented to the Academy of Sciences six years later, in 1849, where he presented a theory of doubly periodic functions in terms of Cauchy's theory of functions of a complex variable. However, his original manuscript was found by Belhoste in 1988 and published some years later (Belhoste 1996).

[2]As Lützen has pointed out out in his analysis of Liouville's work, (1990), Liouville's argument extends readily to any doubly periodic complex function.

[3]See Lützen (1990, Chap. 13).

In his memoir, Hermite attempted to show that every doubly periodic function is the quotient of two singly periodic functions (with the same period) which are defined everywhere, and so the function is of the form

$$\frac{\sum\limits_{m} A_m \exp\left(2m\frac{i\pi x}{a}\right)}{\sum\limits_{m} B_m \exp\left(2m\frac{i\pi x}{a}\right)}.$$

He could make no progress with the idea that the quotient also be periodic with period b except in special cases. So he turned to study doubly periodic functions of the form $F(x) = \frac{\Pi(x)}{\Phi(x)}$, where the functions $\Pi(x)$ and $\Phi(x)$ are periodic with period a.

The values of such a function are known everywhere when they are known on a period parallelogram, and Hermite integrated the function around a period parallelogram to obtain expressions for the Fourier coefficients of F. He found the important result that the sum of the residues of a doubly periodic function in its period parallelogram vanishes. Then he showed that the doubly periodic function F can be written in the form

$$A_0 + \frac{2i\pi}{a}\{r_1\varphi(z - z_1) + r_2\varphi(z - z_2) + \cdots + r_n\varphi(z - z_n)\}$$

where the r_j are the residues of F at the points z_j, $j = 1, 2, \ldots, n$ and the function $\varphi(z)$ is periodic with period a. It is not a theta function, but is required to satisfy $\varphi(z + b) = \varphi(z) - 1$, a condition that makes the function F doubly periodic—that b is a period follows from the fact that the sum of the residues vanishes.

11.3 Liouville on His Theorem for Doubly Periodic Functions

In December 1844 Liouville informed the Académie des Sciences that he had found a general principle that seemed to him to give the theory "an uncommon character of unity and simplicity":

> If a (single-valued) function is doubly periodic, and if one recognises that it never becomes infinite, one can, from this alone, affirm that it reduces to a constant.

But he did not publish a proof of this remarkable fact, nor was he to. Perhaps foolishly, he confined himself to giving private lectures on it. Among his audience when they visited Paris in 1847 were the German mathematicians Carl Borchardt and Joachimstal, who on Liouville's instruction gave copies of their notes to Jacobi and Dirichlet on their return home. Borchardt eventually published his notes of these lectures in 1880, by which time he had become the editor of the *Journal für die reine und angewandte Mathematik*.[4]

Cauchy recognised the generality of the principle at once, and at the very next meeting of the Académie announced this theorem:

> If a single-valued function reduces to a determinate constant F for every infinite value of z, then it will reduce to the same constant value when the variable z has any finite value.

This needs a little interpreting. Cauchy meant that a complex function of a complex variable (a concept he had still not precisely defined) which is defined everywhere and is never infinite (it has no poles), and which it is bounded in a neighbourhood of infinity, reduces to a constant. Such behaviour is a strong and useful constraint on the global behaviour of complex functions that cannot apply in general to similar functions from $\mathbb{R}^2$ to $\mathbb{R}^2$.

Cauchy gave a proof, based on his calculus of residues, and seemed to suggest that he had possessed the means for publishing this theorem since at least 1831. His remarks amount to a claim that this result, today known as Liouville's principle, should really be credited to Cauchy, since Liouville only applied it in the special context of elliptic function theory. In support of this claim, Cauchy went on to give five different proofs of Liouville's principle in under a year.

Thus provoked, Liouville decided to lecture on his work at the Collège de France in 1851. On this occasion he argued that a bounded complex function f is constant by arguing that if f takes its maximum value at a point z_0 say, then the Cauchy Integral Formula implies, on setting $z_0 + re^{i\theta} = z$, that

$$|f(z_0)| = \left|\frac{1}{2\pi}\int_0^{2\pi} f\left(z_0 + re^{i\theta}\right) d\theta\right| \leq \int_0^{2\pi} \frac{1}{2\pi}\left|f\left(z_0 + re^{i\theta}\right)\right| d\theta \leq |f(z_0)|, \tag{11.1}$$

which is impossible.

In the audience were two young French mathematicians in Cauchy's orbit: Charles Briot and Claude Bouquet. They decided that the time had come to write a book on the new theory of complex functions that Cauchy had finally got round to organising and publishing, and to show how the exciting and better-known theory of elliptic functions could be fitted into that framework. Their book appeared in 1859, with generous acknowledgements to both Cauchy and Liouville, and because it was the first textbook in the world to present the new theory, it did much to spread Cauchy's and Liouville's ideas. Liouville was not best pleased, however, and later wrote in his notebooks : "MM. Briot and Bouquet, cowardly thieves, but the most worthy Jesuits. Elected as thieves by the Academy!!!!!" (Quoted in Belhoste and Lützen, (1984, 28)).

---

[4]In an otherwise faithful publication he chose to replace Liouville's original proof with one of his own, based on the calculus of residues, as he said in an introductory footnote.

The controversy continued when Cauchy reported on Hermite's paper on 31 March 1851. He told his audience of the close connection that Hermite had established between elliptic functions and doubly periodic functions, and the result that the sum of the residues of a doubly periodic function inside its period parallelogram is zero. Liouville stood up and protested that he had had such a general theory of doubly periodic functions since 1844. As he explained, this theory was based on the general principle that a doubly periodic function must have at least two poles in its period parallelogram unless it is a constant. (As the lecture notes taken by Borchardt establish, this is equivalent to the principle Liouville had asserted in 1844, but it is not the same.) What Liouville had noticed was that if a given complex function f without poles is composed with the doubly periodic function $\sin am$ then the composite function $f(\sin am(z))$ is constant, and so f itself must be constant. This reduces the general case to that of a doubly periodic function without poles, which was known to be constant.

In conclusion, it would seem that priority in the discovery of the theorem resides with Liouville, who however, to his subsequent distress, did not publish it, so for once, the attribution of a name in the history of mathematics seems reasonably justified. But it was Cauchy who extracted Liouville's principle from its original setting in the theory of elliptic functions, and gave it a much more natural proof within a theory of complex functions. His protestations of priority are also somewhat paradoxical. The more he showed how automatically it followed from his work twenty years before, the odder it looked that he had never noticed it before. One must suppose that a global principle of this kind lay outside his way of conceiving the subject, which was local and rooted in the study of power series.

11.4 Doubly Periodic Functions and Complex Functions

As we saw with the work of Gauss in Sect. 9.5 there would seem to be two theories: one concerns the functions obtained by inverting an elliptic integral with real or complex modulus, the other doubly periodic functions with two arbitrary periods whose ratio is not real. The obvious question is whether these theories are coextensive. But there were difficulties in the way of inverting elliptic integrals with a complex modulus, and equally there were problems in finding a suitable modulus given a pair of periods.

Liouville's theory of elliptic functions, as eventually published by Borchardt and Joachimstal as (Liouville 1880), investigates the problem of showing thatdoubly periodic functions are elliptic functions. In the first half of the memoir Liouville showed that, for a given pair of periods, every doubly periodic function with those periods is a rational function of a doubly periodic function with those periods and exactly two poles. This reduces the problem to that of realising a doubly periodic function with only two poles as an elliptic function. It would then be necessary to show thatthere are doubly periodic functions having any pair of periodsω_1 and ω_2

(assuming only that the quotient of the periods is not real) and that there are elliptic function with the same periods, and to find the corresponding modulus.

However, Liouville did not fully investigate these questions. He began by showing that *if* there is a doubly periodic function $\phi(z)$ with periods $2\omega_1$ and $2\omega_2$, poles α and β, and zeros a and b (when, necessarily, $a + b = \alpha + \beta$) then $\phi(z)$ satisfies the differential equation

$$\{\phi'(z)\}^2 = A\phi^4(z) + B\phi^3(z) + C\phi^2(z) + D\phi(z) + E \tag{11.2}$$

A, B, C, D, and E being constants. If moreover $\alpha = \beta$, then, he said, $A = 0$. To construct a doubly periodic function with arbitrary periods it would then be necessary to examine how the periods of ϕ depend on the constants $A, B, \ldots E$. All that Liouville showed, however, was that if $\phi(z)$ is an even function (that is, $\phi(z) = \phi(-z)$) then the function $\phi(z)$ is a constant multiple of either $\sin am(z)$ or its reciprocal. For in that case, as he showed, the differential equation reduced to

$$\phi'(z)^2 = \left(\phi'(0)\right)^2 \left\{1 - \phi^2(z)\right\} \left\{1 - k^2\phi^2(z)\right\} \tag{11.3}$$

where

$$k = \frac{1}{\phi(\alpha + \frac{\beta-\alpha}{2})} = \frac{1}{\phi(\frac{\alpha+\beta}{2})}.$$

It is true that the connection between the periods $2\omega_1$ and $2\omega_2$ and the constants $A, B, \ldots, E$ would be difficult to elucidate.[5] But Liouville settled for showing that his theory embraced the theory of Abel and Jacobi. Perhaps he took the view that the existence of doubly periodic functions with arbitrary periods was beyond doubt. But while he advanced the subject of doubly periodic functions, he did not unite it with the older theory of elliptic functions.

In this context Jacobi's comments of 1847, made after Borchardt had returned to Germany, are most interesting. He wrote (Jacobi 1847, 519):

> Liouville and Hermite, in works of which only a suggestion has been published, have derived all the results about addition, multiplication, transformation, and division from the purely periodic properties of the transcendents ...One can say that ...the new and true foundations for the theory of elliptic functions have been found. Apart from the simplicity with which the entire structure can now be presented, it accords with the advantage of greater strength and clarity than was possible up to now in those treatments that began with integrals. For the theory of imaginary values of algebraic functions and their integrals is still not built up enough for the consideration of such integrals, over the entire domain of real and imaginary values with both constant and variable magnitudes equally involved, to be taken as basic. This shortcoming is not intrinsic to the method which served me in the *Fundamenta Nova*, but goes back to a gap in the general method of analysis.

---

[5]The mathematician who pioneered this route successfully was Weierstrass in the 1860s.

In the event, it was Briot and Bouquet who brought elliptic function theory and complex function theory together in a series of three papers (Briot and Bouquet 1856a, b, c) and a book (Briot and Bouquet 1859). The papers approached the subject through the theory of differential equations, and the book can count as the first exposition of Cauchy's theory of analytic functions, although it is not without significant mistakes.[6]

[6] See Bottazzini and Gray (2013, 4.6).

Chapter 12
Revision

A good way to begin to think historically is to try to imagine what things looked like at some date in the past without the benefit of our modern knowledge. After all, that is how everyone must proceed in their own present—no-one can know the future with certainty, we must all take risks. I have put a generic question appropriate to such an approach at this stage in the course at the end of this chapter, along with some advice about how to tackle it.

Chapters 1–3

The first three chapters are scene setting for the chapters that follows. To consolidate your study, it is not as important to follow the mathematics in detail as it is to follow the arguments in outline, identifying the key points and assessing their historical significance.

When Lagrange criticised the poor state of the foundations of the calculus this was not the complaint of an outsider such as Bishop Berkeley but the considered opinion of the leading 'pure' mathematician of his day. He followed it with his own account of the foundations, and so gave the problem and his solution a heightened visibility in the mathematical community. There are two things to note about his solution: it is entirely algebraic, making no reference to infinitesimals, limits, or any other intuitive or geometrical argument; and it would suggest that every function is infinitely differentiable and can be given a power series expansion about any point in its domain. No-one objected to this conclusion, least of all Lagrange himself, so, knowing what we know today, we have to wonder why.

Possible answers are overlapping and partly contradictory. It is clear that Lagrange himself admitted that sometimes things can go wrong. The theorem is false for the square root function $y = \sqrt{x}$ at the origin, for example. But, Lagrange would say, of course it is. Theorems are accompanied by small print, their obvious exceptions. We deal in the same way today with statements about the natural or social world and allow that statements are true when they are only true most of the time (the

J. Gray, *The Real and the Complex: A History of Analysis in the 19th Century*, Springer Undergraduate Mathematics Series, DOI 10.1007/978-3-319-23715-2_12

claim that 'Anyone can run a mile in under twenty minutes' is not refuted by the observation that some people cannot run at all). Lagrange's theorem was taken to be true in general, obvious exceptions aside. There may be isolated points where the theorem is wrong, but quite generally the claim is correct.

And indeed, what would you say a typical function was? What would Lagrange and his contemporaries have regarded as a good example to test out a theory on? What sort of object would it be: a graph drawn by 'the free motion of the hand'? A finite algebraic expression? An infinite series, and if so, of powers of x or in terms of sines and cosines? A graph would presumably be capable of being expressed in some formalism, say that of infinite series, so the question becomes: what are they like? And, since the calculus is a local operation, what in particular are they like on very small domains? It is not unreasonable to believe that in fact all functions do have power series expansions, and no counter-examples to this general rule were known.

The same conclusion arises when we look at Fourier's work. There is more geometry here, but the same confidence that the calculus applies everywhere and all the time except where obviously—and trivially—it cannot. That attitude is a feeling that nothing in nature, and nothing in mathematics, obstructs the application of the calculus. It is then a matter of taste what one does with this feeling: for the 'purer' mathematician, Lagrange, it is better to have a proof; for the more 'applied' mathematician, Fourier, is is simply a reliable basis for getting on and doing good solid work.

As for Legendre's work on elliptic integrals, it was cause for an outright scandal that mathematicians could not integrate $\int \frac{dt}{\sqrt{(1-t^2)(1-k^2t^2)}}$, but we shall not be pursuing the resolution of that matter as a piece of real analysis, because that is not how mathematicians took the story up. It is worth noting, however, what Legendre did do with his lengthy investigations over the course of a productive life. He found numerous applications for the new functions he introduced, of which the pendulum and the rotating solid were and are obviously important. And to do so he evaluated tables for his new functions, by an ingenious argument. It is striking how few familiar functions there were at the time (sin, cos, log, exp, and a few more) and to be any use they have to be useful numerically in concrete problems. This raises a valuable question about any question in mathematics: what counts as an answer? One very good way to know a function is to know a table of values. (A graph is another, some general property of it a third, its behaviour for large values of the variable a fourth, and so on.)

In looking at this very important and highly regarded work by three major mathematicians we can also see a very interesting fact about mathematics as it is practised that will help us to understand the subject historically. A possible judgement of this work is that Lagrange was simply wrong, that Fourier was naive, and that Legendre missed the true significance of what he had achieved. Our problem with such a judgement is that it is historically crass, and it is worth thinking why.

First, we must admit its kernel of truth (most dangerous falsehoods have a kernel of truth). Lagrange *was* wrong that the calculus can be given rigorous algebraic foundations along the lines he suggested; Fourier *was* wrong, and certainly naive, in believing that every function equals its Fourier series; Legendre *did* miss the idea of inverting the elliptic integrals he studied and then treating them as functions of a complex variable.

However, a moment's thought reveals how useless such criticisms are. If a mathematician of the calibre of Lagrange, and everyone else of his generation, failed to come up with, say, Cauchy's ideas then this better suggests how difficult it was to carry through Cauchy's programme. As we noted, Grabiner has carefully described how much of the apparatus Cauchy used, in the form of an algebra of inequalities, was used by Lagrange. This did not suggest to people that an arithmetic defence of the calculus in terms of limits could be made to work, despite d'Alembert's (and even Newton's) remarks in that direction, and that says a lot about the magnitude of the change that Cauchy initiated.

Similarly, Fourier promoted a powerful idea without worrying about the finer details. That is not what mathematicians are taught to do in the 21st century—although one can suspect that not every truly original piece of research these days is entirely watertight. Fourier had every reason to believe his approach would work in many cases. What the later work investigating under what conditions, and to what extent, a function does obey Fourier's command tells us is the extent to which Cauchy's introduction of the purely continuous function into mathematics was to change every one's understanding of what a function is—and that was to be a century-long story if indeed it is finished even today.

And yes, as Legendre was the first to admit, Abel and Jacobi found marvellous new things to say about the elliptic functions that can be extracted from the elliptic integrals he spent much of his life studying. But Legendre had done a great deal to create new families of functions, give them both a theoretical basis and a practical one (his tables of their values), and demonstrate how they arise in significant areas of mechanics. There is no reason to expect more of anyone.

Just as we should not treat a famous mathematician from the past as a genius incapable of error, we should not treat them as failures for not seeing what later mathematicians saw, or what we value today. They did not say everything, and not everything they said was right: that is how research gets done.

Chapters 4–6

The three chapters on Cauchy's work in the 1810s and 1820s make an interesting contrast, one that was highlighted by his professional situation. His work on real analysis is very carefully structured, his work on complex analysis rather disorganised. But even to frame the comparison in those terms is misleading. Cauchy knew very well that there was a subject called the calculus that occupied itself with functions, differentiation, and integration. Even by 1830 it was only intermittently clear that there was a distinct, if overlapping subject that could be called complex analysis, and which occupied itself with a distinct class of functions. So he could accept the task of teaching the calculus and the consequent requirement to present his material in an orderly and instructive fashion and allow his research on complex functions to take him where-ever it led.

That said, nothing should obscure the remarkable degree of organisation in Cauchy's two lecture courses of 1821 and 1823 and the high degree of novelty they contain. The traditional presentation of this material put up a defence of the fundamental principles of the calculus and then got down to the proper business of developing methods for solving problems: evaluating integrals, solving differential equations, and generally thinking with the calculus. Cauchy introduced his novel concept of limit, made much more use of the ε, δ, N methodology (if only verbally), defined not only continuity but convergence, differentiable, and integrable, and therefore had much more work to do to substantiate and articulate these new, or newly reformulated, ideas. He was confronted at almost every stage with the need to recast old ideas and techniques in a way that would fit in with the theory that he was creating, before he could get round to solving traditional problems.

In the other half of this work, Cauchy was for much of the time occupied precisely with methods. How can you evaluate double integrals? When and why does the order of integration matter? Why, rather mysteriously, can it help to evaluate a real integral by passing to complex variables? So it was quite natural for him to consider, as he did, the meaning of an integral along a path in the complex plane first when the path is a side of a rectangle with sides parallel to the coordinate axes. In that context, guided by the fact that the value of a repeated integral does not usually depend on the order of integration, it is reasonable to suspect that the integral of a function taken around the whole boundary should vanish. It is also natural to suppose that when the integral round the whole boundary does not vanish (which is precisely the case when the order of integration does matter) that the function being integrated should display some singular behaviour. Put all this together, and it becomes entirely reasonable for Cauchy to think he had made tremendous progress with the problem at hand, that of studying real integrals by possibly complex methods. It is only a later generation, guided in part by Cauchy's own ideas of the 1840s and 1850s, that would see all this work of the 1820s as a set of deep discoveries and missed opportunities.

What of the other unexpected feature of his work, its mistakes? To be sure, they sit oddly with a man who has a reputation for introducing rigorous real analysis, but they sit rather comfortably with a man whose collected works, originally written over a period of some 40 years, fill 31 large volumes. One might suppose that someone writing so fast would occasionally miss important points. But in fact the issue of mistakes by mathematicians is worth thinking about. Some, such as the famous 'mistake' concerning the behaviour of a convergent sum of continuous functions, remind us just how deep and subtle the behaviour of functions can be. It is quite tricky to say exactly what Cauchy failed to see, even in modern terms. This alone should guard us against the foolish idea that the 'great' mathematicians see clearly all the time. Their mistakes can sometimes show that there was still much left to be discovered.

But there is another way in which Cauchy makes mistakes: he is prone to slip into his arguments assumptions that should have entered the formulation of his theorems. One could say that the attentive reader could (should?) simply rewrite the statements of such theorems accordingly, but the effect is disconcerting nonetheless and makes one wonder whether Cauchy was not himself confused. It surely made it harder for Cauchy himself to see clearly what was important.

It is likely that this habit or failing of Cauchy's was tied to an important view about the nature of mathematics, one that we no longer share (at least when we put on our philosophical hats), which is that mathematics is true because it is an idealised description of natural objects. The process of idealisation is unproblematic, it renders to the mind tidied up, simplified versions of real existing things that we can reason about accurately precisely because they have been simplified. The same process we saw at work with Lagrange, whose confidence in the differentiable nature of functions was sustained by his acquaintance with them, applies to Cauchy with his wider acquaintance. Cauchy in the 1810s and 1820s was occupied with the study of real functions, which could be illuminated on occasion by introducing imaginaries. There might then be a need to spell out carefully what the 'passage from the real to the imaginary' involves and say what becomes of a function of a real variable in this process. But there may not be any apparent need to have, invent, or find a theory of complex functions of a complex variable.

On this view one can sustain exceptions to a theorem: they do not refute the theorem until there is a better point of view that gives them greater weight. This explains why a palpable falsehood bothered neither Cauchy nor Abel, but a contrived example produced by Cauchy was destructive for Lagrange's foundations of the calculus. The example, $y = e^{-1/x^2}$, is a strange object. Its definition must of course be augmented by the requirement that $y = 0$ at $x = 0$, which makes it not only continuous but infinitely differentiable at the origin. Such a definition was entirely novel in its day, when all functions were defined by formal expressions. Why was this peculiar object not dismissed as another 'exception', if not indeed banished from mathematics altogether as too contrived?

The most obvious answer, which is not as cynical as it may sound, is indeed correct: it was not marginalised because its discoverer did not want it to be marginalised. Cauchy wanted this function not only to show a weakness in Lagrange's theory but to show a virtue of his own. It became a forceful example, indeed a counter-example, because it is part of another way of thinking about functions that has many other features to commend it. One reason a contemporary could see readily that there was going to be no way this function could be dismissed as belonging to the realm of trivial exceptions was that it belonged very naturally to another substantial theory.

All of which suggests that one helpful way to follow the history of mathematics might be in terms of what mathematicians thought they were talking about, and discovering not only more things to talk about but better things to say about them. Mathematicians who believe, rightly or wrongly, that they know what they are talking about can look at new results and assess their worth as a naturalist does, in terms of their truth, how typical or important they are, and start to assess their implications long before they look at any proof.

Chapters 7–11

It is clear that the work of Abel and Jacobi has its origins in the work of Legendre. Legendre's theory of elliptic integrals had established itself as a new branch in the theory of functions of a real variable, one that enlarged the domain of what could be explicitly integrated and added to the list of problems in mechanics that could be solved. The contribution of Abel and Jacobi was to turn elliptic integrals into elliptic functions and to make the new elliptic functions a rich and interesting collection of complex-valued functions of a complex variable. This was a different, more conceptual, and, if only because complicated, a more abstract kind of knowledge. Why was this regarded as exceptionally good mathematics, worthy of prizes and capable of making people's reputations? Why was the sensible reaction not to say things like this: elliptic functions are useless, it is elliptic integrals that arise in actual problems; complex-valued functions of a complex variable are an abstraction made solely for abstraction's sake and we need not (and therefore should not) go there?

Such questions should always be asked of any new piece of mathematics. It is only healthy to ask, in an open-minded way, what is good and important about any new piece of work. Answers may be hard to come by: it is hard to predict the future and easy to make cheap claims for or against anything while waiting for real evidence to come in. The historian, of course, is more interested in reasons given at the time, not those with hindsight or those that 'ought' to have been given. In the present case, we know that Legendre was delighted to see his elliptic integrals given a new lease of life, and that both Abel and Jacobi had shown how well the new ideas solved problems raised by Legendre's work and extended it. The new functions also had much more comprehensible properties than the corresponding integrals, and not only a richer collection too, but one that was made accessible by the analogy with the familiar trigonometric functions. It helped, of course, that existing and well-regarded problems elsewhere in mathematics were illuminated by the new ideas.

It is harder to say what feelings were about the truly complex (as opposed to real) nature of the new functions. Algebra and analysis in the 18th Century had been no more involved with the complex numbers than was necessary, or perhaps opportune. At the start of the 19th Century there had been a flurry of interest in the philosophical 'meaning' of complex numbers. There was a renewed interest in proofs of the fundamental theorem of algebra, and Gauss, in his *Disquisitiones Arithmeticae*, had shown that there were good reasons for studying what are called cyclotomic integers. These are expressions involving ordinary integers and roots of unity in a way that kept many of the properties of the familiar integers—and of course one could ask: what was good about taking mathematics in that direction? It's an open question as to whether a growing interest in complex variable and complex functions came before the work of Abel and Jacobi, was crystalised by it, or was created by it. In this connection the different attitudes of Gauss and Cauchy are worth thinking about, and then indeed one would want to know how these exceptional mathematicians influenced their contemporaries. Were they listened to, were they regarded as too difficult?

If we look past 1835 we can begin to see what people did in fact do in response to all this mixture of work. We see new developments in the theory of complex functions, complicated by Cauchy's wanderings round Europe, and better and better attempts to use the new theory to make sense of the great examples of complex functions, which were the elliptic functions. The obstacle that they all perceived was the two-valued nature of the integrand in an elliptic integral, which seemed to take it out of Cauchy's domain of ideas entirely; it is worth thinking about why this was so. Do you think this will turn out to be a permanent obstacle, or will it be overcome?

The contrast between the bustling, energetic, and oddly solipsistic Cauchy and the quieter, more socially responsible Gauss is fascinating at every level. Purely metaphorically, I imagine Cauchy as the dinner guest who talks all the time, is frequently brilliant, and moves rapidly from one topic to another and back always with something to say. Gauss I imagine as a person of many fewer words, but, on balance, with the more difficult and deeper things to say. Gauss, more than Cauchy, saw that complex numbers and complex functions were an important new development, even if one wants to say that Cauchy did more for the subject in the end.

But the fundamental question remains: what would you, as an aspiring mathematician in 1830 or 1835 have done with all this new work? Carried on as you were? Learned the new stuff at once? Perhaps it depends on how confident you were of your own merits, your own potential contributions.

A Question

Imagine it is 1835 and you are young British professor finishing a year or two studying mathematics abroad. You are writing a letter to a very good student who will take your place out here, advising them on what they should study when they arrive. Describe recent developments in EITHER the theory of functions of a real variable OR in the topic of complex functions and complex integrals.

Advice

Think hard about what constitutes good advice, as opposed to information.

When you are offering an opinion or judgement (as in 'How important was . . .' or 'why was . . . .') give a brief argument in support of your opinion.

Distinguish between contemporary criticisms of someone's work and your own judgements. Don't be afraid of offering your own judgements, but do not say anything from the perspective of the present day that could not have been said in 1835.

You do not have very many words, so anything you say about personalities must be essential.

Chapter 13
Gauss, Green, and Potential Theory

13.1 Introduction

The basic idea of a potential function is very simple. It applies to problems in dynamics where one is given a force, and the idea is to find a function $f(x, y, z)$ whose partial derivatives give the components of a force. The function, if it exists, is called the potential function for the given force. It was introduced, without a name, by Lagrange in his study of gravitational attraction. The idea became even more important when Coulomb showed that electrostatic forces also obeyed an inverse-square law and then Ampère showed that there is also an inverse-square law in electrodynamics. It is in this fast-moving context that gauss and Green did their important theoretic work on potential functions in their 1830s and 1840s that is the subject of this chapter.

13.2 Potential Theory

First, some remarks about notation (for a revision of the theory, see Appendix C). In the 1830 s and 1840s mathematicians wrote out the corresponding expressions in full. Today we have a good notation using symbols like ∇f and Δ, or terms like div, grad, and curl. These are not only compact, they allow the use of vector operations (the dot and cross product) and the use of vector identities. I have decided to use ∇f and Δ, but purely as abbreviations.

We shall write the three partial derivatives of a function as $\left(\frac{\partial f}{\partial x_1}, \frac{\partial f}{\partial x_2}, \frac{\partial f}{\partial x_3}\right)$, denote it by ∇f, and refer to it as the gradient of f. We also write

$$\Delta V := \frac{\partial^2 V}{\partial x^2} + \frac{\partial^2 V}{\partial y^2} + \frac{\partial^2 V}{\partial y^2}.$$

J. Gray, *The Real and the Complex: A History of Analysis in the 19th Century*,
Springer Undergraduate Mathematics Series, DOI 10.1007/978-3-319-23715-2_13

Given a function f and a surface with an outward pointing normal $\mathbf{n}$ at each point, we write

$$\frac{\partial f}{\partial S} \text{ and } \nabla f.\mathbf{n}$$

for the normal derivative of the function f.

Lagrange introduced the idea of a potential function in 1773. His question was: how to determine the potential function for a given body, and thus its gravitational attraction? After Legendre had opened up the subject, Laplace showed how to find the potential function, V, for the gravitational force due to an arbitrary spheroid in 1782 by assuming that it satisfied the differential equation

$$\Delta V := \frac{\partial^2 V}{\partial x^2} + \frac{\partial^2 V}{\partial y^2} + \frac{\partial^2 V}{\partial y^2} = 0,$$

which is why we speak of Laplace's equation today. In the 1780s Charles Coulomb, a French military engineer, proposed the inverse-square law in electrostatics, which differs from the law of gravitation only because the force is one of repulsion and so a minus sign enters the formulae. In his efforts to produce a mathematical theory of electrostatics in 1813 Poisson was led to generalise Laplace's equation to $\Delta V = -4\pi\rho$, which applies also to points inside a solid (the function ρ is the density of the solid).

These mathematicians had noticed that a force $\mathbf{F}(\mathbf{p})$ exerted on a unit mass at a point $\mathbf{p}$ by a mass distribution $\rho(\mathbf{x})$ is given by the integral

$$\mathbf{F}(\mathbf{p}) = \int \frac{\rho(\mathbf{x})(\mathbf{x} - \mathbf{p})}{|\mathbf{x} - \mathbf{p}|^3} d\mathbf{x}.$$

This is a complicated vector expression, but Lagrange's potential function is the much simpler scalar function

$$V(\mathbf{p}) = \int \frac{\rho(\mathbf{x})}{|\mathbf{x} - \mathbf{p}|} d\mathbf{x}.$$

Direct computation shows that $-\nabla V(\mathbf{p}) = \mathbf{F}$, and for the forces of gravity and electro-statics, as Laplace showed, $\Delta V = 0$ wherever $\rho = 0$ and $\Delta V = -4\pi\rho$ wherever $\rho \neq 0$.

Among the reasons for wanting to study potential functions were the motions of the moon and more generally of the newly discovered asteroids. The moon was Gauss's interest in 1813 when he studied homogeneous ellipsoids. He had other interests that we cannot consider in this course for reasons of space, notably the two-dimensional case when the corresponding potential function is called the logarithmic potential, and he was interested in the problem of the conformal mapping of one region on another in connection with geodesy and terrestrial map making.

But the topic became even more interesting when it was shown that it applied to the new and much more mysterious processes of magnetism and electro-dynamics.[1] The leading figure here was the French mathematician André-Marie Ampère, who responded within days to Oersted's experiments on electro-magnetism with a rush of activity in 1820. In experiments involving the magnetism produced by an electric current, the Earth's magnetic field is a complicating factor. Ampère derived an experimental set up that eliminated this effect, and Oersted's complicated findings were immediately simplified. By 1826 Ampère had written his definitive masterpiece, his *Mémoire sur la théorie mathématique des phénomènes électrodynamiques, uniquement déduite de l'expérience* (or, *Memoir on the mathematical theory of electrodynamics solely deduced from experiment*).

Ampère offered a unified theory of electricity and magnetism in the form of mathematical laws, and seldom ventured into the realm of physical explanations about the nature of electric current. Magnetism he explained in terms of electric currents in the magnetised body, an idea he extended to the magnetism of the Earth. His abstinence deliberately recalled the way Fourier had not offered a physical account of heat and heat diffusion. His mathematical theory fitted in well with the contemporary and brilliant experimental work of Faraday, and for a time seemed to offer a better explanation of the facts than Faraday's, and so they were widely taken up.

In 1828 Gauss visited the newly built magnetic observatory that Alexander von Humboldt had constructed for Berlin University on French principles. Gauss was not impressed; he was used to much higher standards of accuracy in his work on astronomy and land surveying. He went back to Göttingen and three years later was able to team up with the young physicist Wilhelm Weber, who was appointed to the chair of physics at Göttingen in 1831. They were attracted by the opportunities for new mathematics, for a new or at least deepening science of electromagnetism, and for applications to navigation and geodesy. They were successful in all three respects, although we can only follow the first of them here. In this course it is more interesting to pursue the development of the general theory, and, Gauss aside, one name stands out in the 1830s and 1840s, that of the Englishman George Green. Green published first, but his work was not picked up until after Gauss's had been read, so in terms of their reception it is better to take Gauss first.

Before we do, note also one key difference between potential theory applied to the study of mass and applied to the study of charge. The more important departure from gravitation to electrostatics follows from the fact the charge in a body is located entirely on its surface, so it is natural to study the surface distribution of charge. This gives rise to a potential $V(\mathbf{p}) = \int \frac{\rho(\mathbf{x})}{|\mathbf{x}-\mathbf{p}|} dS$, where the integral is taken only over the surface. On the surface it is assumed not only to be continuous but to have tangential derivatives, and two one-sided normal derivatives: the (outward) normal derivative is $\frac{\partial u}{\partial \nu_+} = \boldsymbol{\nu}_+ . \nabla u$ (the limit is taken along the normal going outwards) and the (inward) normal derivative is $\frac{\partial u}{\partial \nu_-} = \boldsymbol{\nu}_- . \nabla u$ (the limit is taken along the normal

[1] For the next few paragraphs this treatment follows Darrigol's *Electrodynamics from Ampère to Einstein*, (2000).

going inwards). These differ because there is mass in the surface, and their difference is:

$$\frac{\partial u}{\partial \nu_-} - \frac{\partial u}{\partial \nu_+} = 4\pi\mu,$$

where μ is the mass distribution on the surface.

Because of the importance of the subjects to which potential theory could be applied, there developed a prodigious industry of finding classes of functions specially adapted to particular cases of these equations. For example, if the Earth is the solid of interest, it might be considered as a sphere, a spheroid (rather less likely as an ellipsoid), or as almost spheroidal, each change bringing greater accuracy and relevance at the cost of greater complexity. For example, to this day the vibrations of the earth caused by earthquakes are studied using functions called spherical harmonics. We do not have the time to go in this direction, which requires a certain virtuosity, but it should not be forgotten.

13.3 Gauss

Gauss's most important work on the subject is his *Allgemeine Lehrsätze* (*General Propositions*) of 1840. In it he first gave rigorous proofs of results already published by the French—although Gauss for some reason did not mention them by name. Then he set out some new results of his own. The first is a *reciprocity theorem*: If μ and μ' are two mass distributions on regions vol and vol' respectively, and V and V' are their associated potential functions, then $\int_{vol} V'\mu = \int_{vol'} V\mu'$. His proof for this is very straight forward. He started with the simplest system, one consisting of a single mass M at the point $\mathbf{P}$ and with a potential function V, the other of a single mass M' at the point $\mathbf{P}'$ with a potential function V'. In this case

$$MV'(\mathbf{P}) = M\frac{M'}{|\mathbf{P} - \mathbf{P}'|} = M'V(\mathbf{P}').$$

So a similar expression holds for two mass distributions each consisting of a finite number of points, in the form of a sum

$$\Sigma_i M_i V_{i'}(\mathbf{P}_i) = \Sigma_j M_{j'} V_j(\mathbf{P}_{j'}),$$

where the sums are taken over points in each system (to get started, suppose one consists of one point and one of two, and write out the corresponding expressions.)

Finally, for continuous mass distribution on curves, or in surfaces or in solids, the sum is replaced by an integral. Gauss also paid attention the case when the two mass distributions lie on the same surface, which he dealt with as the limiting case of mass distributions on two surfaces that are brought together.

It is sometimes easiest to regard the reciprocity theorem as the equality $\int V'\mu = \int V\mu'$ where the two integrals are taken over the whole of space, and a non-zero contribution to an integral is made only where there is mass.

From this Gauss derived an expression for the potential $V(0)$ at the centre of a sphere S of radius R due to a mass distribution μ outside the sphere and having a potential V. This is his Theorem of the arithmetic mean

$$V(0) = \frac{1}{4\pi R^2}\int_S V \,.$$

Gauss gave this proof. Let S denote the sphere of radius R, B its interior, and X its exterior, $X = \mathbb{R}^3 \setminus (B \cup S)$. We shall compare three mass distributions. The first is given by a function μ in the region X that lies outside the sphere S; this mass distribution has a potential V. The second mass distribution, μ_1, is a uniform distribution on the sphere of mass $4\pi R^2$. The potential function associated with this, V_1', is a constant, $4\pi R$, inside the sphere, and at a distance $r > R$ outside the sphere the potential is $4\pi R^2/r$. Notice that these two functions for the potential associated to the second mass distribution agree on the sphere. It is a not entirely trivial fact that in regions outside the sphere the potential is the same as the potential of a mass of $4\pi R^2$ at the centre of the sphere, and this leads us to our third mass distribution, μ_2, which is indeed a mass of $4\pi R^2$ at the centre of the sphere, which is the origin, O, with the corresponding potential function $V_2(r) = 4\pi R^2/r$.

Now, the reciprocity theorem applied to the first and second of these says

$$\int_{\mathbb{R}^3} V\mu_1 = \int_{\mathbb{R}^3} V_1\mu.$$

We write each integral as a sum of three: $\int_{\mathbb{R}^3} = \int_B + \int_S + \int_X$, and we then find that some of the contributions are zero:

$$\int_B V\mu_1 + \int_S V\mu_1 + \int_X V\mu_1 = 0 + \int_S V\mu_1 + 0,$$

because the mass distribution μ_1 is concentrated on the sphere S. So the left hand side reduces to $\int_S V\mu_1 = \int_S V$, because the mass distribution is uniform. The right hand side, when similarly written as a sum, becomes

$$\int_B V_1\mu + \int_S V_1\mu + \int_X V_1\mu = 0 + 0 + \int_X V_1\mu,$$

because the mass distribution μ is concentrated in the region X. Furthermore, $\int_X V_1\mu = \int_X \frac{4\pi R^2}{r}\mu$. So the first use of the reciprocity theorem says that

$$\int_S V = \int_x \frac{4\pi R^2}{r}\mu \,.$$

To deal with the second of these integrals, we apply the reciprocity theorem to the first and third integrals. It now says

$$\int_{\mathbb{R}^3} V\mu_2 = \int_{\mathbb{R}^3} V_2\mu.$$

We write each integral as a sum of three: $\int_{\mathbb{R}^3} = \int_B + \int_S + \int_X$, and we again find that some of the contributions are zero:

$$\int_B V\mu_2 + \int_S V\mu_2 + \int_X V\mu_2 = \int_B \mu_2 + 0 + 0,$$

because the mass distribution μ_2 is concentrated at the centre of B. So the left hand side reduces to $\int_O V\mu_2 = 4\pi R^2 V(0)$. The right hand side, when similarly written as a sum, becomes

$$\int_B V_2\mu + \int_S V_2\mu + \int_X V_2\mu = 0 + 0 + \int_X V_2\mu,$$

because the mass distribution μ is concentrated in the region X. So this use of the reciprocity theorem says that

$$4\pi R^2 V(0) = \int_X \frac{4\pi R^2}{r}\mu\,.$$

But the first use said that $\int_S V = \int_X \frac{4\pi R^2}{r}\mu$. So we deduce that $\int_S V = \frac{1}{4\pi R^2} V(0)$, which establishes Gauss's Theorem of the arithmetic mean.

Gauss then showed that if the given mass distribution also has a total mass M^0 inside the sphere then

$$\int_{S'} V = 4\pi R^2 V(0) + 4\pi R M^0.$$

From these results we can deduce (although Gauss did not) that this implies that the potential cannot have a maximum or a minimum outside the mass distribution.

Gauss then proved what is today called Gauss's law or Gauss's theorem:

If V is a potential due to a mass distribution and M the total mass inside a closed surface S then

$$\int_S \frac{\partial V}{\partial S} = -4\pi M.$$

The theorem has this corollary: If there is no mass inside then

$$\int_{surf} \nabla V.\mathbf{n} = 0.$$

Gauss then proved that

$$\int_{vol} \nabla V.\nabla V = -\int_{surf} V(\nabla V.\mathbf{n}).$$

It will be convenient to follow Green's proof a little later on, which is not that different from Gauss's anyway. For a sketch proof and a critical comment, see the remarks at the end of this chapter.

From this Gauss deduced that if a potential due to a charge distribution on a surface is constant then it is zero everywhere if and only if the total charge is zero, and the potential is smaller than this constant value away from the surface if and only if the total charge is different from zero. In particular, the potential function itself cannot be zero everywhere in a region without vanishing altogether, as we shall shortly see.

13.4 Green

The English mathematician George Green was born on 14 July 1793, in the town of Nottingham. He inherited a successful milling and bakery business from his father, and on the basis of his own reading in libraries and with a little help from local mathematicians who had passed through Cambridge he set to work on his own in mathematics. He read what he could of the relevant literature, which was not much: Laplace's *Mécanique Celeste*, some papers by Poisson, Biot, and Coulomb (more, by the way, than he could have learned if he had gone to Cambridge). Then, in 1828 he wrote the remarkable work entitled *An Essay on the Application of Mathematical Analysis to the Theories of Electricity and Magnetism*, which we shall consider below and which brought him to the attention of Sir Edward Ffrench Bromhead, a local dignitary and Cambridge graduate. In due course Green himself went to Cambridge, where he was a student and then, for a brief time, a Fellow. At Cambridge he wrote his few other papers, which deal with magnetism, electricity, and hydrodynamics, but finding little to stimulate him there he returned to Nottingham, where he died a few weeks short of his forty-eighth birthday, not even, it would seem, a local celebrity.[2]

Green had become interested in the hugely popular topic of the emerging new physics of electricity and magnetism. People were excited by dramatic public displays of experiments, and less flamboyant scientists discovered new phenomena for which there was no satisfactory theory. Green set himself the task of facilitating, as he put it, "the application of analysis to one of the most interesting of the physical sciences". The new physics was to call for new mathematics, and was based on intuitions of a non-Newtonian kind. This was to prove the secret of their lasting impact. The very word 'analysis' in the title reminds us that Green was self-taught, for it refers to the calculus as it had become in France, not to the sterile exercises in Newtonian methods preferred in England.

[2]We know remarkably little about Green, and much of what we do know was discovered by the indefatigable Mary Cannell and can be found in her fine book on Green: *George Green, Mathematician and Physicist, 1793–1841* (1993).

To Green is due the convenient term 'potential function' that describes the effect upon a point of forces coming from a system of bodies. The introduction of the name helped move the concept to centre stage. Green had learned from Laplace and Poisson that the potential function satisfies two partial differential equations: one for points outside the body, and another for points inside. He then formulated what today is called Green's theorem, which can be thought of as a reciprocity theorem, and which generalises the fundamental theorem of the calculus to several variables.

Green's reciprocity theorem concerns a closed surface S enclosing a volume V, and two functions, U and U', of the 3 variables x, y, z. We may write dv for the volume element in V and ds for the area element of S, although I shall usually suppress that part of the notation. We also write $f_{\mathbf{n}}$ for the normal derivative of a function f—Green wrote δ where we write Δ, w for what we write as $\mathbf{n}$, and $\left(\frac{df}{dw}\right)$ for what we write as $f_{\mathbf{n}}$.

Green's reciprocity theorem says that

$$\int_V U\Delta U' dv - \int_V U'\Delta U dv = \int_S U' U_{\mathbf{n}} - \int_S U U'_{\mathbf{n}}.$$

One way to understand this is to take U' to be the constant function 1, when the theorem reduces to

$$\int_V \Delta U dv = -\int_S U_{\mathbf{n}},$$

when it now says Gauss's law.

It seems that Green initially found his reciprocity theorem difficult to prove, but he eventually came to this elegant proof. He considered the integral $\int_{vol} \nabla U.\nabla V$, where U and V are arbitrary functions, so $\nabla^2 U \neq 0$. Green, more or less as Gauss was to do, integrated this expression by parts, taking each of the three parts of the sum separately, which gave him

$$\int_{vol} \nabla U.\nabla V = \int_{surf} V(\mathbf{n}.\nabla U) - \int_{vol} V\Delta U.$$

This reduces to Gauss's law when U is a potential function and there is no mass in the region being integrated over, but in the general case that Green was working with the final term does not vanish. However, Green noted that U and V can be exchanged and the integral does not change, so it is also the case that

$$\int_{vol} \nabla V.\nabla U = \int_{surf} U(\nabla V.\mathbf{n}) - \int_{vol} U\Delta V.$$

On subtracting this from the first equation, he obtained the desired result:

$$\int_{vol} V\Delta U - \int_{vol} U\Delta V = \int_{surf} [U(\mathbf{n}.\nabla V) - V(\mathbf{n}.\nabla U)] \qquad (13.1)$$

It remained for Green to show how to solve Laplace's and Poisson's equation and so determine the potential function for a wide variety of distributions of mass or charge. He had the remarkable idea of solving these equations in extreme cases, where the solution is easy to discover, notably the case where the potential was caused by a single charge at an isolated point. In this case the potential function therefore satisfies Laplace's equation in any region that does not contain the point. Green assumed that the potential increased like $1/r$ as one neared the point charge, and showed how to solve the equation for such a curious function by an attractive use of the consequences of Green's theorem that he had developed in the earlier part of his *Essay*. These functions are called Green's functions today, and they are a standard part of the repertoire in potential theory.

To pass from the solution to the problem with these extreme boundary conditions to more plausible ones where the electricity is distributed through an entire body, Green argued first that a problem with finitely many point charges is solved simply by adding the solutions, and second that infinitely many point charges distributed in some way are handled by replacing addition with integration, on the assumption that the potential function vanishes on the boundary of the body.

Green's reciprocity theorem is a piece of rigorous mathematics, whether pure or applied, but his analysis of potential functions in terms of 'Green's functions' was not, as he was well aware, although he did give a plausible limiting argument to show how such potential functions can be defined. Amusingly, naive arguments involving Green's functions have tended to strike mathematicians as likely to make sense on physical grounds, and physicists (such as Maxwell) as likely to be amenable to rigorous mathematics.

However, there are many other situations where a point source of influence is a natural object to study. To mathematicians among Green's eventual readers the example of a complex function was irresistible. Cauchy had shown during the 1820s that much of their theory follows from the distribution of their poles, and later writers, such as Riemann, saw Green's theorem as the natural way to prove the Cauchy integral theorem. To a physicist a single force concentrated at a point might represent an impulse. A load on a beam might be concentrated at a point. These extreme situations have the advantage over more general ones of simplifying the attendant mathematics, which is why Green's functions have become such a powerful idea.

Green's *Essay* was not appreciated in his lifetime, although his later papers earned him a modest reputation. That the *Essay* became known at all was the work of William Thomson, who had picked up a stray reference to it and finally tracked the *Essay* down the day before taking his degree and leaving for Paris in January 1845. Thomson was entranced. He took the *Essay* to Paris, and showed it to Joseph Liouville and his friend Charles Sturm. They too were excited by it, and so too was Crelle, who immediately accepted it for publication in his journal, the leading mathematical journal of the day. It was published in three installments between 1850 and 1854.

The combination of ideas, their elegance, and their clear presentation even when rigour lay out of reach, impressed mathematicians, and rapidly ensured that Green's name was henceforth securely attached to his discoveries, even when some had by then been discovered by others. From Crelle, news of the essay passed to Dirichlet, the leading mathematical analyst of his day, and from him to Bernhard Riemann, and William Thomson remained a staunch advocate of them, as was Maxwell.

The standard equations of mathematical physics can always be treated using Green's functions. The reason is that the equations are linear, and so may be said to have a family of solutions, every one of which is a sum (in a suitable sense) of what may be called the basic solutions. If a basis of solutions can be found by solving extreme cases (such as concentrating the electricity at a point) then the method of Green's functions will work, and this is usually the case. If the equations are not linear, their solutions may still fill out a cone, in which case the ones arising from the extremal problems will lie on the boundary of the cone.

Gauss's Theorem (After Feynman)

Given a vector field $\mathbf{V}$ and a volume Ω enclosed by a surface S, first consider an infinitesimal cube in the volume, of sides dx, dy, dz. The flow along the vector field out of the upper face of the cube is given by multiplying the vertical component of $\mathbf{V}$, which we write $\mathbf{V}_3(upper)$ by the area of the face, $dxdy$: $\mathbf{V}_3(upper)dxdy$. As for the lower face, the flow there is $\mathbf{V}_3(lower)dxdy$, so the flow through these faces is $\mathbf{V}_3(upper)dxdy - \mathbf{V}_3(lower)dxdy$. This is $(\mathbf{V}_3(upper) - \mathbf{V}_3(lower))dxdy$. Now, the difference $(\mathbf{V}_3(upper) - \mathbf{V}_3(lower))$ can be evaluated by taking the first term in the Taylor expansion: $(\mathbf{V}_3)_z dz$. So the flux through these faces is $(\mathbf{V}_3)_z dxdydz$. Therefore the flux through the cube (by looking at the two other pairs of faces) is $((\mathbf{V}_1)_x + (\mathbf{V}_2)_y + (\mathbf{V}_3)_z)dxdydz$. This can be written as $\nabla.\mathbf{V}dv$. On the other hand, the flow is also the sum $\mathbf{n}.\mathbf{V}$, the sum of the normal directions.

We pass from the infinitesimal cube to the finite solid body by integration. The adjacent faces of neighbouring cubes contribute equal and opposite amounts to the sum, because what leaves one enters the other, and so the contribution here is entirely from the surface, and the result is $\int_V \nabla.\mathbf{V}dv = \int_S \mathbf{n}.\mathbf{V}$. Note that the RHS can be written as a normal derivative.

It is worth pointing out that this informal argument has one interesting flaw: it is unclear how the infinitesimal cubes can fill out a body whose sides are not parallel to the coordinate planes. When we recall the familiar paradox in two dimensions that 'shows' that the two shorter sides of a right-angles triangle have a length equal to the hypotenuse we can begin to see the need for infinitesimal solids with sides parallel to arbitrary planes.

In this fallacious argument, the sides AB and BC of a right-angled triangle (see Fig. 13.1) are repeatedly bisected to produce a sequences of curves

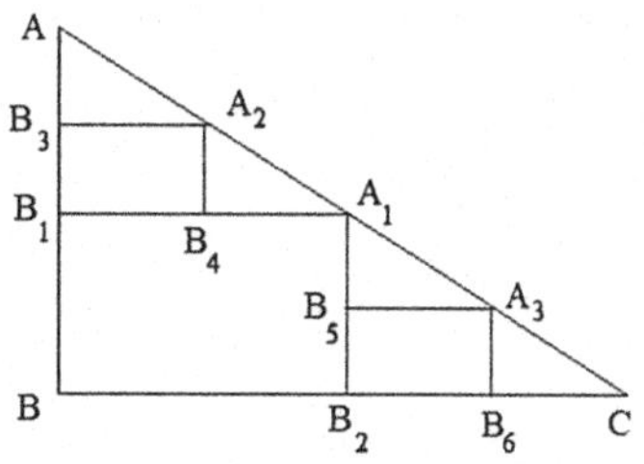

Fig. 13.1 Two sides of a triangle successively approximating a third

$$ABC,\, AB_1A_1B_2C,\, AB_3A_2B_4A_1B_5A_3B_6C, \ldots$$

then tends pointwise to the edge AC. Each curve has length $AB + BC$ and the lengths of these curves do not, of course, tend to the length AC.

Chapter 14
Dirichlet, Potential Theory, and Fourier Series

Gustav Lejeune Dirichlet (1805–1859)

14.1 Introduction

Dirichlet is often regarded as the man who, more than any other, brought rigour to mathematical analysis. Here we look at his career and then at two of his contributions: the Dirichlet principle in potential theory and his study of when a Fourier series converges to the function that it represents.

J. Gray, *The Real and the Complex: A History of Analysis in the 19th Century*,
Springer Undergraduate Mathematics Series, DOI 10.1007/978-3-319-23715-2_14

14.2 Dirichlet's Career

Gustav Peter Lejeune Dirichlet is one of the most important mathematicians of the 19th Century. He was born in Düren, near Cologne in Germany in 1805, which at the time was occupied by the French under Napoleon (it became part of Prussia after 1815). Finding that there was not much opportunity to learn advanced mathematics where he was, at the age of 17 he went to Paris, for Paris was then by far the leading place for mathematics in the world. There he made many good social contacts, including Fourier, and picked up a strong interest in Fourier series and mathematical physics. However, his consuming interest was in number theory, and when he was 20 he presented a memoir to the French Academy of Sciences in which he showed that many cases of the equation $x^5 + y^5 = z^5$ cannot be solved in integers. The remaining cases were then dealt with by Legendre, but unhappily Legendre's publication unduly minimised Dirichlet's contribution. Even so, Dirichlet had almost completely resolved the first new case of Fermat's Last Theorem since Euler had dealt with cubes. Dirichlet returned to Germany in 1826 where his acquaintance with Fourier brought him an introduction to Alexander von Humboldt, whose brother Wilhelm had founded the new University of Berlin in 1810, and in 1828 Dirichlet was appointed a Professor of Mathematics there. In 1831 he married Rebecca Mendelssohn-Bartholdy, the sister of the famous composer; Dirichlet was a good pianist in his own right.

Dirichlet went on to make many contributions to number theory, one of the most notable being his proof that every arithmetic progression of the form $\{a + bk \mid k = 1, 2, \ldots\}$ where a and b are relatively prime contains infinitely many primes. Through his papers and his lectures, he explained Gauss's work to the rest of the world so successfully that he helped make number theory into a major branch of pure mathematics, and almost the leading branch in Germany. This work led directly to the work of Richard Dedekind in the 1860s and beyond, and thence to Hilbert.

More important here was Dirichlet's attitude to mathematical rigour, captured marvellously in the remark his life-long friend Jacobi made in a letter to Alexander von Humboldt:

> If Gauss says he has proved something, it seems very probable to me; if Cauchy says so, it is about as likely as not; if Dirichlet says so, it is certain. I would gladly not get involved in such delicacies.[1]

And Dirichlet's wife tells us that Jacobi would spend hours with Dirichlet

> Being silent about mathematics. They never spared each other, and Dirichlet often told him the bitterest truths, but Jacobi understood this well and he made his great mind bend before Dirichlet's great character.[2]

[1] Jacobi to A. von Humboldt, 21 December 1846, quoted in Biermann (1973, 53).

[2] Quoted in Scharlau and Opolka (1985, 148).

It can be argued that Dirichlet. more than Cauchy or even Gauss, is the man who brought rigorous mathematics into the world. Two examples, one more successful than the other, will illustrate what that means. The first is his work on potential theory, the second his work on the convergence of Fourier series.

14.3 Dirichlet and Potential Theory

Green had claimed in 1833 that one can solve this problem: given a solid body and a continuous function V defined on its boundary, there is a function f which is harmonic on the interior of the body and agrees with V on the boundary of the body. His argument was as follows: the successful function minimises the integral

$$\int_{vol}\left(\left(\frac{\partial f}{\partial x}\right)^2+\left(\frac{\partial f}{\partial y}\right)^2+\left(\frac{\partial f}{\partial z}\right)^2\right)dxdydz.$$

Thomson made the same claim a few years later, in 1847, but the assertion has been attached for many years to the name of Dirichlet because of lectures he gave in 1856–1857 at Göttingen that were published posthumously in an edition by Grube in 1876.

According to Grube, what was known as 'Dirichlet's principle' was the claim that there is always a unique function on an arbitrary bounded domain which, with its first partial derivatives, is continuous, satisfies Laplace's equation everywhere in the domain, and reduces to a given value at every point of the boundary of the surface. So in Grube's version the Dirichlet Principle was this bold existence and uniqueness theorem.[3]

In the lectures, Dirichlet observed that the problem of finding such a function explicitly "cannot be solved; we can only speak of an existence proof for it. The latter presents no difficulty." He now stated the 'Dirichlet principle':

> For every bounded connected domain T there are clearly infinitely many functions u continuous together with their first-order derivatives, for x, y, z which reduce to a given value on this surface. Among these functions there will be at least one which reduces the following integral $U=\int_{vol}\left((\frac{\partial u}{\partial x})^2+(\frac{\partial u}{\partial y})^2+(\frac{\partial u}{\partial z})^2\right)$ extended over the domain T, to a minimum; it is evident that this integral has a minimum since it cannot become negative. We can now show the following:
>
> 1. Every such function u which minimizes U, satisfies the differential equation $\frac{\partial^2 u}{\partial x^2}+\frac{\partial^2 u}{\partial y^2}+\frac{\partial^2 u}{\partial z^2}=0$ everywhere in the domain T. This already makes it clear that there always exists a function u having the desired property, namely that function for which U becomes a minimum.

[3]This remark, and the quote that follows, are taken from Dirichlet (1876, 127–128), quoted in Bottazzini, *The* Higher Calculus, 300.

2. Every function u which satisfies the [above, JJG] differential equation within the domain T, minimizes the integral U.
3. The integral U can have only one minimum. It follows from 2 and 3 that there is only one function u with the desired property.

In short, the 'Dirichlet problem'—find a function satisfying the above conditions—can be solved by an appeal to the 'Dirichlet principle'. The principle provides a function that minimises a certain integral, and this function solves the problem.

When this 'principle' is applied in a genuine physical setting it seems entirely reasonable: nature herself tells you there is an equilibrium distribution of charge, and so forth. But there is a question when the matter in hand is entirely mathematical. Proofs are required for each of the next three assertions:

there are functions that extend the function defined on the surface to the whole of the domain in question;
there are functions of this kind for which the integral is finite; and
there are functions of this kind that minimise the integral. Unless these conditions can be met the class of functions Dirichlet wished us to consider is empty and the argument stops right there. When Dirichlet said "It is evident that this integral has a minimum since it cannot become negative", the only proper reply is "Not exactly, Lord Copper". The function $f(x) = x$ defined on the open interval $(0, 1)$ is never negative but it never has a minimum. The function $F(x) = \tanh(x) + 1$ defined on the whole of $\mathbb{R}$ is never negative but it too never has a minimum. It is surprising, and a little disappointing, that Dirichlet would make such naive remarks, especially given the sophistication of his earlier work on Fourier series to which we turn next.

The fate of the Dirichlet principle will detain us later, when we look first at the work of Dirichlet's greatest student, Bernhard Riemann, and then at the criticisms his work engendered among his followers and in the rival school of complex function theorists, led by Karl Weierstrass in Berlin.

14.4 Dirichlet and Fourier Series

Dirichlet came to the topic of Fourier series knowing only, he said, of one paper by Cauchy on the subject. There had, in fact, been quite some discussion of Fourier's ideas in print, and one can suppose that Dirichlet heard them discussed during his stay in Paris. For example, Poisson had commented on Fourier's methods for finding the coefficients in a Fourier series as follows: "It seems to me that the formula for the expansion of functions in series has not in fact been demonstrated in a precise and rigorous manner",[4] a comment that Bottazzini says seems to have been widely shared. Later, Poisson wrote (1835, 186) "the determination of the coefficients essentially supposes that one knows, in addition to the form of the functions, that they are expandable"—here I take it that it is the expandability of an arbitrary given function

[4]See Poisson (1823a, 46), quoted in Bottazzini, *The* Higher Calculus 188, as is the next quotation.

that is at stake. Bottazzini notes that Sturm endorsed this shrewd observation: "Fourier and other geometers seem to have misunderstood the importance and the difficulty of this problem, which they have confused with that of determining the coefficients".[5] The distinction here is worth noting. It might be that the coefficients could all be evaluated and the corresponding Fourier series shown to converge—does it follow that the Fourier series is equal to the function for every value of the variable (in the given interval)?

Cauchy took up the subject in 1827, dismissed Poisson's approach as unconvincing, and wrote[6]

> In series of this type, the coefficients of the different terms are ordinarily definite integrals that include sines and cosines; and when the integrations can be made, by reason of the particular form attributed to the function that it is necessary to expand, it is easily seen that the series obtained are convergent. Nevertheless it is always desirable that this convergence be demonstrated in a general manner, independently of the values of the functions.

Cauchy's 'proof' is interesting for a variety of reasons. It uses his new theory of complex integration and his theory of residues, it produces a convergence criterion that turns out to be flawed, but most of all because it gives us a vivid picture of a mathematician at work, pulling ideas from unexpected places and trying anything that might work. As Bottazzini points out (p. 190) the resulting paper looks more like experimental science than the logical hypothetico-deductive structures we expect meta-mathematics to be. Here I pass over the use of complex function theory to get to the convergence question.

Cauchy wanted to investigate the convergence of a Fourier series of a function $f(x)$ on the interval $[0, a]$. He took the series, and transformed it into another (using his complex methods) and then argued that the convergence question reduced to establishing the convergence of a series $\Sigma_n v_n$ by comparison with one whose general term is w_n. In particular, he claimed that if the series $\Sigma_n w_n$ converges, and $w_n \to v_n$ as n tends to infinity, then $\Sigma_n v_n$ also converges. (In the case at hand, Cauchy set $w_n = -\frac{1}{2n\pi}(f(a) - f(0)) \sin(\frac{2n\pi x}{a})$).

Dirichlet began his paper (see the translation in Appendix A.2) by observing that the fact that a Fourier series expansion of an arbitrary function converges is remarkable (note that he did not doubt the convergence). But, he said, no-one had given a satisfactory proof, and so far as he knew only Cauchy had really tried, and that proof was flawed by fallacious convergence argument. Moreover, the use of complex methods seemed foreign to the topic at hand, and he noted that Cauchy elsewhere also spoke of the need for caution when making real functions into complex functions.

As for the convergence 'criterion', that was refuted by this example. Let

$$w_n = \frac{(-1)^n}{\sqrt{n}} \text{ and } v_n = \frac{(-1)^n}{\sqrt{n}}\left(1 + \frac{(-1)^n}{\sqrt{n}}\right).$$

---

[5]See Sturm (1836, 400), quoted in Bottazzini, *The* Higher Calculus, 188.

[6]Quoted in Bottazzini, *The* Higher Calculus, 188.

The first series is convergent, the second one is not because $v_n = w_n + \frac{1}{n}$, so the difference of the second from the first is the well-known divergent series $1 + \frac{1}{2} + \frac{1}{3} + \cdots$. However, both w_n and v_n tend to zero as n tends to infinity.

Dirichlet therefore proposed to establish the convergence of a Fourier series directly, and also that the series and the function agreed. The Fourier series associated with a function $\phi(x)$ on the interval $[-\pi, \pi]$ Dirichlet wrote as

$$\frac{1}{2}b_0 + b_1 \cos x + b_2 \cos 2x + \cdots + b_m \cos mx + \cdots$$
$$+ a_1 \sin x + a_2 \sin 2x + \cdots + a_m \sin mx + \cdots$$

where the coefficients are given by

$$b_m = \frac{1}{\pi}\int_{-\pi}^{\pi} \phi(\alpha) \cos m\alpha d\alpha,$$
$$a_m = \frac{1}{\pi}\int_{-\pi}^{\pi} \phi(\alpha) \sin m\alpha d\alpha.$$

The series is therefore composed of terms of the form

$$\cos mx \cos m\alpha + \sin mx \sin m\alpha = \cos m(\alpha - x),$$

by a standard trigonometry identity. The first n terms of the Fourier series can then be summed, and the sum is:

$$\frac{1}{\pi}\int_{-\pi}^{\pi} \frac{\sin(2n+1)\frac{\alpha-x}{2}}{2\sin\frac{\alpha-x}{2}}\phi(\alpha)d\alpha.$$

So the summation of the Fourier series is established if and only if this integral, today known as a Dirichlet integral in Dirichlet's honour, converges to a finite sum as n tends to infinity.

Dirichlet first considered the case where the arbitrary function $\phi(x)$ is continuous, positive, and monotonic decreasing in $[0, h]$ with $0 < h < \frac{\pi}{2}$. He looked at the convergence of the integral $\int_0^h \frac{\sin(2n+1)\beta}{\sin\beta}\phi(\beta)d\beta$ as n tends to infinity. The argument is delicate and original. The integrand is 'infinite' when the denominator vanishes, which happens when $\beta = 0$ and (outside this range of integration, when $\beta = 2k\pi$. The integrand changes sign whenever $\sin(2n+1)\pi\beta$ vanishes, which happens much more often, whenever $(2n+1)\beta$ is an even integer, that is, whenever $0 < \beta = \frac{2k}{2n+1} < 1$. For future use, let us temporarily call an interval any of the regions between the points where the integrand vanishes (and so where the integrand has a single sign).

An aside Fig. 14.1, which is the graph of $\frac{\sin(41x)}{\sin x}$ in the range $[-\pi, \pi]$, shows clearly that the function takes its maximum values at $-\pi$, 0 and π, and otherwise

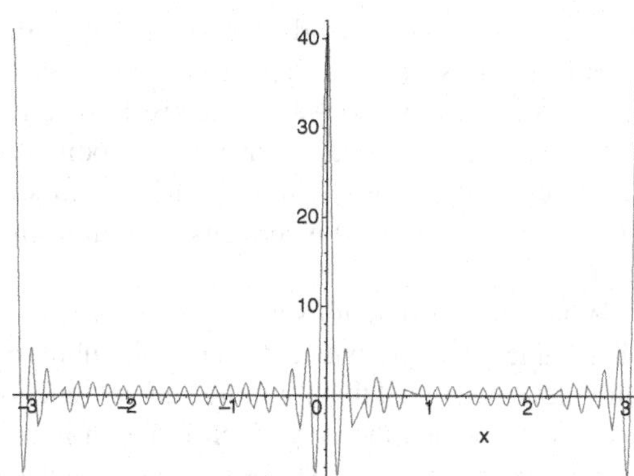

Fig. 14.1 The function $\frac{\sin(41x)}{\sin x}$ in the range $[-\pi, \pi]$

contributes small alternating amounts to the integral $\int_{-\pi}^{\pi} \frac{\sin(41x)}{\sin x} dx$. Remarkably, $\int_{-\pi}^{\pi} \frac{\sin(41x)}{\sin x} dx = 2\pi$.

Now, the absolute value of the integrand shrinks as one goes from each interval to the next, because the function $\phi(x)$ is monotonic decreasing. So the value of the integral is a finite sum of an alternating series of positive and negative terms that shrink in absolute value, and the question is to establish the convergence of this sum as the number of terms included increases without limit. By the mean value theorem applied to the function $\phi(x)$ this alternating series is a sum of terms ρ_k times an integral of the form $\int_{(k-1)\pi/n}^{k\pi/n} \frac{\sin n\beta}{\sin \beta} d\beta$. Because the absolute values on each interval are shrinking, the absolute values of the ρ_k are all less than the absolute value of the first one, ρ_1. So the problem is reduces to the study of sums of terms of the form

$$\int_{(k-1)\pi/n}^{k\pi/n} \frac{\sin n\beta}{\sin \beta} d\beta.$$

As n tends to infinity the integral $\int_{(k-1)\pi/n}^{k\pi/n} \frac{\sin n\beta}{\sin \beta} d\beta$ tends to

$$\int_{(k-1)\pi}^{k\pi} \frac{\sin \gamma}{\sin \gamma/n} d\gamma/n$$

—put $\gamma = \beta n$. Happily, $\int_0^\infty \frac{\sin \gamma}{\gamma} d\gamma$ is finite, in fact equal to $\frac{\pi}{2}$—A non-trivial result first obtained by Euler. So the alternating series converges as n tends to infinity, and with it the integral in β.

What then of the Fourier series? Dirichlet now understood the value of the integral in β and therefore the integral in $\alpha - x$, which I repeat here for convenience:

$$\frac{1}{\pi} \int_{-\pi}^{\pi} \frac{\sin(2n+1)\frac{\alpha - x}{2}}{2 \sin \frac{\alpha - x}{2}} \phi(\alpha) d\alpha.$$

His arguments apply to two parts of this integral, the integral from $-\pi$ to x and from x to π, as can be seen from the substitution $x - 2\beta = \alpha$ in the first case and

$x + 2\beta = \alpha$ in the second case. Note that when you make these substitutions the variable x turns up in the upper end points of the integrals, which now run from 0 to $\frac{\pi+x}{2}$ and from 0 to $\frac{\pi-x}{2}$ respectively. The result of this is that the conclusions of the preceding arguments have to be modified somewhat, and they now give that the limit of the integral from $-\pi$ to π (and so the sum of the Fourier series) is $\frac{1}{2}(\phi(x+\varepsilon) - \phi(x-\varepsilon))$ where ε is an infinitesimal. So, if ϕ is continuous, the limit is $\phi(x)$.

What class of functions has Dirichlet shown to have convergent Fourier series? Those functions ϕ for which $\phi(x)$ is continuous, positive, and monotonic decreasing. Dirichlet now argued that for more general functions ϕ one can consider the different intervals on which the new given function is continuous, positive, and monotonic decreasing or increasing. The original argument works in each of these settings, and so convergence is again assured on the whole interval—provided that the decomposition of the function ϕ into monotonic regions requires only finitely many intervals. For the same reason, the condition that the function ϕ be continuous everywhere can be dropped to the extent that the function may be allowed to jump at finitely many points (which produce a further decomposition of the original interval). Dirichlet spoke of the function having sudden changes at isolated points.

The convergence of the Fourier series was therefore established by Dirichlet for functions (on the interval $[-\pi, \pi]$) with finitely many points where they have jump discontinuities and which have a finite number of maxima and minima. The value of the Fourier series at the jump was shown to be the average of its values, one might say, on either side of the jump.

Dirichlet then added a famous remark. The necessity of imposing the condition that the function have only finitely many jumps was felt, he said, by contemplating a function $\phi(x)$ that takes one value c when x is rational and another value d when x is irrational. This strange function, never previously considered in mathematics, was introduced to show the value of a certain restriction needed for the proof of the theorem. It could be, of course, that the restriction could be partly lifted; Dirichlet's function showed however that some restriction was necessary.

Dirichlet's proof had also imposed some other conditions on the functions that could be shown to agree with their Fourier series. They must have only finitely many maxima and minima in any interval. Dirichlet noted that this condition too required both comment and further investigation, and expressed the hope that he would return to the topic in a later note, but he was never to do so. There is some evidence that he was optimistic on this point, but if so he was to be proved wrong in 1876, long after his death, as we shall see.

Dirichlet's work shows a shift of concern, from the description of expected properties ('every function has a convergent Fourier series to which it is equal') to the perceived need to establish these properties together with the observation that they cannot be expected to hold in general.

A Trigonometric Argument

We want to evaluate $\frac{1}{2} + \sum_{k=1}^{n} \cos(kt)$, where $t = x - \alpha$. Consider

$$
\begin{aligned}
1+\sum_{k=1}^{n}(\cos(kt)+i\sin(kt)) &= 1+\sum_{k=1}^{n}e^{ikt} \\
&= \frac{1-e^{i(n+1)t}}{1-e^{it}} = \frac{e^{-i(n+1)t/2}-e^{i(n+1)t/2}}{e^{-it/2}-e^{it/2}}\frac{e^{i(n+1)t/2}}{e^{it/2}} \\
&= \frac{\sin((n+1)t/2)}{\sin(t/2)}e^{int/2} \\
&= \frac{\sin((n+1)t/2)}{\sin(t/2)}(\cos(nt/2)+i\sin(nt/2)).
\end{aligned}
$$

So

$$
\frac{1}{2}+\sum_{k=1}^{n}\cos(kt) = \frac{\sin((n+1)t/2)}{\sin(t/2)}\cos(nt/2)-\frac{1}{2}\,.
$$

To bring this into line with the neater expression that Dirichlet used, we write

$$
\frac{\sin((n+1)t/2)}{\sin(t/2)}\cos(nt/2)-\frac{1}{2} = \frac{\sin((n+1)t/2)\cos(nt/2)-\frac{1}{2}\sin(t/2)}{\sin(t/2)}\,.
$$

Then we use the trigonometric identity $2\sin A\cos B = \sin(A+B)+\sin(A+B)$ to rewrite this as

$$
\frac{\frac{1}{2}\sin((2n+1)t/2)+\frac{1}{2}\sin(t/2)-\frac{1}{2}\sin(t/2)}{\sin(t/2)} = \frac{\sin((2n+1)t/2)}{2\sin(t/2)}\,,
$$

which is the expression Dirichlet used.

Chapter 15
Riemann

Bernhard Riemann (1826 –1866)

15.1 Introduction

This is the first of four chapters in which we look at various aspects of Riemann's work and the way it changed the theory of real and complex functions. In the short space of fifteen years—1851 to 1866—Riemann rewrote both subjects as he did differential geometry and analytic number theory. Here we look at how his introduction of trigonometric series opened the way for mathematicians to study the properties of non-differentiable functions.

J. Gray, *The Real and the Complex: A History of Analysis in the 19th Century*,
Springer Undergraduate Mathematics Series, DOI 10.1007/978-3-319-23715-2_15

15.2 Lipschitz

Among those who responded to Dirichlet's paper on Fourier series was Rudolf Lipschitz in his doctoral dissertation (Lipschitz 1864). Lipschitz considered the ways in which a function could fail to satisfy the sufficient conditions that Dirichlet had established for the convergence of a Fourier series. For example, the function f might be discontinuous at an infinite number of points, although otherwise monotonic and piecewise continuous in the interval $[-\pi, \pi]$. But the difficulty here was that Dirichlet's concept of the integral did not apply in this case, and so Lipschitz wondered if the integral could be suitably redefined. He came to the conclusion that it could, at least if the set of discontinuities was nowhere dense.[1] Here his argument was incorrect, but this was not noticed for a time, which shows how difficult this question really is.

Lipschitz's fallacious argument was that if D, the set of discontinuities, was nowhere dense, then D', the set of limit points of D, must be finite. In this case an argument about singular integrals can be mounted that extends Dirichlet's theorem appropriately. For example, if D' consists of the single point c, then f has at most finitely many points of discontinuity in any interval $(-\pi, c - \varepsilon)$ and in $(c + \varepsilon, \pi)$, for any $\varepsilon > 0$. We shall recall later the significance of looking topologically at the point set D and not in a way that illuminated the concept of integrability.

Once he was past this mistake, Lipschitz then argued correctly that Dirichlet's monotonicity condition could be replaced by a condition on the function that is stronger than piecewise continuity but weaker than differentiability—the original Lipschitz condition.

15.3 Riemann

Riemann was the archetype of the shy mathematician, not much drawn to topics other than mathematics, physics, and philosophy, devout in his religion, conventional in his tastes, close to his family and awkward outside them.[2] As a child, he was taught by his father, a pastor, and then for some years at school before going to Göttingen University. There he had initially intended to study theology, in accordance with his father's wishes—Göttingen was the only university in Riemann's native Hanover with strong links to the Hanover church—but his remarkable ability at mathematics led him to switch subjects. He was always inclined to the conceptual side of things, rather than the computational or algorithmic, and interestingly enough never learned to write his native German eloquently, and his Latin (required for academic purposes) was even worse.

---

[1]In modern terms, a set A is nowhere dense in another set B if the closure of the interior of A is empty. We shall see that this concept was not reliably understood in the period before it was made precise by the techniques of 20th century topology.

[2]For a biography of Riemann, see Laugwitz, *Bernhard Riemann* (2000).

Although Gauss was in Göttingen, where he taught statistics, the level of instruction at Göttingen was not high, and Riemann transferred to Berlin in 1847; German students were allowed to study at any university. There he found his ideal instructor and guide in Dirichlet, from whom he learned about potential theory and partial differential equations, number theory and theory of integration. He also took Jacobi's classes in analytical mechanics and higher algebra, and it has been suggested that these lectures might have stimulated Riemann's tendency to think in an abstract and sophisticated way about the relation of mathematics to physics and the real world.[3]

Riemann returned to Göttingen in 1849 and attended Wilhelm Weber's lectures of mathematical physics and for a while he devoted himself to studies in physics and Naturphilosophie. He met Richard Dedekind, who was greatly influenced by him throughout his long mathematical career, and in 1855 Dirichlet came to Göttingen as Gauss's successor. Dirichlet was a major influence not only on Riemann's work on trigonometric series and number theory, but also on his general, abstract approach. However, Dirichlet died in 1859. Riemann was quickly appointed his successor, but as it was to turn out he had by then published nearly all of his major papers, for he contracted pleurisy in 1862. The only treatment was to spend as much time as possible in the South, and whenever he could he travelled to Italy, where on 20 July 1866 he died near Lake Maggiore, where he is buried. His friends arranged for his published papers, several unpublished ones that were found to be in a good enough state, and some that could be edited from notes, to be published, and the first edition of his *Werke* (edited by Dedekind and Heinrich Weber) appeared in 1876.

No-one disputed Riemann's brilliance, but he had few students, and among them Hankel and Roch died young, and Hattendorff at 48. His successor at Göttingen was Alfred Clebsch, who was joined by Ernst Schering, who was not a truly creative mathematician. Clebsch stepped forward to take up Riemann's ideas, but died, aged 39, in 1872, and eventually Felix Klein took up the task, but by then Berlin had become the dominant centre for mathematics in Germany, and Weierstrass and Kronecker much preferred algebra to geometry, let alone topology, and Riemann's deepest ideas fell into suspicion. His preference for highly abstract arguments over long calculations and explicit methods was alien to them, and rather than make Riemann's insights rigorous on, as it were, Riemann's own terms Weierstrass and his former student Hermann Amandus Schwarz frequently sought to rewrite them in their own, very different, manner.

15.4 Riemann's Publications

The first of Riemann's publications was his doctoral thesis on the foundations of a theory of functions of a complex variable. In it he presented the idea of a 'Riemann surface' as a branched covering of the complex plane, and outlined a proof that any two disc-shaped regions of the plane are not only topologically equivalent but

[3]See Pulte in Jacobi (1996).

analytically so (there is a 1–1 analytic map from each one to the other). The thesis was privately distributed, and had no impact until it was reprinted 25 years later in his Collected Works.

Riemann could now work for the crucial post-doctoral qualification the Habilitation (a necessary and sufficient condition to be allowed to lecture at a German University). In 1854 he successfully presented his Habilitation essay, on trigonometric series and the theory of functions, and gave his Habilitation lecture, on the foundations of geometry. Whole branches of mathematics stem from each of these.

In the lecture he argued that mathematics is about 'n-fold extended quantities' ('n-tuples of real numbers' is the equally unattractive modern term) to which is added some appropriate extra structure, such as a metric to measure distance. This view of differential geometry gradually allowed mathematicians to propose many different descriptions of physical space, because any set of n-fold extended quantities with a metric will do, and as a result Euclidean geometry became just one possibility among many. In particular, complex numbers are pairs of real numbers; some extra structure (given by the Cauchy–Riemann equations) enables one to define analytic functions.

15.5 Riemann's Paper on Trigonometric Series

One of the papers Riemann wrote for his Habilitation in 1854 was undoubtedly the one he wrote under the influence of Dirichlet.[4] Its subject is the representation of a function by its Fourier series. It opens with a remarkably thorough account of the history of the topic to date, going back as far as Euler and d'Alembert, and which also contains information obtained directly from Dirichlet himself. This account forms the kernel of every historical account written since. Turning to the mathematics, Riemann conceded that functions that do not satisfy Dirichlet's conditions are unlikely to occur in nature, but he still felt that they were worth studying for two reasons (see Riemann 1867a, §3). One was that Dirichlet had argued that "the topic has a very close connection with the principles of infinitesimal calculus, and can serve to bring greater clarity and rigour to these principles.". The other was that Fourier series were useful in pure mathematics, including number theory, where precisely the sorts of functions that Dirichlet had excluded were important.

Riemann began his investigation by redefining the Cauchy integral (see the extract in Appendix A.4). Cauchy had considered only those subdivisions of an interval into subintervals of equal size. Riemann proposed to let the intervals vary, and said that a function f on a given interval was integrable if the area given by the obvious rectangles tended to a limit as the widths of the intervals in the partition tended to zero. This is not much of a departure from Cauchy's idea. But then Riemann argued that if the oscillation of the function on each interval is also bounded and the

[4] The entire paper has been translated into English and published in Bernhard Riemann, *Collected papers*, translated by Roger Baker, Charles Christenson and Henry Orde, Kendrick Press, 2004; see pp. 219–256.

contributions of the 'tops' of the rectangles is bounded as the intervals tend to zero, then the function will still be integrable. His idea was that each partition presents an approximation to the function as a step function. If the successive step functions do not differ uncontrollably as the intervals shrink, then the function should have an area.

So Riemann defined integrability without reference to continuity. Next he turned his attention to functions with infinitely many points of discontinuity. As he put it (1867a, §6) "Since these functions have never been considered before, it will be good to give a definite example". He first defined the function $r(x)$ as follows.[5] For each real number x,

$$r(x) = \begin{cases} x - m(x) & \text{if } x \neq n/2 \\ 0 & \text{if } x = n/2 \end{cases}$$

where $m(x)$ is the integer such that $|x - m(x)|$ is a minimum, and n is odd. So the function $r(x)$ takes values between $-1/2$ and $1/2$, and is zero where $x = n/2$ with n an odd integer. Its graph is discussed and illustrated in Sect. 15.6.

Riemann then defined the function

$$f(x) = r(x) + \frac{r(2x)}{2^2} + \cdots + \frac{r(nx)}{n^2} + \cdots.$$

This function is discontinuous at every point $x = m/2n$ where m and $2n$ are relatively prime; this is a dense set of rational numbers. At these points, the right and left hand limits of the function f are

$$f(x+) = f(x) - \frac{1}{2n^2}\left(\sum_k \frac{1}{(2k+1)^2}\right) = f(x) - \frac{\pi^2}{16n^2}$$

and

$$f(x-) = f(x) + \frac{1}{2n^2}\left(\sum_k \frac{1}{(2k+1)^2}\right) = f(x) + \frac{\pi^2}{16n^2}.$$

Nonetheless, the function f is integrable, because the contribution of jumps exceeds a given bound at only finitely many points.

A good way to think of Riemann's fundamental question about integrals, which led him to redefine the Cauchy integral, is to think of the integral $\int_a^x f(t)dt$ as x goes from a to, say, b as calculating a moving average. To be precise, by the intermediate value theorem, $\int_a^x f(t)dt = (x-a)g(x)$, where $g(x)$ is the average value of $f(x)$ on the interval (a, x). So Riemann was looking for functions which may oscillate wildly and even be discontinuous, but which always have an average value at every point of an interval on which they are defined. From this perspective it is not only reasonable that an integrable function can fail to be defined at some points (the average of

[5] Riemann's notation for this function was (x), but this causes problems by disappearing when the function is composed with others, so I have introduced this notation.

infinitely many values should not be affected if we lose finitely many values) it is quite plausible that many points of discontinuity and some degree of oscillation in the values of the function should not affect its integrability. In the example of the case just considered each jump affects the average value by a finite amount, but the function is integrable because the contributions of the jumps are negligible except at finitely many points.

Commentators agree that Riemann's deepest insight came with his radically new point of departure. He began with a trigonometric series, which he wrote

$$\Omega(x) = A_0 + A_1(x) + A_2(x) + \cdots + A_n(x) + \cdots, \tag{15.1}$$

where

$$A_0 = a_0/2,\ A_n(x) = a_n \cos nx + b_n \sin nx.$$

In §9 he assumed that $A_n(x)$ converged to 0 as n increased—in fact, he treated this as uniform convergence throughout the paper: the convergence is for all values of x under consideration. Next, he formally integrated the series $\Omega(x)$ twice to obtain the series

$$F(x) = C + C'x + A_0x^2/2 - \sum_n \frac{A_n(x)}{n^2} \tag{15.2}$$

The hypothesis on the coefficients $A_n(x)$ ensure that the series $F(x)$ converges for every value of x and represents a continuous function of x. In fact, the function $F(x)$ has the important properties that (Lemma 1):

$$\mathcal{D}^2F(x) := \lim_{\alpha,\beta\to 0} \frac{F(x+\alpha+\beta) - F(x+\alpha-\beta) - F(x-\alpha+\beta) + F(x-\alpha-\beta)}{4\alpha\beta}$$

exists and equals $\Omega(x)$ whenever Ω converges.
Also (Lemma 2):

$$\lim_{\alpha\to 0} \frac{F(x+\alpha) - 2F(x) + F(x-\alpha)}{\alpha} = 0 \tag{15.3}$$

for all x.

Riemann then proved the important result (Lemma 3): today known as Riemann's Lemma.

Let $b < c$ be arbitrary real numbers, $\lambda(x)$ a function with continuous first differential quotient within $[b, c]$, and which vanishes at b and c, and whose second differential quotient has only finitely many maxima and minima in $[b, c]$. Then

$$\lim_{\mu\to\infty} \mu^2 \int_b^c F(x) \cos \mu(x-a)\lambda(x)dx = 0.$$

The proof is astute manipulation of series, rather than difficult.

Using these results, in §9 Riemann derived necessary and sufficient conditions for a function to be represented at a point by a trigonometric series Ω. These were that

1. For a function f to be representable on $[0, 2\pi]$ by a Fourier series, there must be a continuous function F, derived from f by double formal integration, such that Lemmas 1 and 3 hold.
2. Conversely, if these conditions hold, f is representable by the Fourier series everywhere the Fourier series converges.

The proofs of these conditions made heavy use of the Dirichlet integral, and will not be discussed here. Nor were they found to be convincing, although they are correct: Smith and du Bois-Reymond both disputed the sufficiency.

In §10 Riemann made the remarkable observation that if the coefficients of Ω become infinitely small then the convergence at a value of x depends only on the immediate neighbourhood of x. This is in marked contrast with the idea that the series is defined on a whole interval, and if it were a Fourier series its coefficients would be determined as integrals over that interval. But it is absolutely in line with Dirichlet's observation in the much simpler case that he had considered.

Thus far, Riemann has assumed that the coefficients of the trigonometric series $F(x)$, Eq. (15.1), all become arbitrarily small for large enough n whatever the value of x. He now started to weaken this assumption—I omit a discussion of this section.

Thus far he had largely considered functions of the kind that Dirichlet had looked at. In §12 Riemann began a study of the representation of functions which become infinite at finitely many points but have only finitely many maxima and minima, by trigonometric series whose terms become arbitrarily small. Then he considered the representation of functions which have only finitely many maxima and minima, by series whose terms do not become infinitely small.

Then in §13 he opened the door to all sorts of unexpected, and largely unknown, functions. A list of his results must suffice:

1. Functions with infinitely many maxima and minima can be integrable without being representable by a Fourier series. For example,

$$\frac{d(x^{\nu}\cos(1/x))}{dx}$$

on $0 \leq x \leq 2\pi$, with $0 < \nu < 1/2$. His proof involves showing that the integral that would define the Fourier coefficients produces a quantity that does not satisfy the conditions laid down in the above lemmas.

2. Conversely, there are functions which cannot be integrated but can be represented by a trigonometric series. For example, the function

$$\sum_{k=1}^{\infty} \frac{r(kx)}{k}$$

exists for every rational value of x and is representable by

$$\sum_{1}^{\infty} \frac{s(x)}{n\pi} \sin(2n\pi x),$$

where $s(x) := \sum \sigma^d - (-1)^d$ and is taken positive or negative according as n is odd or even, and σ ranges over all the divisors of n. This series is not bounded in any interval, no matter how small, and consequently is not integrable.
Another example is

$$\Sigma c_n \cos n^2 x,$$

where $c_n \to 0$ but $\Sigma c_n = \infty$. As Riemann explained, "For, if $x/2\pi$ is rational with denominator m in lowest terms, then clearly the series either converges or diverges to infinity according as

$$\sum_{n=0}^{m-1} \cos n^2 x, \ \sum_{n=0}^{m-1} \sin n^2 x$$

are zero or not. Both cases arise, by a well-known theorem on partitioning the circle, for infinitely many values of x between any two bounds, no matter how close." (The reference was to Gauss, *Disquisitiones Arithmeticae* §356.)

3. Likewise the trigonometric series Ω can converge without the series obtained from it by term by term integration being integrable on any interval, however small. For example

$$\Sigma \frac{1}{n^3} \left(1 - q^n\right) \log \left(\frac{-\log(1 - q^n)}{q^n} \right),$$

where the logarithms are chosen so as to vanish for $q = 0$. By Riemann's lemma, this gives an example of a function representable by a trigonometric series that is not a Fourier series.

4. Finally a trigonometric series can converge for infinitely many x in any interval and yet the coefficients not tend to zero. For example

$$\Sigma \sin(n!\pi x),$$

which converges not only for every rational value of x but also some irrational numbers, such as $\sin 1$, $\cos 1$, $\frac{2}{e}$ and its multiples, etc.

Figure 15.1 shows the first four terms of Riemann's series (4), which converges at some points even though the coefficients do not tend to zero.

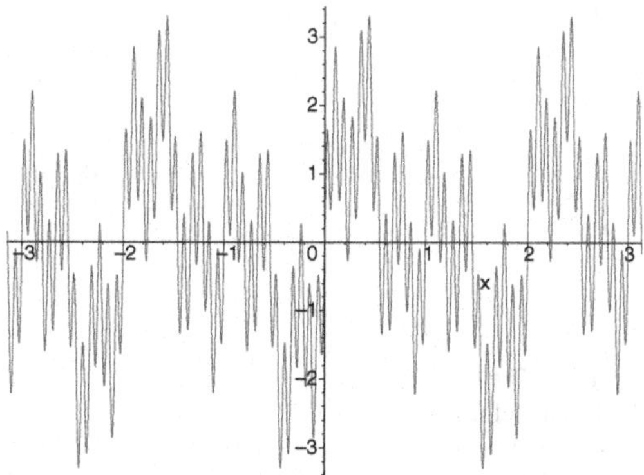

Fig. 15.1 The first four terms of $\sum_1^\infty \sin(n!\pi x)$

Summary

Riemann's key insight was the introduction of the trigonometric series. No assumption was made about the coefficients a_n and b_n, in particular they need not be the coefficients of a Fourier series. This distinction between a Fourier series and a trigonometric series was introduced by Riemann, and he gave examples of trigonometric series that are not Fourier series.

Riemann took some steps towards investigating functions that do not satisfy Dirichlet's conditions for representation by their Fourier series, but historians agree that he found more problems than he could solve. He was not to be the first! But in his case it may have inclined him to keep the paper back from publication, so it was only published for the first time in 1867, after his death the previous year, and only then did it begin to influence the theory of real functions and their Fourier series.

Strange functions, however, continued to occupy him, and Weierstrass was later to recall that he first heard in 1861 or even earlier from students who had attended Riemann's lectures in Göttingen that the function represented by the trigonometric series $\sum_k \frac{\sin(n^2x)}{n^2}$ is a continuous function that is nowhere differentiable. This statement is not quite accurate mathematically—I shall correct it in due course (see p. 204).

15.6 Riemann's Function

Riemann's strange function is very nearly the same as the function $g(x) = x - 1 - \lfloor (x - 1/2) \rfloor$. Both have period 1 and the same graph except that Riemann's function vanishes at the points $n/2$ where n is an odd integer and the new function g takes the value $-1/2$ there. For graphing purposes these differences are insignificant, because we may change the value of a function at finitely many points without changing its integral. But these differences need to be remembered when taking an infinite sum

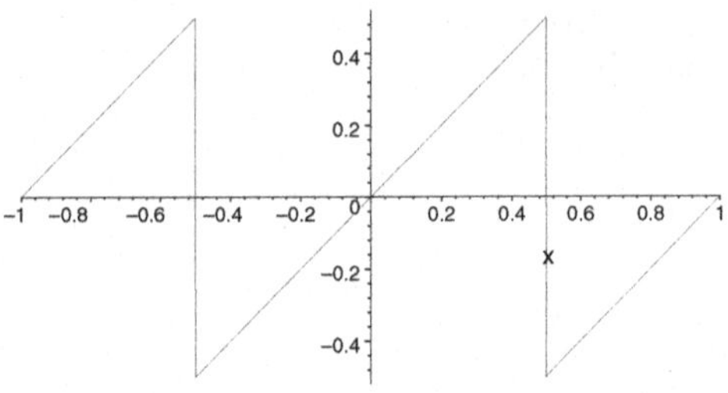

Fig. 15.2 Riemann's function $g(x)$

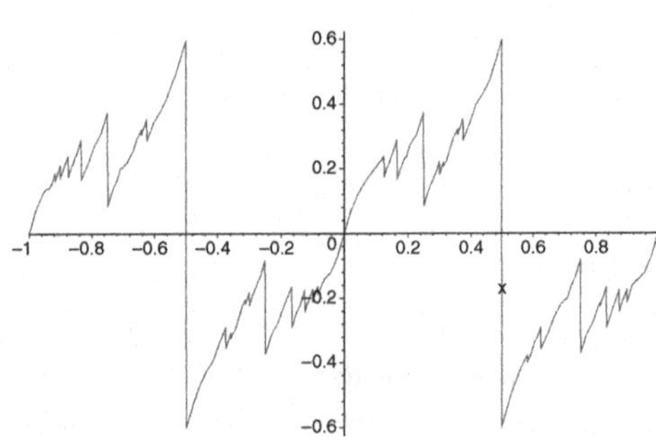

Fig. 15.3 The first 15 terms of the sum defining Riemann's function

(which generates an infinite set of discrepancies). We can study finite approximations to Riemann's function

$$\sum_{1}^{\infty} \frac{r(kx)}{k^2}.$$

as if they are

$$\sum_{1}^{N} \frac{g(kx)}{k^2}.$$

The function $g(x)$ has a graph like this (see Fig. 15.2; the verticals are an artefact of Maple and are incorrect mathematically—the function has a jump discontinuity at those points).

Figure 15.3 shows the first 15 terms of the sum defining Riemann's function.

The next graph, Fig. 15.4, shows the difference between the first 15 and the first 5 terms of the sum defining Riemann's function. It shows how crinkly the successive graphs are becoming.

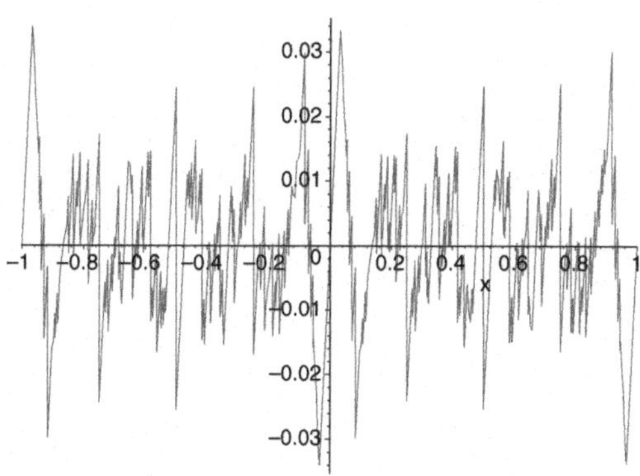

Fig. 15.4 The crinkliness of terms 6 to 15 of the sum defining Riemann's function

15.7 Exercises

1. What value could be attached to unrigorous mathematics in the 1840s? Give examples of mathematics showing various levels of rigour.
2. Does mathematics have to be more rigorous than physicists require?
3. How can a continuous function fail to be differentiable? Can you give any examples of a function that is continuous on an interval but fails to be differentiable at infinitely many points? What properties would a function have to have for it to be continuous on an interval and differentiable nowhere?

Chapter 16
Riemann and Complex Function Theory

16.1 Introduction

In his doctoral dissertation of 1851 Riemann gave a complete, clear introduction to complex function theory as an autonomous domain in mathematics. He gave the subject a clear starting point in the Cauchy–Riemann equations, brought in geometrical ideas that he may have learned from Gauss, and developed a profound connection with the theory of harmonic functions. He presented a theory, with uncertain levels of rigour, that applied not only complex-valued functions defined on the plane of complex numbers but to complex-valued functions on any two-dimensional domain, and outlined a proof that all simply-connected domains can be mapped one onto another by complex analytic maps. For a translation of the start of this paper, see Appendix A.3.

16.2 The Doctoral Dissertation of 1851

Riemann saw, with remarkable clarity was that what was required of a function f of x and y for it to be a complex-valued function of a complex variable $z = x + iy$ was that it could be treated as a function of a single variable, specifically, that it be a differentiable function of z. This requires that the expression $\frac{df}{dz}$ make sense, which means that the expression

$$\lim_{\delta\to 0} \frac{f(z+\delta) - f(z)}{\delta}$$

must be independent of the direction in which δ tends to zero. For, if there were to be a sense of direction for δ it would have to be in a space of more than one dimension, contradicting the basic idea that the variable z is a single thing and not a compound thing. He wrote

J. Gray, *The Real and the Complex: A History of Analysis in the 19th Century*,
Springer Undergraduate Mathematics Series, DOI 10.1007/978-3-319-23715-2_16

$$f(z) = f(x+iy) = u + iv = w \quad \text{and} \quad \delta = dx + idy,$$

supposed that $dx + idy = \varepsilon e^{\varphi i}$, and wrote down the corresponding expression for

$$\frac{du + idv}{dx + idy},$$

namely

$$\frac{du + idv}{dx + idy} = \frac{1}{2}\left(\frac{\partial u}{\partial x} + \frac{\partial v}{\partial y}\right) + \frac{1}{2}i\left(\frac{\partial v}{\partial x} - \frac{\partial u}{\partial y}\right)$$

$$+ \frac{1}{2}\left(\left(\frac{\partial u}{\partial x} - \frac{\partial v}{\partial y}\right) + i\left(\frac{\partial v}{\partial x} + \frac{\partial u}{\partial y}\right)\right)e^{-2\varphi i}.$$

For this to be independent of direction, it must be that the term in φ vanishes, so

$$\frac{\partial u}{\partial x} = \frac{\partial v}{\partial y} \quad \text{and} \quad \frac{\partial v}{\partial x} = -\frac{\partial u}{\partial y}. \tag{16.1}$$

These are the celebrated Cauchy–Riemann equations, and Riemann rightly took them as fundamental. In § 4 of his paper he rederived them by writing

$$\frac{\left(\frac{\partial u}{\partial x} + i\frac{\partial v}{\partial x}\right)dx + \left(\frac{\partial v}{\partial y} - i\frac{\partial u}{\partial y}\right)idy}{dx + idy}$$

and observing that this expression takes the same value for all values of dx and dy when Eq. (16.1) hold. He now added that the functions u and v satisfy

$$\frac{\partial^2 u}{\partial x^2} + \frac{\partial^2 u}{\partial y^2} = 0, \quad \frac{\partial^2 v}{\partial x^2} + \frac{\partial^2 v}{\partial y^2} = 0. \tag{16.2}$$

He did not call them harmonic functions, but he appreciated very keenly the connection to potential theory.

Riemann also observed that if the derivative does not vanish then the map is conformal. He may well have picked this idea up from a conversation with Gauss, or by reading and appreciating Gauss's work on conformal maps (see Gauss 1822); in any case it was not an idea that Cauchy had used.

To handle the case of a surface spread out over another so that several points on the upper surface correspond to just one point below, Riemann explained what a branch point is, using the example of the function $w = z^m$, which has a branch point of order m when $z = 0$. Here there are m values of w for every nonzero value of z, and they are cycled around as z is taken on a circle around the origin.

Riemann deduced something else from his definition of a complex function: it need only be defined on a two-dimensional patch. It might be defined on only a part of the complex plane, or perhaps on some other surface altogether. In this way he freed the complex variable to roam over a large class of surfaces; his theory of complex functions is not about functions defined on the plane of complex numbers; rather, it is about them and their generalisations defined on any two-dimensional patch whatever.

To study a complicated surface he outlined a procedure whereby it could be cut into manageable pieces. He said a piece of surface was connected when any two points in it can be joined by a curve.[1] He defined a boundary cut (*Querschnitte* in German) as a simple (i.e. non self-intersecting) curve joining two points on the boundary. At every stage existing cuts were counted as parts of the boundary, and to get started, if the surface had no boundary at all, one was allowed to delete a point. The cutting process would end if any further cut divided the surface into two pieces; such surfaces he called simply connected.

To investigate how the nature of the surface was related to the number of cuts needed to make it simply connected, he considered a surface T cut up in two separate ways. He supposed that it fell into m_1 simply connected pieces via a system of n_1 boundary cuts, and into m_2 simply connected piece after n_2 boundary cuts, and looked at the result of superimposing one system of cuts upon the other. By counting carefully he concluded that $n_1 - m_1 = n_2 - m_2$. It follows that if a surface is cut into m simply connected pieces by n boundary cuts, then the number $n - m$ is a constant, and Riemann called this number the order of connectivity of the surface. On this definition, a surface given in one piece is n-fold connected if a system of $n-1$ boundary cuts is required to cut it into one simply connected piece. The term 'genus' was introduced later in Clebsch (1864): a surface is of genus g if it is $2g + 1$-fold connected.

16.3 Harmonic Functions

Riemann's very general idea of the possible domains of a complex function made it difficult for him to define them by power series expansions, but not impossible. This was, indeed, to be Weierstrass's approach later, but Riemann did not want to define complex functions by their representations as a power series. Rather, he wanted to define them abstractly in a way that should minimise the otherwise lengthy manipulations that would be involved. His preferred way to obtain existence theorems for complex functions was to exploit the fact that their real and imaginary parts are harmonic. The crucial theorem relates the integral of a function taken over a region and the integral of another function taken over the boundary of that region. It was known already to Cauchy, Green, Gauss and Dirichlet, and it says:

$$\int \left(\frac{\partial X}{\partial x} + \frac{\partial Y}{\partial y} \right) dT = \int (X \cos \xi + Y \cos \eta) ds \tag{16.3}$$

[1]This is the modern definition of path connected.

where the first integral is taken over the surface and the other along the boundary, and the angles ξ and η are the angles the normal makes with the x- and y- axes respectively.

Using this theorem, Riemann obtained these essential results:

- a harmonic function defined on a surface T that covers the plane once possesses all its derivatives.
- if both the harmonic function u and its normal derivative $\frac{\partial u}{\partial p}$ vanish along a curve, then u vanishes everywhere, and so if the values of u and its normal derivative are prescribed along a curve they are determined everywhere.
- a harmonic function u cannot be constant on a piece of surface, and cannot attain a maximum or a minimum at a point inside T.

To obtain these results, Riemann first proved the theorem (16.3) and then applied it in the special case when the surface integral vanishes and the surface T has been cut into a simply connected surface T^*. The theorem now says that the integral along the boundary is a single-valued function of its upper endpoint (the lower endpoint being assumed fixed) provided that the upper endpoint does not cross the boundary cuts. If the upper endpoint moves across a cut, then the value of the integral increases by a constant quantity that depends on the cut in question but not on the position where the cut is crossed[2]), and each cut determines an independent quantity.

Now, to apply this result to complex functions, Riemann had to allow that there will be points where the function is infinite. He extended his analysis to deal with functions that are harmonic except at a finite set of points, and showed that such a function u possesses all its derivatives. To do this he estimated the contribution of the 'bad' points to the value of the integral along the boundary, and hence deduced the existence of all the sought-for derivatives. The other results followed swiftly.

Riemann now turned to the study of a complex function $w = u + iv$ defined on a simply connected region that covers (a part of) the plane just once. Since the real and imaginary parts have now been shown to be infinitely differentiable, the same is true of the complex function w. To study the case where a function becomes infinite at some interior point z' of its domain of definition, Riemann introduced polar coordinates centred on the singular point. If the function behaves like the μth power of ρ, where ρ is the radius, Riemann let m be the least integer greater than or equal to μ. Then $\left(z - z'\right)^m w$ tends to zero with ρ, and so $\left(z - z'\right)^{m-1} w$ is a function of z. He denoted its value at z' by a_{m-1}, and by repeating the argument $m - 2$ times showed that the function w can be changed into a finite, continuous one by subtracting an expression of the form

$$\frac{a_1}{z - z'} + \frac{a_2}{(z - z')^2} + \cdots + \frac{a_{m-1}}{(z - z')^{m-1}}.$$

So, as he remarked, a function that becomes infinite to some finite order does so to an integral order.

---

[2]To see this, apply the Cauchy integral theorem to the loop obtained from two different crossings; see § 17.4 below.

16.4 Riemann's Defence of the Dirichlet Principle

Riemann did not mention the Dirichlet principle by name in his paper of 1851, and moreover he did not uncritically rely on the principle but attempted to prove it. As he explained in next general paper on complex function theory, the monumental paper on Abelian functions (1857a), he was concerned in his (1851) to generalise this principle to cases where prescribed discontinuities are allowed.

He argued as follows in his (1851, § 16). He defined an integral over a surface T of the form

$$L(\alpha, \beta) = \int \left(\frac{\partial \alpha}{\partial x} - \frac{\partial \beta}{\partial y}\right)^2 + \left(\frac{\partial \alpha}{\partial y} + \frac{\partial \beta}{\partial x}\right)^2 dT. \tag{16.4}$$

It is required to be finite for arbitrary functions $\alpha(x, y)$ and $\beta(x, y)$; we note that this integral will vanish if the functions α and β satisfy the Cauchy–Riemann equations. Riemann then argued that by varying α by a continuous function that is at worst discontinuous only at single points and that is zero everywhere on the boundary of T, the integral can be made to attain a minimal value. Once that is established—and this is the contentious part—then it is straight-forward to show that the minimising function is a unique function (apart from any isolated the points of discontinuity), and moreover that this unique minimizing function is harmonic.

To prove the first claim, Riemann considered a function λ that vanishes on the boundary, that could be discontinuous at isolated points, and for which the integral

$$L(\lambda) = \int \left(\left(\frac{\partial \lambda}{\partial x}\right)^2 + \left(\frac{\partial \lambda}{\partial y}\right)^2\right) dT$$

is finite. He let $\alpha + \lambda = \omega$, and considered the integral

$$\int \left(\frac{\partial \omega}{\partial x} - \frac{\partial \beta}{\partial y}\right)^2 + \left(\frac{\partial \omega}{\partial y} + \frac{\partial \beta}{\partial x}\right)^2 dT.$$

which he denoted Ω. He then asserted (1851, § 16) that

> The totality of these functions represents a connected domain closed in itself, in which each function can be transformed continuously into every other, and a function cannot approach indefinitely closely to one which is discontinuous along a curve without $L(\lambda)$ becoming infinite.

So for each λ, Ω only becomes infinite with L, which depends continuously on λ and can never be less than zero; consequently has at least one minimum.

This argument is too naive to pass as a proof, but it is an attempt. It proceeds by generalising the fact that a continuous function defined on a closed interval is bounded and attains its bounds to the situation of a function L defined and continuous on a 'closed' set of functions, and claims that in this situation the function L likewise attains its lower bound. It even contains an argument that nothing goes wrong in the passage to the limit. If it is to be rejected, the grounds are that the proof is too weak, not that the claim was simply accepted.

It is interesting to note that when Riemann's student Gustav Roch published his account of Riemann's lectures of 1861/62 (Roch 1863, 1865) he avoided mention of Dirichlet's principle here, although Riemann had mentioned it the lecture course for 1855/56 which his friend Dedekind attended.[3] Roch's account does go on to include the Riemann mapping theorem.[4] So did repent of its proof? According to Klein (see Klein 1894–95, 492), Weierstrass once told him that "Riemann had never laid any particular value on finding his existence proofs with Dirichlet's principle", so perhaps he did. But perhaps he merely withdrew it as being unsuitable for an elementary lecture course.

The two remaining claims follow less problematically. Uniqueness followed an examination of functions of the form $u + h$ near to a minimum of u, which shows that Ω can be written in the form $M + 2Nh + Lh^2$. Then, for u to be a minimum it must be the case that $N = 0$. For the same reason, any two minimising functions u must give the same value of M and can only differ by having different isolated points of discontinuity that do not alter the value of the integral. These arguments depended on showing that the limit function cannot be discontinuous along a curve, and Riemann established this in § 17.

Then in § 18 Riemann applied a Green's function argument to the integral for N to show that the vanishing of N implied that the minimising function u was harmonic.

Riemann then deduced that if a complex function is defined on a connected surface T for which the integral

$$L(\alpha, \beta) = \int \left(\frac{\partial \alpha}{\partial x} - \frac{\partial \beta}{\partial y}\right)^2 + \left(\frac{\partial \alpha}{\partial y} + \frac{\partial \beta}{\partial x}\right)^2 dT \tag{16.5}$$

is finite, and if T can be cut up into a simply-connected region T^*, then there is always a unique function $\mu + \nu i$ of z such that μ vanishes at all but finitely many points on the boundary, ν takes an arbitrary value at an arbitrary point, and isolated discontinuities of μ in T or ν in T^* leave $L(\mu)$ and $L(\nu)$ finite and the values of ν differ only by a constant at corresponding points on the sides of the boundary cuts.

This result brings us to the heart of Riemann's conceptual approach to complex function theory and to the split that was to develop in the subject because Weierstrass could not accept it at all. In Riemann's opinion (see Riemann 1851, § 19):

> [These] principles open the way to the study of definite functions of a complex variable independent of an expression for it

With these words Riemann offered an approach to the study of functions that did not depend on any particular expression for it, and accordingly played down the role of long complicated manipulations of such expressions (such as Jacobi or Kummer had developed). It moves the study of complex functions firmly over to the study of

---

[3]See Göttingen, Akt Nr. 37.

[4]See Roch (1863, 19–28). Roch's lectures are considered in Sect. 17.2.

harmonic functions, which are determined by their boundary values, but now in a context where functions may have some singular points.

At this time, and arguably for many years afterwards, it was usual to allow complex functions to be many-valued, as the functions that take the nth root or the logarithm of a complex variable are. Riemann dealt with this by extending the domain of such a function to region T that is multiply connected and then passing to a simply connected region T^* by making boundary cuts. In this way a many-valued function is obtained which jumps as one crosses the cuts, and a branch of an n-valued function is extended to a function on a larger domain.

Riemann also spelled out that he considered that his claim about the existence and uniqueness of functions, given suitable information about their values on the boundary of their domain, applied to arbitrary simply connected domains and not merely regions of the plane.

In Riemann's opinion, his approach greatly simplified such tasks as deciding when two expressions represent the same function. Riemann did consider whether the class of complex functions as he defined them agreed with the class of functions definable by a finite or infinite number of operations of addition, multiplication, subtraction and division, and he claimed that they did. Had that been true his programme might have had an easier time, but Weierstrass's later refutation of that claim was to be one of his reasons for disputing the fundamental validity of Riemann's approach (see below, § 20.4).

But even on its own terms Riemann's approach raises an urgent uniqueness question that his mapping theorem was designed to address.

16.5 The Riemann Mapping Theorem

As we have seen, Riemann's approach to the study of the possible domains of complex functions was to divide a domain into simply connected pieces. His defence of the Dirichlet principle allowed him to assert the existence of complex functions on simply-connected domains, although must be said that few mathematicians at the time, and even for half a century afterwards, were persuaded by this existence claim, and under the influence of Weierstrass and people around him, it came to be believed that the argument rested on an appeal to Dirichlet's principle and had therefore to be rejected.

But from Riemann's perspective, it was natural, and perhaps essential, for him to see if, from the point of view of his theory of complex functions, there was essentially only one simply connected domain or if there were many. This is the topic he addressed in the second part of his paper of 1851, where Riemann offered an argument to show that[5]

[5]See Riemann (1851, § 21).

> Two given simply connected plane surfaces can always be mapped onto one another in such a way that each point of the one corresponds to a unique point of the other in a continuous way and the correspondence is conformal; moreover, the correspondence between an arbitrary interior point and an arbitrary boundary point of the one and the other may be given arbitrarily, but when this is done the correspondence is determined completely.

This important result, in a somewhat more precise form, is nowadays called the Riemann mapping theorem. It establishes not only that there are complex functions defined on any simply connected region, and indicates how they can be defined, but also that any two such regions are equivalent for the purposes of Riemannian complex function theory.

The claim amounts to showing—as an example of the application of his general results, as he put it (1851, § 21)—that any two simply connected planar regions can be mapped conformally onto one another. To do this, Riemann began by noting that it was enough to show that any such region T can be mapped conformally onto the unit disc. To obtain the conformal map of T to the unit disc, and to show that it extended to the boundary, Riemann took local coordinates $z - z_0$ defined on a suitably small disc around an arbitrary interior point of T, cut this disc along a radius, and defined the map $f(z) = \log(z - z_0) = \log(r) + \varphi i$ on that disc. He then extended the cut to a specified point on the boundary of T, and the function f to a continuous function $\alpha + \beta i$ defined on the whole of T that agreed with f on the boundary of the disc and had these properties:

1. α vanished on the boundary of T,
2. $\alpha + \beta i$ jumped across the cut by $2\pi i$ (like log).

He then argued that, by his general principle, there is a complex function $u + iv$ with suitable jumps. The function u will take every value from $-\infty$ (at z_0) to 0 (on the boundary of T). Then, because T was simply connected and the function u was harmonic, the level sets $u^{-1}(a)$ are single simple closed curves for every value of a. Therefore, he concluded, the sought-for function was

$$e^{u+vi} : T \to D$$

After a few remarks indicating that the entire theory could readily be extended to regions which are not simply connected, this remarkable paper came to an end.

16.6 Reactions to the Paper of 1851

Riemann's thesis was not published well enough for it to have a significant impact. As was the custom at the time it printed as a separate thesis and distributed to a number of German universities, but it was not reprinted in a journal such as Crelle's *Journal für die reine und angewandte Mathematik*. This is the main reason his approach did not spread in the early 1850s. It is also clear that anyone who read the paper would have encountered many troubling novelties that show up in its vagueness at certain crucial points. There is the striking and emphasis on arbitrary (and above all non-planar) domains, and topological concepts such as cuts and connectivity. A complex function is both a single object with characteristic properties (complex differentiability, the Cauchy–Riemann equations, conformality) and the sum of its real and imaginary parts which, being harmonic, are reduced to their boundary behaviour and their singular points. Dirichlet's principle, whether assumed or 'proved', lurks at the heart of the paper. It is entirely right that Lars Ahlfors, a leading complex analyst of the 20th century, was to say that Riemann wrote "almost cryptic messages to the future" and that his mapping theorem is in a form that "would defy any attempt at proof, even with modern methods".

The contrast with the approach of the previous generation was stark. In the hands of a master such as Jacobi, complex functions were given, indeed defined, by certain formal expressions: infinite series, infinite products, implicit functions, and integrals. His theory of elliptic functions, although daunting, had a directness about it: expressions for functions yielded information about the functions. Riemann's presentation was meant to change the type of questions one could ask about a complex function, away from the computational and towards the conceptual.

16.7 Riemann Surfaces and Conformal Maps

Riemann Surfaces

Riemann surfaces were greeted as a very difficult idea, as we shall discuss in Chap. 18. Here a picture by Gustav Holzmüller, an otherwise minor mathematician who excelled at providing pictures to illustrate theorems, has a go at explaining what two branch points of order three might look like (see Holzmüller 1882, Fig. 49) (Fig. 16.1).

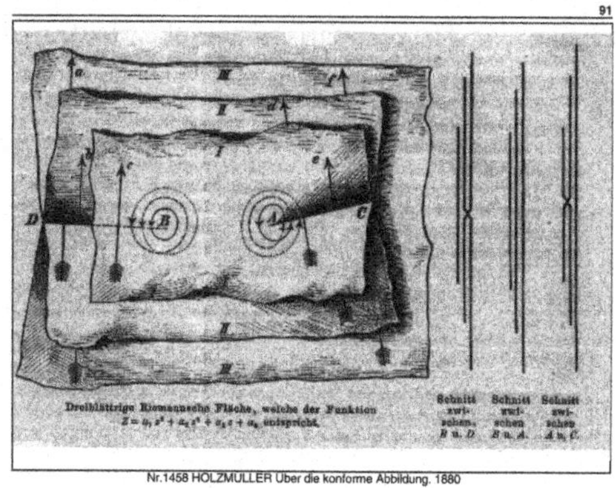

Fig. 16.1 Two branch points of order 3

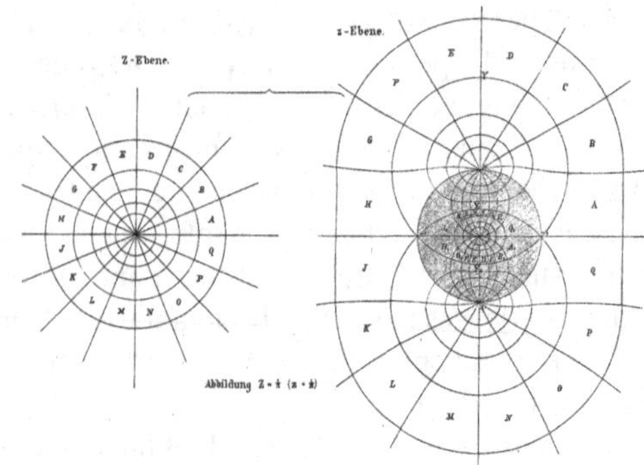

Fig. 16.2 A Holzmüller picture of $\frac{1}{2}\left(z+\frac{1}{z}\right)$

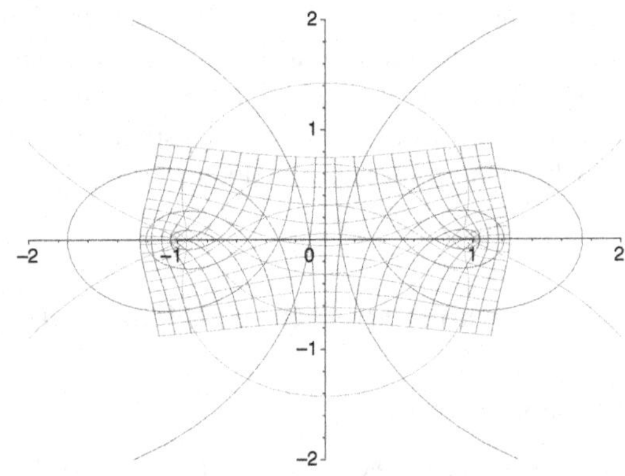

Fig. 16.3 A Maple picture of $\frac{1}{2}\left(z+\frac{1}{z}\right)$

Conformal Maps

Holzmüller also enjoyed depicting the effect of various maps on the complex plane. This one (Holzmüller 1882, Fig. 40) is an attempt to describe the map (later used in the Zhukovskii aerofoil) $Z = \frac{1}{2}\left(z+\frac{1}{z}\right)$. The map is 2–1 except at ± 1. The points $\pm i$ map to 0. The comparison with the Maple map is instructive (Figs. 16.2 and 16.3).

Chapter 17
Riemann's Later Complex Function Theory

17.1 Introduction

Riemann continued to develop his own ideas, extending them to multiply-connected domains defined by algebraic curves. He presented them in public when he lectured on complex functions, in particular elliptic and Abelian functions, in 1855/56 and again in 1861/62, and he published them in his remarkable paper on abelian functions in 1857.[1]

17.2 Riemann's Lectures from 1855/56 to 1861/62

A number of sources have survived that illuminate how Riemann lectured on complex function theory during his years as a Professor in Göttingen, and they make for an interesting comparison with his published papers. Some extracts were published in the first edition of the *Werke*, and in the *Nachträge* that accompanied the second edition. In 1896 and 1899 Hermann Stahl published his own notes on the Riemann's lectures on the theory of Abelian functions and the theory of elliptic functions respectively.

Riemann's lecture course of 1861/62 was also written up by Roch and published in 1863 and 1865, making them one of the few sources for his ideas that were available in his lifetime. In it, complex numbers are first explained geometrically, as corresponding to points in the plane, and then algebraically as the end result of a process of inverting the operations of arithmetic (for example, inverting addition to obtain subtraction). One begins with the natural numbers and proceeds to the integers, the rational numbers, and finally the complex numbers. The interpretation

[1]Riemann also applied these ideas to the theory of the hypergeometric equation; I plan to discuss this in a later course of lectures. See also Gray (2000a).

J. Gray, *The Real and the Complex: A History of Analysis in the 19th Century*, Springer Undergraduate Mathematics Series, DOI 10.1007/978-3-319-23715-2_17

that goes back to Wallis[2] of $\sqrt{-1}$ as a mean proportional between $+1$ and -1 was given, and Gauss's views of 1831 were supported, and then it was conceded that the geometric was more important for making complex quantities intuitive than for establishing their reality.

The lectures then presented the definition of a complex function w as a complex-valued function that is complex differentiable, and therefore satisfies the equation $\frac{\partial w}{\partial y} = i\frac{\partial w}{\partial x}$. This ensured that its derivative depended only on the position of z but not on the direction of dz. Roch (but not Riemann in 1861) then observed that a complex function is conformal except where the derivative vanishes; Riemann also pointed out that the real and imaginary parts of the function are harmonic, although he did not use the word.

The integral of a complex function along a path was then defined, which led to the question of when the integral around a closed path vanished. Riemann, as he had in his (1851), took a Green's theorem approach, proving that under suitable conditions on functions X and Y

$$\iint \left(\frac{\partial Y}{\partial x} - \frac{\partial X}{\partial y}\right) dxdy = \int (Xdx + Ydy)$$

where the first integral is taken over a plane region and the second integral over its boundary. When $X = w$ and $Y = iw$, $\frac{\partial Y}{\partial x} - \frac{\partial X}{\partial y} = 0$ and so Riemann concluded that the integral of a complex function around a closed contour vanished when the function w was everywhere finite and continuous inside the contour.

What happens when w becomes infinite at a point in the region? The example $\int\limits_1^z \frac{dz}{z} = \log z$ shows that in such cases the integral around a closed contour need not vanish. Riemann showed that if a function $f(z)$ defined in a region is such that there is a point a_i in the region where $\lim_{z\to a_i} f(z)$ is infinite as $z \to a_i$ but $\lim_{z\to a_i} (z - a_i) f(z)$ is finite and equal to c_i, say, and the function is finite everywhere else in the region, then the integral of taken clockwise around, is $2\pi c_i$. From this, Riemann deduced that if a function f is analytic everywhere in the interior, and t is an arbitrary interior point, then

$$f(t) = \frac{1}{2\pi i} \int \frac{f(z)}{z-t} dz,$$

and the nth derivatives satisfy:

$$f^{(n)}(t) = \frac{1.2....n}{2\pi i} \int \frac{f(z)}{(z-t)^{n+1}} dz.$$

These are the Cauchy integral formulae, which Riemann could have read—along with the strange term "finite and continuous"—in Cauchy's *Exercises de*

[2]See Wallis, *Algebra*, 1673, Vol. 2, Ch. 66, and the English translation in Smith, *Sourcebook*, 46–54.

mathématiques, which he borrowed from the Göttingen University Library when he was a student in 1847.

These formulae allowed Riemann, as they had Cauchy, to show how a complex function can be developed as a convergent power series on a suitable disc, and as a convergent Laurent series inside an annulus. Furthermore, if a function vanishes along a curve all the coefficients in its power series expansion vanish, so the only complex function that vanishes along a curve is the zero function, and any identity between functions that holds for real values of their arguments holds also for complex values.

Now that Riemann had shown that the complex functions are the ones that, at least locally, have power series expansions, he could investigate properties of complex functions by looking at their power series representations, and in this way he began the investigation of the extent to which a function is known up to a constant when functions are given which have the same infinities.

Riemann said that a function f is infinite of order m at $z = a$ if it is the case that $f(z)(z-a)^{m-1} = \infty$ when $z = a$ but $f(z)(z-a)^m$ does not (later, people said that such a function has a pole of order m; the term was introduced in Neumann (1865, 38)). Riemann also similarly defined what it is for a function to be infinitely small of order m (i.e. have a zero of order m). This allowed him to show that if a function has only finitely many points, including $z = \infty$, where it is infinite, then it is a rational function, and if it is never infinite even at $z = \infty$ then it must be a constant. Riemann gave two proofs of this result, which he did not call Liouville's theorem: one by looking at the power series expansion, and another using the fact that under the stated conditions the integral

$$f(t) = \frac{1}{2\pi i}\int \frac{f(z)}{z-t}dz$$

taken on a circle centre the origin approaches a constant value as the radius tends to infinity. This is because it can be regarded as the integral around $z = \infty$ and that is finite by earlier results.

Riemann finished this part of his course by giving a proof of the fundamental theorem of algebra. He first looked at the integral of the logarithmic derivative of a function f and showed that

$$\frac{1}{2\pi i}\int \frac{f'(z)}{f(z)}dz = \frac{1}{2\pi i}\int \frac{d\log f(z)}{dz}dz \tag{17.1}$$

evaluated on the boundary of a region, is equal to the number of zeros minus the number of poles inside the region, counted according to multiplicity.

Now, a polynomial of order n is infinite to order n at $z = \infty$, but (17.1) vanishes for such functions, so the single nth order pole of the polynomial at infinity, which is its only pole, must be balanced by a total of n zeros.

Riemann also gave another proof, which, as he remarked, is essentially Gauss's third proof of the fundamental theorem, and proceeds by evaluating (17.1) for the function $\frac{f(z)}{z^n}$, and interpreting the answer, n.

Riemann finished the course by showing how some real definite integrals can be evaluated by contour integration, and then giving a brief description of the branch points of a many-valued function.

17.3 Riemann's Theory of Elliptic Functions

As one might expect, Riemann's geometrical theory of functions was well suited to deal with elliptic functions, so much so that one wonders at what stage Riemann had that aim in mind. Unhappily, there is no evidence to illuminate that question, so we must set it aside.

His lecture courses on elliptic and algebraic functions were usually offered after the courses on complex function theory, and the lectures for 1861 are interesting because in them he showed for the first time how the geometry of the complex curve (which is a real two-dimensional surface) defined by the function $y^2 = f(x)$, where $f(x)$ is a cubic or quartic in x, illuminates the whole study of elliptic integrals and elliptic functions.[3] Riemann's grasp of the surface as a branched covering of the sphere enabled him to do what both Jacobi and Cauchy had been unable to do and to keep track of the values of the integrand (which, being a square root, is a two-valued expression). This was the first time that Cauchy's approach to complex function theory was made to apply to many-valued integrands apart from Puiseux's (1850) in which had investigated integrals of algebraic functions.

In these lectures Riemann confined himself to the traditional cases when the roots of the quartic are either real or come in complex conjugate pairs, but there is no reason to suppose his analysis was restricted to these cases—the most likely reason is pedagogical.

Riemann took the surface T corresponding to the curve, which is a double cover of the sphere branched over the zeros of $f(z)$ and at $z = \infty$ when $f(z)$ is a cubic, and cut it into a simply connected region T' by two cuts a and b. Then he showed that the elliptic integral in its normal form maps T' onto a copy of a rectangle in the complex plane. The sides of the rectangle are determined by the periods of the integral, and the rectangles tile the plane. The value of the integral depends on the path and the end point of the integral determines the value of the integral only up to a sum of periods. This illustrates the many-valued nature of the integral. The net of rectangles to the plane displays the doubly periodic nature of the inverse map. The periods arise as integrals along the cuts, and Riemann had now shown how they arise by integrating along non-contractible closed curves on a torus.

[3]For the published course see Stahl (1899); the chapter on the presentation of the theory of theta functions was taken from the course of 1856.

Riemann's account of Jacobi's theory of theta functions also displays a novel use of his theory of functions. He defined a theta function by a power series. To get the Jacobian elliptic functions as quotients of theta functions he wrote down a quotient that had the correct zeros and poles and then multiplied by an exponential factor to get the periods right, and used a little trick to get the constant multiple right. The result that the elliptic function equals this quotient then followed from Liouville's theorem.

In the course of 1856 he had given an alternative account: he defined the period parallelogram, $P = \{0, 1, \tau, 1+\tau\}$ for the theta function, pointing out that it was not, strictly, a periodic function. Integration around the boundary showed that it had exactly one zero inside P, which could be found explicitly from the power series to be at $\tau/2$. Thence, following Jacobi, the other three theta functions, and a suitable quotient ($x := a^2 \left(\frac{\theta_1(v)}{\theta(v)}\right)^2$) was then doubly periodic with a double pole at $\tau/2$. From which it followed by counting poles that $\frac{dv}{dx}$ was the square root of a cubic, and so $\sqrt{x} = sn(2Kv)$.

The lecture course of 1856 is therefore more traditional in its approach, but it is also interesting because it was a try out for some of the material that was later published in the great memoir of 1857 on Abelian functions, to which we shall turn after quickly looking at a Riemannian treatment of elliptic integrals and functions that was given in 1864 and which helps us understand what Riemann was doing.

Durège on Elliptic Integrals

In his book of 1864, Heinrich Durège took 14 pages (pp. 187–201) to connect elliptic functions back to the corresponding elliptic integrals. It is interesting to follow his account, because it illustrates what had to be done to make Riemann's geometric ideas comprehensible to an audience for whom they were unfamiliar.

Durège considered the elliptic integral

$$w = \int_0^z \frac{dz}{\sqrt{(1-z^2)(1-k^2z^2)}}.$$

As he observed, there are two values of w for each value of z except where $z = +1, -1, +\frac{1}{k}, -\frac{1}{k}$. He expressed this awkwardly and arguably misleadingly by saying the z-surface always has two leaves and has four points of discontinuity, which are called its branch points. These points are not to be excluded, because at them the integrand is only infinite with order $\frac{1}{2}$. Nor is the point ∞ to be excluded, because there $zf(z) = \lim_{z\to\infty} \frac{z}{\sqrt{(1-z^2)(1-k^2z^2)}} = 0$.

To render the surface simply connected, Durège had to choose cuts in the surface that go between the branch points; he chose to join $+1$ and -1 and to join $+\frac{1}{k}$ and $-\frac{1}{k}$. This led him to Fig. 17.1.

The transition from a continuous to a dotted line is made because a loop such as q_2 crosses from one sheet to the other, whereas a loop such as q_1 stays entirely in one sheet of the surface. He then explained the many-valued nature of the elliptic integral

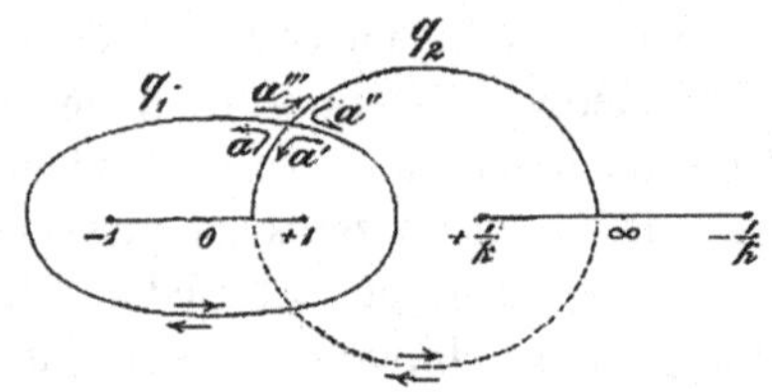

Fig. 17.1 Durège's figure of a simple elliptic function

by tracking the values of the integral along the loops q_1 and q_2, and he depicted the infinitely many values of the integral by means of a parallelogram lattice.

Then, he said (on p. 193) "We also want to follow Riemann and make the connection between the doubly periodic function and the elliptic integral in the opposite direction, namely starting from the doubly periodic function." To this end he supposed he had a doubly periodic continuous function ϕ, so

$$\phi(w + A_1) = \phi(w + A_2) = \phi(w),$$

for some complex numbers A_1 and A_2. He argued that the ratio of these numbers cannot be real, or the function would either be simply periodic or constant. By an earlier result—Liouville's theorem although Durège gave it no special name—the function must have a pole somewhere. On integrating the function around a period parallelogram and applying the Cauchy residue theorem the sum of the residues must vanish, so such a function cannot have just one simple pole and must have at least two. If it has two they have equal and opposite residues, and in this, the simplest case, the function can be written as

$$\phi(w) = \frac{c}{w - r} - \frac{c}{w - s} + \psi(w),$$

where the poles are at $w = r$ and $w = s$ and ψ is a function that is not infinite at any of the points $r + m_1 A_1 + m_2 A_2$ or $s + m_1 A_1 + m_2 A_2$. Durège then showed that $\phi(w) = \phi(r + s - w)$. Moreover,

$$\phi'(w) = \frac{dw}{dz} = -\frac{c}{(w - r)^2} + \frac{c}{(w - s)^2} + \psi'(w).$$

This function has two double poles and so, as a function of w, it takes every value four times in each period parallelogram. The derivative is a single-valued function of w but, Durège then showed, a two-valued function of z, so $\left(\frac{dz}{dw}\right)^2$ is a single-valued function of z. Whence $\left(\frac{dz}{dw}\right)^2$ is of the form $C(z - \alpha)(z - \beta)(z - \gamma)(z - \delta)$ and one has returned to the original elliptic integrals.

17.4 Riemann's Paper of 1857 and Complex Functions

Riemann's paper of 1857 on Abelian functions is undoubtedly one of the most important papers on mathematics published in the 19th century. Unlike the (1851), Riemann's (1857c) was published in the principal mathematical journal of the day, the *Journal für die reine und angewandte Mathematik*, which explains why Riemann repeated many of his earlier ideas. Its importance was appreciated at once by Weierstrass who, as we shall see in Chap. 19 had already been working on the topic for some time, and who, on reading Riemann's account withdrew a paper he was writing on the subject from publication. He was not respond in print until 1869.

The topic is the integrals of arbitrary algebraic functions. As we saw in Chap. 3, the subject of elliptic functions grew out of the study of integrals of the form

$$z = \int \frac{dx}{\sqrt{g(x)}},$$

where the function $y = g(x)$ is a cubic or quartic in x. Geometrically, this can be considered as studying integrals of the form

$$z = \int f(x, y)dx$$

on the complex curve $y^2 = g(x)$, although investigations mostly avoided this interpretation in favour of strictly analytic methods. By the early 1850s attempts had been made to go beyond the quartic case and to contemplate integrals of the form $z = \int f(x, y)dx$ where there is an algebraic relation between x and y of the form $g(x, y) = 0$.

In two papers of 1854 and 1856 Weierstrass made his name by resolving a special case of integrals of the form $\int \frac{w dx}{y}$ on a curve with equation $y^2 = g(x)$, where $g(x)$ is of degree $n > 4$. This, the so-called hyper-elliptic case, had been studied by some earlier writers, including Hermite and some students of Jacobi. But in his (1857) Riemann dealt with integrals of the form $\int f(x, y)dx$ on any plane algebraic curve $g(x, y) = 0$ whatever.

To do so, Riemann pushed the ideas in his (1851) from being a good indication that complex function theory can be done on any 'surface' to showing in detail how it could be done on complex algebraic curves described by equations $g(x, y) = 0$, although he only gave details for curves having at most some double points in the complex plane and no more complicated singularities.

Riemann called the space of points (x, y) such that $g(x, y) = 0$ a surface T. He described how to dissect T by either closed curves or curves with end points on the boundary of the (partially dissected) surface until a simply connected surface T' is obtained. He explained how to think of integrals on T, and discussed when an integral may be independent of the choice path joining its end points. To introduce the idea of

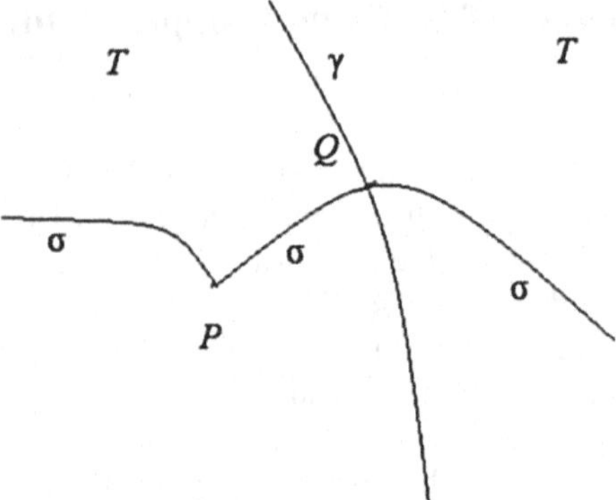

Fig. 17.2 A loop from P to P along a curve σ that crosses a cut γ

complex functions defined on T he summarised his earlier use of Dirichlet's principle and concluded that there is a unique function on T' with prescribed singularities in T' and whose real part may be defined arbitrarily along the cuts that form the boundary of T'. The whole aim of the paper was to show how to study the integrals of complex functions on T, and the use of geometry was wholly new.

Riemann supposed that $g(x, y) = 0$ was irreducible, and of degree n in x and m in y, and regarded it as representing as an n-sheeted covering T of the y-plane; there are, in general, n values of x for each value of y. Points at which there are fewer than n values of x are called branch points.

Riemann remarked that rational functions of x and y (which, by definition, are the quotient of one polynomial in x and y by another), are single-valued functions on T that are branched like x, and he proved the converse later in the paper. The integral of a rational function yielded a many-valued function, because the integral may be conducted along inequivalent paths that wind around the branch points. However, these different analytic continuations differ only by constants, because their derivatives at the same point necessarily agreed. It was this system of functions that Riemann studied.

Suppose that P is point inside T' and that Q_1 and Q_2 are corresponding points on two parts γ_1 and γ_2 of the boundary of T' (so γ_1 and γ_2 are the two sides of a cut γ on T and Q_1 and Q_2 correspond to the same point on T, as shown in Figs. 17.2 and 17.3). Consider an integral of a rational function $f(x, y)dx$ on T along a simple loop σ starting at P and passing through $Q_1 = Q_2$. This path appears on T' as a path from P to Q_1 and a path from Q_2 to P. We do not expect $\int f(x, y)dx$ taken along σ from P to P to vanish, rather, it should be a non-zero period of the integral. This means that the values of the integrals $\int_P^{Q_1} f(x, y)dx$ and $\int_P^{Q_2} f(x, y)dx$ (where the second integral is taken along σ in the reverse direction) will differ, by an amount that Riemann called the jump of the integral at $Q_1 = Q_2$.

Now consider a similar figure in which the loop σ through P deviates along the path σ_2 and crosses γ at another point R, say, as shown in Fig. 17.4 and suppose that

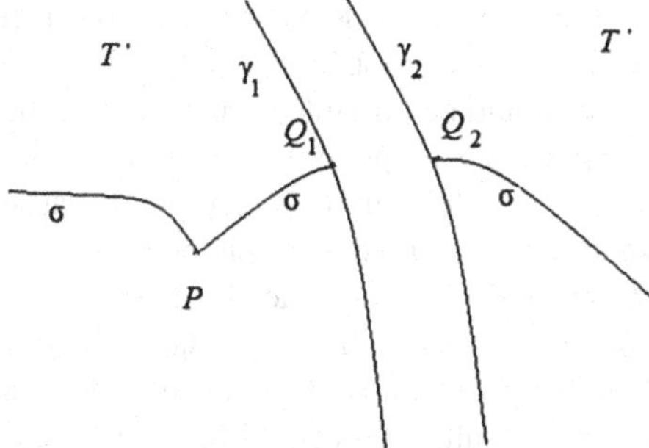

Fig. 17.3 The loop σ as it appears in T'

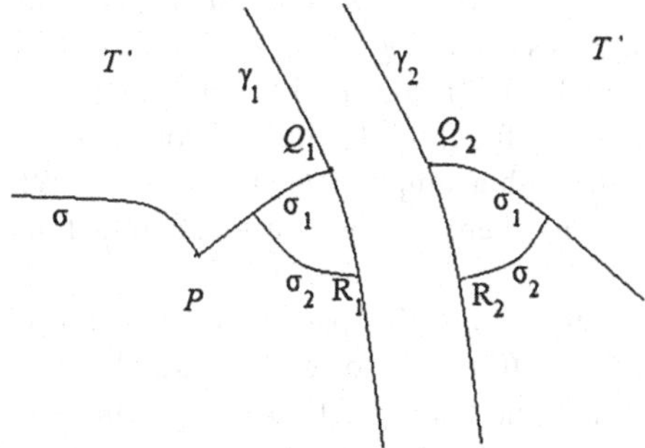

Fig. 17.4 The loops σ_2 and σ_2 as they cross the cut γ

no branch points of T are contained in the region bounded by σ_1 and σ_2. Then, by what Riemann regarded as a consequence of Green's theorem[4] the integral taken in the clockwise direction around σ_1 and the anticlockwise direction around σ_2 is zero, so the contributions of the integrals along σ_1 and σ_2 to the integral along the loop that starts at P are the same, and therefore the jump of the integral at $Q_1 = Q_2$ is the same as jump of the integral at $R_1 = R_2$ (the points on the boundary of T' that correspond to R).

So the conclusion is that the jump of the integral of a rational function across a cut is constant along a cut. It will, of course, vary from cut to cut.

Now, a complex function that is bounded on all of T' and does not jump across any cut will correspond to a complex function that is bounded on all of T, and by Liouville's theorem (not that Riemann gave it a name) such a function must be constant. But an integral with a non-zero jump on at least one cut corresponds to a many-valued function on T, the different values being determined by the path of integration. So Riemann now had a way of generating non-trivial functions on T, at the price of allowing them to be many-valued.

---

[4] And we might regard as the Cauchy integral theorem.

To produce single-valued functions, Riemann used the fact that integrating a function around a point where the function has a simple pole produces a many-valued function. He argued that it should be possible to make the contributions balance, so that the integral of a function with simple poles has jumps of zero across every cut. There will be no contradiction with Liouville's theorem, because the integral is now infinite at the points where the function has simple poles.

The surface T will be made simply connected by a system of, say, $2p$ cuts. These render T into a disc-shaped surface with a boundary of $4p$ sides that are to be identified in pairs, and therefore $2p$ jumps. It remained for Riemann to show how the two ways of introducing multi-valuedness could be made to cancel each other out, and his conclusion was that there was a system of single-valued complex functions on T having simple poles at m points of dimension at least $m - p + 1$.[5] When $m - p + 1 = 1$, or $m = p$, the corresponding functions are all constant, and there are non-constant functions as soon as $m > p$.

This happily agrees with what is already known. When T is a sphere, for which $p = 0$, there are non-constant functions having only one pole; they are the functions of the form $f(z) = \frac{az+b}{cz+d}$, with a simple pole at $z = -d/c$. When $p = 1$, which is the case for elliptic functions, there are non-constant elliptic functions with precisely two poles.

In the second part of this remarkable paper Riemann showed how to generalise the theory of theta functions from one to several variables, which involved him in discussing functions in not one but $p > 1$ variables, This topic remains graduate work to this day and is out of reach here.

[5]This is the so-called Riemann inequality. It was refined to an equality in Roch (1865).

Chapter 18
Responses to Riemann's Work

18.1 Introduction

The impression is given in several places that few people responded quickly to Riemann's work, and those that did so with a view to finding fault with it, even arguing that it should be dismissed. It is true that some mathematicians began their papers by lamenting that nothing seems to have been done, but in fact his papers, with a few notable exceptions, drew a considerable amount of attention, as this chapter discusses.

18.2 Critical Acceptance

Elementary complex function theory on Riemannian principles was presented in Durège's book of 1865, the first book on the subject at all in German, and it was echoed in a long article by Oscar Schlömilch (1866). In 1868 Casorati published his explicitly Riemannian presentation of complex function theory for an Italian audience. More advanced topics were also treated on Riemannian lines: Clebsch (1864) discusses the concept of genus and explains Abel's theorem; Neumann (1865) is about hyperelliptic integrals; Clebsch and Gordan wrote their (1866) on Abelian function theory; in the late 1860s Lazarus Fuchs (1865, 1866) generalised the hypergeometric equation to a class of complex ordinary differential equations all of whose solutions are meromorphic everywhere (now called equations of the Fuchsian type). Riemann's work was corrected and extended by several authors: by Prym, Christoffel and Schwarz on the Riemann mapping theorem in the 1860s, and by Schottky in his (1877), who extended the Riemann mapping theorem to non-simply connected domains. Yet other authors took up the topological aspects of Riemann's function theory, his introduction of trigonometric series, and his differential geometry, and in

J. Gray, *The Real and the Complex: A History of Analysis in the 19th Century*,
Springer Undergraduate Mathematics Series, DOI 10.1007/978-3-319-23715-2_18

the end only three of Riemann's papers were left unattended: the one on the prime number theorem, the one on shock waves, and the one on Bessel's equation.

Moreover, Weierstrass had proposed Riemann in strong terms to the Berlin Academy of Sciences in 1859. When they met they apparently got on well together, although it is true that in later years Weierstrass, egged on by his acolyte Schwarz, was more and more opposed to Riemann's way of thinking. The importance attached to producing the first edition of Riemann's *Werke*, notwithstanding its many delays, and editing his unpublished papers, lecture notes, and fragments, also suggests a real willingness to keep his ideas alive after his premature death.

People who supported Riemann's approach to complex function theory emphasised the importance of considering the global nature of the domain of a complex variable, and at least in simple cases gave a topological analysis of the domain in terms of cuts in the Riemann surface. They allowed the appeal to Dirichlet's principle to establish the existence of complex functions on various domains and having prescribed singularities, although they stopped short of attempting a proof of that principle. But by and large, and certainly in books, they did not follow him into the deeper reaches of Abelian function theory.

Even these matters raised problems. For example, Clebsch's seven distinguished obituarists (Von der Mühll et al. 1874) listed as one of the three hardest problems confronting anyone who approached Riemann's work the problem of understanding the topology of an algebraic curve and, conversely, of passing from an equation to the form of the corresponding Riemann surface. In the 1860s Clebsch and Lüroth described how a Riemann surface may be cut up along loops that encircle the branch points in particularly simple ways, and further contributions were made in Clifford (1877).

As for the contentious Dirichlet principle, when Klein set about implying that he was Riemann's natural successor he suggested that Riemann had advocated Dirichlet's principle on physical grounds, but as Bottazzini has shown in his (1977) former students of Riemann's, notably Prym, disagreed. In fact mathematicians all agreed with Riemann that the principle needed a proof—only physicists such as Helmholtz and Maxwell disagreed. The problem was, as we shall now see, that a proof seemed to be possible only under restrictions that Riemann's gestures had not indicated.

18.3 Prym Refutes the Dirichlet Principle

Ironically, Riemann's former student Friedrich Prym was to be the first person to refute Dirichlet principle. His example (Prym 1871) is interesting. He began by observing that Dirichlet's principle implies that an arbitrary continuous function, u, defined on the boundary of a disc extends to a finite and continuous harmonic function defined on the whole of the disc. As he said, the weak point, in all known proofs of this result could be found by exploiting other of Riemann's ideas, for they relied on the claim that the function u was identical with its Fourier series, and it was a misapprehension to think that Riemann had proved that an arbitrary function is representable by its Fourier series.[1] In fact the best result was still due to Dirichlet and applied to functions having only finitely many maxima and minima.

Prym then dug deeper, and showed that even when Dirichlet's problem can be solved, Riemann's use of the Dirichlet principle could be invalid. He considered a branch of the complex function

$$u + iv = i\sqrt{-\ln(R + x + iy)}$$

defined on a disc of radius $R < \frac{1}{2}$, and introduced polar coordinates ρ and τ centred on the point $(-R, 0)$. In the disc ρ took every value from 0 to $2R < 1$ and τ every value from $-\pi$ to π. The branch of logarithm taken was to satisfy $-\ln(R + x + iy) = -\ln\rho - i\tau$.

The explicit form for u and v in terms of polar coordinates shows that the functions u and v are everywhere defined and single-valued, even on the boundary of the disc, and since the function u is the real part of a complex function it is certainly harmonic. So the Dirichlet problem is solved in this case. However, as Prym then showed, Dirichlet's integral $L(u, 0)$ is infinite. The reason, as his formulae make clear, is that the function u oscillates infinitely often in any neighbourhood of the point $\rho = 0$. Consequently there is no hope that step 2 of Riemann's argument ($L(\mu) < \infty$) could be made to work. So Dirichlet's principle can fail, even when Dirichlet's problem can be solved.

Although Weierstrass's many students and more and more of his contemporaries preferred to use Weierstrass's example of an integral bounded below and that does not attain its bounds as the way to show that Dirichlet's principle is unreliable, the first direct refutation of it came from within the group of Riemann's students. Whether it was decisive is hard to say, but widespread use of the principle in Riemann's work meant that once it was called into question much else of what Riemann had done was also put in doubt.

Weierstrass's example was set out in one of the few papers that Weierstrass published—he generally preferred to confine himself to lectures in Berlin—and it is likely that each of these papers was intended to shake mathematicians out of their complacency. In his (1872) he raised the subject of Dirichlet's principle,explained that if the Dirichlet integral exists and attains its minimum then the minimising function is harmonic and unique, and then he turned to the existence question. Here he offered what he called a simple example to show the inadmissability of Dirichlet's reasoning. He observed that

$$J = \int_{-1}^{1} \left(x \frac{d\varphi}{dx} \right)^2 dx,$$

[1] Prym cited Hankel, another of Riemann's students, as having made this mistake.

where $\varphi(-1) = a \neq b = \varphi(1)$ is always positive and can take any non-zero value however small, but cannot take the value zero unless $\frac{d\varphi}{dx}$ vanishes on the interval $[-1, 1]$, which is ruled out by the boundary conditions.

An impression of how things stood by 1900 is given by the remarks David Hilbert made in what proved to be—as it was intended to be—a speech that set the agenda for the opening decades of the 20th century.

At the end of the 19th century, two mathematicians stood as intellectual leaders: Henri Poincaré in France and the younger figure of David Hilbert in Germany. Hilbert was invited to be one of the keynote speakers at the second International Congress of Mathematicians, to be held in Paris in August 1900. His close friend Minkowski urged him to seize the day, writing to him:

> Most alluring would be the attempt to look into the future, in other words, a characterisation of the problems to which the mathematicians should turn in the future. With this, you might conceivably have people talking about your speech even decades from now. Of course, prophecy is indeed a difficult thing.[2]

After delaying so long that the invitation was nearly withdrawn, Hilbert wrote was has become probably the most cited single paper in mathematics, his address on the Problems of Mathematics. In it he presented two fruitfully contrasting views on mathematics: on the one hand, mathematics for its own sake; on the other mathematics in the service of science. He argued that high standards of rigour and understanding brought these strands harmoniously together, and then offered some 23 problems, several in several parts, as good problems to work on. Such was Hilbert's prestige as the years went by that mathematicians have felt especially honoured when they solved one of the problems.[3] Few remain unsolved today; one of the few to do so is the Riemann hypothesis.

Among the problems Hilbert singled out was the Dirichlet problem and the Dirichlet principle.

> 20. THE GENERAL PROBLEM OF BOUNDARY VALUES.
> An important problem closely connected with the foregoing is the question concerning the existence of solutions of partial differential equations when the values on the boundary of the region are prescribed. This problem is solved in the main by the keen methods of H. A. Schwarz, C. Neumann, and Poincaré for the differential equation of the potential. These methods, however, seem to be generally not capable of direct extension to the case where along the boundary there are prescribed either the differential coefficients or any relations between these and the values of the function. Nor can they be extended immediately to the case where the inquiry is not for potential surfaces but, say, for surfaces of least area, or surfaces of constant positive Gaussian curvature, which are to pass through a prescribed twisted curve or to stretch over a given ring surface. It is my conviction that it will be possible to prove these existence theorems by means of a general principle whose nature is indicated by Dirichlet's principle. This general principle will then perhaps enable us to approach the question : *Has not every regular variation problem a solution, provided certain assumptions regarding the given boundary conditions are satisfied* (say that the functions concerned in

[2] See Rüdenberg and Zassenhaus (1973, 119).

[3] See Gray (2000b) and Yandell (2002).

> these boundary conditions are continuous and have in sections one or more derivatives), *and provided also if need be that the notion of a solution shall be suitably extended?*[4]

Hilbert did not entirely establish a solution in his own later work, but his bold idea was to bridge the gap between mathematicians and physicists by delineating a class of boundary conditions which were general enough to cover the physical situations and yet narrow enough to finesse the mathematical difficulties. And if pressed, Hilbert admitted, one might weaken what was meant by a solution a little. In short, Hilbert raised the issue of establishing, for a large class of equations including several that arise in various physical problem, theorems that guarantee the existence of solutions in a possibly novel sense to be made precise by future research. Much of the research agenda in partial differential equations to this day was thus established for the first time.

18.4 Other Responses

When Dirichlet died in 1859 Rudolf Clebsch was appointed as his successor in Göttingen, and he took the opportunity to see what could be done to promote Riemann's ideas (by this time Riemann was already very ill and was spending as much time as he could in Italy to preserve his failing health). Clebsch had graduated from Jacobi's school in Königsberg and initially worked on elasticity theory and hydrodynamics. His first major paper after his turn to Riemann is his (1864), where he endeavoured to find applications for abelian functions in the study of algebraic curves that could match those already well known and understood for elliptic functions. As he put it (1864, 189):

> The such applications have not yet been sought, although we have had Riemann's theory of these functions for six years, can without doubt be put down in large part to the difficulties in the way of understanding the work under consideration, and which the recent efforts of younger mathematicians have not quite alleviated.

In the 1860s Clebsch emerged as the leader of a strong group of algebraic geometers who saw themselves as leading a wave of new mathematics in just those areas neglected by Gauss. In this spirit they tried to rewrite Riemann's theory of algebraic curves in the language of plane projective geometry, and to find a largely algebraic path from there to analysis. This appalled Riemann's surviving pupils. Clebsch and his younger colleague Paul Gordan set out their approach in detail in their book (1866), and Prym wrote to Casorati in 1866 that "The attempt to base function theory on algebra is completely useless" and that they would never have dared to publish the foreword in Riemann's lifetime.

What enraged Prym was the preference for algebra over topology. However imperfectly, Riemann had attempted to get away from defining a Riemann surface by

[4]By a regular problem he means one such as the Dirichlet problem. Here he referred to his own lecture on Dirichlet's principle: (Hilbert 1900, 1904).

an equation, and had attempted to prove that any finite Riemann surface could be described by (an equivalence class of) equations. This separated the surface, thought of abstractly, from any embedding it might have in some complex projective space; the Clebsch-Gordan treatment collapsed that distinction. The surface was regarded as a complex curve in the complex plane, and the genus was defined in terms of the degree of the equation defining the curve and the number of its double points and cusps (curves having worse singularities were for the time being excluded). Even though they could show that birational transformations of the surface did not change the value of the genus (at least for curves that were not too singular) their definition was evidently not intrinsic.

The function-theoretic side of the book was more successful. Whereas even Riemann's followers had to admit that his theory of theta functions was not intimately connected to the rest of the theory, the book spelled out at length the connection between theta functions and integrals of the third kind, which Riemann had only sketched.

However, even Clebsch was not long enlisted in Riemann's cause. He was to have edited Riemann's *Werke*, but he died unexpectedly of diphtheria in 1872 when only 39. The task passed to Riemann's friend Dedekind, and a younger mathematician, Heinrich Weber, who was more sympathetic to Riemann's ideas.

The people Clebsch left to pursue his ideas continued to prefer an algebraic geometry based on a study of equations, and while they named theorems after Riemann, it was only after giving them entirely different proofs. They came to regard Riemann's reliance on the 'misleading' Dirichlet's principle as the problem, as Brill and Noether's monumental historical survey of algebraic function theory, written in 1894, makes clear. Definitions become indefinite, they wrote (1894, 265), and

> At such a level of generality the idea of a function, incomprehensible and evaporating, no longer delivers reliable conclusions. In order to delineate precisely the domain of validity of the theorems under consideration, people have recently abandoned Riemann's path completely.

Riemann had also had some students of his own at Göttingen; we have already mentioned Roch, Hattendorff, Hankel, and Prym. There was also Enrico Betti, who was a very valuable Italian contact. But as we noted earlier Roch died in 1866 age 26 and Hankel in 1872 age 34; Hattendorff died in 1882 age 48. Had Klein arrived in Göttingen in 1872 to study with a 45 year-old Riemann recovered from pleurisy, how different the growth of complex function theory would surely have been.

The response from Berlin, by far the dominant university for mathematics at the time, was different. Weierstrass and Kronecker were the most influential analysts, and until their personal relations became clouded in acrimony in the 1880s they were often engaged together in trying to sort out fundamental points in both real and complex analysis. This opened the way for Schwarz, initially with Weierstrass's support, to examine more and more of what Riemann had done with a view to making it not only more rigorous but also more orthodoxically Weierstrassian. This resulted in the Schwarz–Christoffel approach to the Riemann mapping theorem and Schwarz's alternating method for solving a modified form of Dirichlet's problem (Schwarz

1870a, b). Riemann's insights were reworked at a price it will be instructive to examine when we have considered what these alternative, Weierstrassian principles were.

In any case, Riemann's topological way of thinking was not one Weierstrass wished to advance, and because he liked to present function theory entirely as he wished to see it and gave very few references to other literature the relatively large audiences he attracted would not necessarily know when they were entering Riemannian territory. Instead power series methods were given a central place; even the Cauchy integral formula was eliminated from his presentation of complex function theory. One unexpected result was that few of Weierstrass's former students shared his preferences and he left less of a school behind him than he wished.

Italian mathematicians proved the most sympathetic to what Riemann had been trying to say. Apart from Casorati, who never met Riemann, there was also Riemann's friend Betti, who ran the Scuola Normale in Pisa for 30 years, The Scuola Normale was a small, high quality institution where most of the next generation of Italian mathematicians trained. Betti was faithful to the breadth of Riemann's interests, and that influenced the work that came out of that school, for example in the papers they submitted in due course to Klein when he was establishing the *Mathematische Annalen*—the journal founded by Clebsch that he now edited—and trying to outflank Berlin's *Journal für die reine und angewandte Mathematik*.

On the other hand, Riemann's influence was negligible in France for a generation, and this, together with the increasingly ambiguous attitude to it in Berlin, led to a justifiably profound sense of a rediscovery of Riemann's ideas that attended the rehabilitation of his ideas in Göttingen in the 1880s. Klein began this process when he arrived in Göttingen in 1886, Hilbert took it up in due course, and so very markedly did Hermann Weyl in the next generation.

Klein had become a professor in 1872 at the remarkably early age of 23, and swiftly attached himself to Clebsch, whom he later described as a divinely inspired teacher. He was highly talented, very ambitious, and saw an opportunity for himself in mastering what Riemann had done, not least because he shared some of Riemann's traits, a rich intuition and a liking for geometry among them. But in 1881 the young French mathematician Henri Poincaré began to publish a stream of discoveries involving non-Euclidean geometry and automorphic functions, topics close to Klein's interests but treated from a different perspective (oddly enough, a more Riemannian one) and Klein's efforts to match this work, as he was later to say, cost him his health.

Part of Klein's programme to recover was to lay ever greater claim to the Riemannian mantle. His energies diversified out of research, which he came to delegate to his best students, and into building up mathematics at Göttingen in direct competition with Berlin. He sought money from the state of Prussia, brought in industrial money for work on applied mathematics, and set about creating a tradition of mathematics at Göttingen that embraced Gauss, Riemann, and his own day.[5] He helped edit the many unpublished papers of Gauss; he wrote a richly interesting history of

[5] See Tobies (1981).

mathematics in the 19th century; and finally he broke with the successors of Clebsch. Clebsch is the only mathematician Klein criticised in his popular lectures in Evanston in 1893, and Brill and Noether repaid him by slighting his contributions in their own historical survey (1894).

Textbooks

If we count the book (Briot and Bouquet 1859) as a book on complex function and elliptic function theory, then the first book on complex function theory proper is the one by Durège, a former colleague of Riemann's. As one would expect, it is along the lines of Riemann's lectures. It ran to four editions, was translated into English for use in American universities, and was used by those in the Weierstrassian school when they found themselves outside Berlin. For a time Teubner editions carried an endorsement of it by Fuchs, for example. In due course there was a polarisation of German textbooks into either Riemannian or Weierstrassian styles, and then there was an attempt to combine the best of both. But curiously, Weierstrass himself never published his version (or, better, versions) and the result was to be a failure to spread his ideas abroad, where the Cauchy–Riemann style prospered.

In 1865 the first edition of Carl Neumann's *Vorlesungen über Riemann's Theorie der Abel'schen Integrale* came out. Neumann confined his analysis to the hyperelliptic case, and with that restriction aimed to make Riemann readable. In the second edition Neumann did the general case of Abelian functions, but now he retreated from his earlier endorsement of Dirichlet's principle. The first edition was disparaged by Roch, who, although he found it praiseworthy, found it too long to be lectures, often verbose, and to differ too much from Riemann's own presentation of these ideas. The second edition, when it was re-read in Göttingen later, was said by Klein to have made every thing look so easy it was insulting, which shows how far things had swung Riemann's way.[6]

Mention should also be made of the book by Schlömilch (who, by the way, was the person who encouraged the young Roch to go to Göttingen and study with Riemann). His (1866) contains enough material on the subject to count as only the fourth book on complex function theory to be published, and it ran to several editions. Pages 35 to 111 cover functions of a complex variable, and further chapters look at elliptic integrals and elliptic functions. Like Durège he used the theory of Riemann surfaces to deduce the properties of elliptic functions from those of elliptic integrals.

Casorati's remarkable book of 1868 is one of the best textbooks. It opens with a reliable and insightful historical account of almost 200 pages, and then devotes over 250 pages to describing the current state of the art. The ideas and techniques of Cauchy, Riemann, and Weierstrass, are all presented, and the book did a lot to establish Riemannian complex function theory in Italy. Oddly enough, the first part about Riemann surfaces and Abelian functions is more difficult than the second, which goes over ground that Riemann himself had covered in his lectures.

[6]See Klein *Entwickelung* (1926–1927, 1, 273).

18.5 Complex Algebraic Geometry

It has become entirely natural for mathematicians today to regard the equation $f(x, y) = 0$, where f is a polynomial in x and y, as defining a curve in the Cartesian plane with coordinates x and y, and to think of the curve as the set of points (x, y) in the plane for which $f(x, y) = 0$. If not from the time of Descartes then certainly since Euler it has seemed natural to treat these curves algebraically by analysing their equations. Thus the fundamental theorem of algebra says that a polynomial equation of degree n has as n solutions, and geometrically that says that a curve of degree n will meet a straight line in n points.[7] A more complicated argument establishes Bezout's theorem, that two curves of degrees k and m respectively will meet in km points.

There is a price to pay. A line should meet a conic section in two points, according to the algebra, but it may not meet it at all (in such cases the algebra gives complex points of intersection). The line may be a tangent to the curve, in which case algebra suggests counting the point twice. And the line $x = a$ meets the parabola $y = x^2$ in the point (a, a^2)—but where is the second point of intersection? All of these questions can be answered by moving to the setting of complex projective geometry. In 19th century language, projective space contains points at infinity, and this allows mathematicians to give a rigorous account of Bezout's theorem once complex coordinates are introduced. Which raises the question: who first endowed a curve with all its complex points?

It became gradually accepted that two real curves of degrees k and m respectively will meet in km points, provided some of the 'points' are complex. But this was variously regarded as an artefact of the algebra, or as an indication that the curves possessed a symmetry of some kind (a map from the curve to itself that fixed the real points and switched each complex one with its conjugate, for example). The issue is when a plane curve given by a polynomial equation $f(x, y) = 0$ was taken to be a subset of $\mathbb{C} \times \mathbb{C}$.

One partial answer is provided in a paper of Cauchy's on elliptic function theory that he published in 1846, in which he spoke of the "complex variable y [that] satisfies the finite equation $y^2 = \left(1 - x^2\right)\left(1 - k^2x^2\right)$ [and] varies with x by insensible degrees". What this involves was soon addressed by Puiseux in his long paper (1850), in which, inspired by Cauchy's work, he showed how to regard a function $f(u, z) = 0$ of degree m in u as an m-sheeted covering of the z-plane branched over a number of points.[8] These examples suggest that complex analysts were moving towards a geometric appreciation of some functions, but they fall far short of Riemann's vision.

Riemann was clear on every occasion that he regarded an algebraic curve with equation $f(s, z) = 0$ of degree n in s and m in z, as an equation in complex variables that should be thought of providing an n-fold covering of the complex z-plane

[7]The degree of a polynomial $f(x, y)$ is defined to be the highest degree of its monomials.

[8]Although we know from Laugwitz (1999, 92) that Riemann had read Cauchy's report on Puiseux's memoir by December 1851 it seems unlikely that Riemann had anything to learn from Puiseux by the time his was writing his doctoral thesis.

(or, better, z-sphere). He then explained how such a covering surface could be regarded (in many different ways) as an algebraic curve in $\mathbb{C} \times \mathbb{C}$. From his standpoint a curve simply had complex points on it.

Such an idea is implicit in the work of Abel and Jacobi on elliptic functions, but they never presented it geometrically. Indeed, Jacobi found the idea of a two-valued integrand in an elliptic integral sufficiently complicated to motivate his move to base the theory of elliptic functions on the theory of theta functions and so avoid the integrals altogether. Riemann's vision of complex algebraic curves was another barrier to the reception of his ideas, as Cayley's comments in a paper of 1878 indicate. He distinguished carefully between Riemann's approach and the simpler idea that a complex curve is a set of points in $\mathbb{C} \times \mathbb{C}$, and then of the second idea "I was under the impression that the theory was a known one; but I have not found it anywhere set out in detail". It would seem that only somewhere between Clebsch's turn to algebraic geometry in 1864 and Cayley's remarks 14 years later did curves acquire their full complement of complex points.

18.6 Exercises

1. What is the value of the Riemann mapping theorem?
2. Illustrate, with examples, the value in being able to define complex functions on a variety of surfaces other than the plane. How does Riemann's framework help explain the double periodicity of elliptic functions?
3. What aspects of Riemann's work were adequately established by the time he died? What problems confronted mathematicians wishing to take up Riemann's work after his death in 1866?

Chapter 19
Weierstrass

Weierstrass (1815–1897)

19.1 Introduction

A powerful, algebraic, alternative to Riemann's geometric complex function theory was developed by Weierstrass. In this chapter we look at his early work, his successful arrival in Berlin, and his first lecture courses. Weierstrass was to be a powerful influence on the development of mathematics in Germany, and very often his influence was indirect, because he frequently confined his discoveries to his lectures in Berlin.

J. Gray, *The Real and the Complex: A History of Analysis in the 19th Century*,
Springer Undergraduate Mathematics Series, DOI 10.1007/978-3-319-23715-2_19

19.2 Weierstrass's Early Years

If the generic myth of the great mathematician is someone who comes from nowhere and dazzles immediately (as, for example, Galois did) then Karl Weierstrass is the antithesis, in that he published his first paper in 1854 when he was almost 40. Even so, it must be admitted then he then rose very rapidly to prominence, because in this paper he solved the Jacobi inversion problem for hyperelliptic integrals, a fact that startled his contemporaries, and he went on to dominate the mathematical world for the next 40 years.

Much of the mystery surrounding his late appearance dissipates, only to be replaced by another, when we learn that Weierstrass had been conducting important research for several years by 1854 but had not bothered to publish it. A deeper mystery is that he had a lifelong aversion to seeing his work appear in print, and published very little, confining himself lectures.

Weierstrass was born on October 31, 1815 in Ostenfelde near Münster, and spent four years, from 1834 to 1838 at the university of Bonn supposedly studying courses in jurisprudence and administrative matters that would lead to a career in the public services but without getting a degree or even trying to pass any examinations. Karl's brother Peter told Gösta Mittag–Leffler that Karl was active in the Jugendabenteuern, joined the Korps Saxonia and spent his time more in cellars and student fights than in studying.[1] On the other hand, other sources suggest that Weierstrass had begun to read Crelle's *Journal* while still at school, and he himself claimed in 1840, in a report on his education, that he had studied mathematics intensively while at Bonn. Perhaps both accounts are true. At all events, Weierstrass does seem to have studied Laplace's *Mécanique céleste* and Jacobi's *Fundamenta Nova* on his own, and at some stage he must have realised that his future lay with mathematics.

Weierstrass entered the Theological and Philosophical Academy in Münster in 1839 to obtain a teacher's diploma. The decisive event for his future career was Christoph Gudermann's lecture course on the theory of elliptic functions. Gudermann was a friend of Jacobi and he based his course on his friend's *Fundamenta Nova*, developing the elliptic functions as power series. There are three fundamental elliptic functions, for which Gudermann introduced the convenient notation $\mathrm{sn}(x)$, $\mathrm{cn}(x)$, and $\mathrm{dn}(x)$. As with the sine and cosine functions in trigonometry, these functions are connected by differential equations that can be made to yield the corresponding power series by the method of undetermined coefficients, but the argument is considerably more complicated.

Weierstrass took up the subject, which was his introduction to the formal theory of power series, published a thesis that Gudermann thought very highly of, and after passing the examination, moved to the Gymnasium Paulinum in Münster.

Here he wrote his second paper (Weierstrass 1841a), although he did not attempt to publish it.[2] This is odd, because, as Mittag–Leffler (1923, 27) was to remark, in this paper.

[1] See Mittag–Leffler (1923, 14–15).

[2] It was published for the first time in his *Werke* towards the end of his life.

> one can find, as far as is known, the first rigorous proof of Cauchy's theorem on the integration between two given complex limits without any application of double integrals or surface integrals; this is given in such a way that one obtains at the very same time the Laurent theorem, which became known only two years later through a note by Cauchy.

It is all the more remarkable, to quote Mittag–Leffler again, when one notes that "Weierstrass first became acquainted with Cauchy's works in 1842".

As Bottazzini was the first to observe (see Bottazzini 2002), the emphasis on power series methods is characteristic of Weierstrass's approach to complex analysis because it allowed him to develop uniform methods that apply to functions of both one and several variables, and complex functions of several variables were intended by him to open up the study of hyper-elliptic and abelian functions. This commitment is apparent in his (1841b), the second paper Weierstrass wrote in 1841. In it he stated and proved three theorems about power series expansions (by which he meant both 'ordinary' power series and Laurent series expansions). Theorems (A) and (B) show how to evaluate the coefficients of a Laurent series in one and resp. several complex variables without the use of integrals. Weierstrass did not know of Cauchy's Turin 1831 Mémoire on the 'calculus of limits', and derived the relevant inequalities for Taylor series independently.

These theorems run as follows:

Theorem (A): Let $F(x) = \sum_{\nu=-\infty}^{\nu=+\infty} A_\nu x^\nu$ be a power series of the complex variable x with given coefficients and r 'any determinate, positive magnitude which lies within the domain of convergence of the series' (Weierstrass 1841b, 67). Let $\sup |F(x)| = g$ for $|x| = r$. Then,

$$\left|A_\mu\right| \leq gr^{-\mu}$$

for every integer value of μ.

Weierstrass's proof is based on his assumption that the series is uniformly convergent for $|x| = r$.

Theorem (B) extended this result to functions $F(x_1, \ldots, x_\rho))$ of several complex variables. Let $F(x_1, \ldots, x_\rho) = \Sigma_\nu A_{\upsilon_1, \nu_2, \ldots, \nu_\rho} x_1^{\nu_1} x_2^{\nu_2} \ldots x_\rho^{\nu_\rho}$ be a convergent series, where $x_1, x_2, \ldots, x_\rho$ are complex variables, $A_{\upsilon_1, \nu_2, \ldots, \nu_\rho}$ are given constants and $\nu_i \in \mathbb{Z}$, and let $r_i (1 \leq i \leq \rho)$ be a system of positive quantities such that the point $(x_1, x_2, \ldots x_\rho) = (r_1, r_2, \ldots r_\rho)$ belongs to the domain of convergence. In addition, let $\sup \left|F\left(x_1, \ldots, x_\rho\right)\right| = g$ for $|x_i| = r_i$ for any i. Then

$$A_{\upsilon_1, \nu_2, \ldots, \nu_\rho} \leq gr_1^{-\mu_1} r_2^{-\mu_2} \ldots r_\rho^{-\mu_\rho}$$

for every system of integer values of μ_i.

The proof is similar to the first one, and is based on the assumption that the series for values x such that $|x_i| = r_i$ for any i converges uniformly.

Weierstrass's Theorem (C) is the double series theorem nowadays named after him: Given an infinite sequence of functions $F_\mu(x_1, \ldots, x_\rho)$ defined by 'ordinary' power series

$$F_\mu(x_1, \ldots, x_\rho) = \sum_\nu A^\mu_{\nu_1, \nu_2, \ldots, \nu_\rho} x_1^{\nu_1} x_2^{\nu_2} \ldots x_\rho^{\nu_\rho}$$

such that each of the functions (and their sum) is 'unconditionally and uniformly' convergent in a determinate neighbourhood G of the origin $(0, 0, \ldots, 0)$, then

$$\sum_{\mu=0}^{+\infty} F_\mu(x_1, \ldots, x_\rho) = \sum_\nu A_{\nu_1, \nu_2, \ldots, \nu_\rho} x_1^{\nu_1} x_2^{\nu_2} \ldots x_\rho^{\nu_\rho}$$

where $\sum_\mu A^\mu_{\nu_1, \nu_2, \ldots, \nu_\rho} = A_{\nu_1, \nu_2, \ldots, \nu_\rho} < +\infty$

In other words, under the given hypotheses about convergence, the sum of the series of functions F_μ can be expressed by a power series in $x_1, \ldots, x_\rho$ whose coefficients are given by the sum of the corresponding coefficients of the power series which represent the functions F_μ.

Weierstrass proved this theorem by using the Cauchy inequalities for Taylor coefficients that he had already obtained in Theorem (B). His arguments show that he now had a clearer understanding of the phenomenon of the uniform convergence of series, as far as the interchanging of two infinite series was concerned.[3]

The importance of these theorems, and particularly the last, is that "for Weierstrass the double series theorem was the key to convergence theory", to quote (Remmert 1989, 251). From Theorem (C) Weierstrass obtained the theorem on differentiating uniformly convergent series. He developed each F_μ into its Taylor series around any point $a_1, \ldots, a_\rho$ of G and observed that, as a consequence of Theorem (C), for fixed $a_1, \ldots, a_\rho$ they converge in some fixed disk centred at $a_1, \ldots, a_\rho$ and lying in G. Therefore the series

$$\sum_{\mu=0}^{+\infty} F_\mu(a_1, \ldots, a_\rho)$$

represents an analytic function there, and has derivatives of any order which are given by

$$\frac{\partial}{\partial x_\lambda} \sum_{\mu=0}^{+\infty} F_\mu(x_1, \ldots, x_\rho) = \sum_{\mu=0}^{+\infty} \frac{\partial}{\partial x_\lambda} F_\mu(x_1, \ldots, x_\rho), \quad \lambda = 1, \ldots, \rho.$$

As with his previous papers, even this paper remained unpublished until 1894, when the volumes of Weierstrass's *Werke* began to appear. However, Weierstrass

[3]However, Weierstrass made the unnecessary hypothesis that the series converges unconditionally, i.e. absolutely.

sometimes presented Theorem (A), and even also Theorems (B) and (C), in his Berlin lectures on the introduction to analytic function theory.[4]

A striking feature of these early papers, and it persisted throughout Weierstrass's career, is his avoidance of methods that rely on the Cauchy integral theorem. Bottazzini has suggested that Theorem (B) suggests that Weierstrass's drive for a theory of several variables may provide the reason. To be sure, Weierstrass could have proved Theorem (A)—which is about functions of a single variable—by introducing an integral taken around the boundary of a disk of a suitable radius. But for functions of several variables case there was no way then to handle the integral of a function of several complex variables taken over the boundary of a polydisk. However, the method of power series applied with not much extra work to several complex variables as well as one. For this reason, Bottazzini argues, Weierstrass seems to have turned against integrals ('transcendental' methods, as he later called them) in favour of power series as early as 1841.

In the spring of 1842 wrote a third paper that he also withheld from publication. In it he established the existence of solutions of a system of n algebraic ordinary differential equations, unaware of what Cauchy had done in a series of papers starting with (Cauchy 1842).[5] These three papers gave Weierstrass the foundations of all of his later theory of analytic functions.

In autumn 1842 Weierstrass obtained a position as teacher at the Katholisches Progymnasium in Deutsch Krone in West Prussia. He had to teach a wide range of subjects, but in spite of this he was able to continue his research. In 1848 he got a position at the Gymnasium in Braunsberg. The next year he published a draft of a theory of Abelian functions in Programmheft of the Gymnasium—academic schools in Prussia were proud of the qualities of their teachers and encouraged them to do some sort of research. Weierstrass then developed the basic ideas in this paper in the two papers that appeared in 1854 and 1856 in Crelle's *Journal* and made his name.

In the first Weierstrass solved the Jacobi inversion problem in the hyper-elliptic case, a problem that had reduced Jacobi, on his own admission "almost to despair" until he had found the key in a paper of Abel's. Even so, understanding of the problem remained limited until Weierstrass solved it by his completely different route. His 1854 paper "caused a degree of surprise all over the mathematical world, which remains almost unique in the history of our science" (Killing 1897, 718). Borchardt, the editor of Crelle's journal, went to Braunsberg to meet the author of the paper. Liouville wrote a complimentary letter to Weierstrass, and published a French translation of his paper in the *Journal de mathématiques* that he edited. Soon thereafter Weierstrass was awarded by a doctor degree *honoris causa* from Königsberg University, and a delegation led by Friedrich Richelot went to Braunsberg

[4]Bottazzini has considered Killing (1868, 66–68), Hettner (1874, 278–294) and Hurwitz (1988a 99–100, 123 and 155), see Bottazzini and Gray, (Chap. 7).

[5]Weierstrass considered n equations of the form $\frac{dx_j}{dt} = G_j(x_1, \ldots, x_n)$, for n unknown functions $x_1, \ldots, x_n$, where the functions G_j are polynomial functions in $x_1, \ldots, x_n$.

to hail Weierstrass as "our teacher".[6] The paper (1856a) is an expanded and detailed version of the first one, and was also published in Crelle's *Journal.*

The publication of his 1854 paper opened the way to Weierstrass's academic career. In January 1855 Crelle wrote to the Prussian Minister of Education to ask them to see that Weierstrass could be found an academic position in order to continue his research. The Ministry naturally consulted Dirichlet, who confirmed Crelle's views and hopes while noting, as elder generations of mathematicians are prone to do, that Weierstrass "has given only partial proofs of his researches [in his 1854 paper], where the intermediate explanations are lacking" (in Dugac (1973, 52)). But in 1855 Gauss died, and this provoked a round of academic hirings and promotions. Initially, Weierstrass was an unsuccessful candidate for a chair in Breslau, but this failure seems to have shaken up the Berlin mathematicians, for very soon, thanks to the joint efforts of Alexander von Humboldt and Richelot, Weierstrass was appointed professor at the Berlin Gewerbeinstitut (the Industry Institute, later Gewerbeakademie, today the Technische Universität). In September 1856 he became a professor at Berlin University. There he joined Kummer and the privately wealthy Leopold Kronecker, a former student of Dirichlet's and Kummer's and (after 1860) a member of the Berlin Academy of Sciences. So by the mid-1850s the celebrated 'triumvirate' of Kummer, Kronecker and Weierstrass, which was to dominate the German mathematical milieu for some 25 years, had settled in the Prussian capital. For the former teacher at Braunsberg, at the age of 40 a new life began.

19.3 Weierstrass in Berlin

Weierstrass threw himself into teaching. Among his early courses was one on 'The applications of elliptic functions to geometry and mechanics' (Winter Semester, WS, 1857/58). In winter 1859/60 he lectured for the first time on 'Introduction to analysis', and in summer 1860 he gave a course on integral calculus. Then he came back to his favourite subject, elliptic functions, and devoted a course to a selection of problems to be solved by means of these functions. In WS 1861–62 he announced a course on 'The general theory of analytic functions', but a serious breakdown caused by overwork forced him to stop lecturing for one academic year. Once he recovered he resumed teaching elliptic functions and their applications, and in summer 1863 delivered his first course on the introduction to the theory of Abelian functions. Eventually, in WS 1863/64 he was able to teach the promised course on analytic function theory.

A Course on Differential and Integral Calculus

We get a glimpse of Weierstrass's early lectures on analysis by looking at a set of (unpublished) lecture notes taken by Schwarz from a course on 'Differential and Integral Calculus' given by Weierstrass in summer 1861 at the Gewerbeinstitute in

[6] See Killing (1897, 718).

Berlin. These are at times rather sloppy, and not always any better than Cauchy had already done. That said, the last topic Weierstrass presented was the differentiation of infinite series. He began by raising the question about the conditions one has to impose so that a series of continuous functions, each of which is convergent for all x within two given limits, represents a continuous function. He answered the question precisely, saying that the series is convergent "in equal degree".[7]

> when the sum of r terms of the series following the nth (r being an arbitrary positive whole number) for increasing n can be made smaller than any arbitrarily given magnitude, and this such that one may take the same n for all values of x and the property still holds true when instead of n one takes $n+m$, m being any positive whole number.

Nowadays this form of convergence is called uniform convergence. Weierstrass's clumsy expression is the result of using words rather than symbols, but it also reflects the difficulty of the concept that Weierstrass is trying to convey.

By means of an $\varepsilon - n$ argument Weierstrass then proved rigorously that the uniform convergence of the series

$$\sum_{n=1}^{\infty} \varphi_n(x)$$

of continuous functions in a given interval (a, b) ensures the continuity of the sum of the series.[8]

Weierstrass then investigated the term by term differentiability of a series of functions. He considered two series of functions $\sum \varphi_n(x)$ and $\sum \varphi_n'(x)$, that are continuous on a given interval and such that $\varphi_n'(x)$ is the derivative of the corresponding term $\varphi_n(x)$ for each n. Under the hypothesis that the series converge uniformly in the interval, he proved that the function

$$\varphi'(x) = \sum \varphi_n'(x)$$

is indeed the derivative of the function

$$\varphi(x) = \sum \varphi_n(x)$$

there; that is, term by term differentiation is valid. Now Weierstrass could use the differential calculus to obtain the series expansion of elementary transcendent functions such as the exponential and the trigonometric functions. Finally, he determined the series expansion of a function given the series expansion of its derivative.

This course exerted a great influence on Schwarz and, in turn on Eduard Heine who was a professor at Halle when Schwarz began teaching there in 1867. From

---

[7]In German 'Convergenz in gleichem Grade', a term he took from Gudermann, as we discuss below, see § 22.4.

[8]Schwarz mistakenly wrote that the theorem states a necessary and sufficient condition, but the proof clearly shows that it is a sufficient condition.

Heine himself we learn that Weierstrass's ideas were spread in conversations with Weierstrass, Schwarz, and Cantor as well.[9]

19.4 Weierstrass's Programme of Lectures

As his lecture courses developed over the years, and especially after July 1864 when he was appointed *Ordinarius* at the University, and gave up teaching at the Gewerbe-institut, Weierstrass became more and more deeply concerned with rigour in analysis. In particular, he rightly realised that the theory of analytic functions required rigorous foundations, and in his eyes these functions were to provide the foundations of the whole building of the theory of elliptic and Abelian functions. Poincaré caught this well in his obituary of Weierstrass (Poincaré 1899) when he summarised Weierstrass's work in this way:

1. To deepen the general theory of functions of one, two and several variables – this was the basis on which the whole pyramid should be raised.
2. To improve the theory of transcendental and elliptic functions and to put them into a form which could be easily generalised to Abelian functions, the latter being a 'natural extension' of the former.
3. Eventually, to tackle Abelian functions themselves.

Unlike other mathematicians, who preferred papers or books, Weierstrass chose to realise this programme through his university lectures. Over the years one of his major concerns became the establishing of analysis, and the foundations of analytic function theory in particular, with absolute rigour. Until the end of his teaching career he presented a course in analysis in a cycle of lectures over four consecutive terms, that ran as follows:

1. Introduction to analytic function theory
2. Elliptic functions
3. Abelian functions
4. Applications of elliptic functions or, sometimes, the calculus of variations.

Most of his original discoveries were presented first, and on occasion only, in his lectures, and the steadily evolving lecture cycle was often the only way to find out what his opinions and insights were. For example, when Weierstrass proved that a convergent power series is uniformly convergent within the domain of convergence, he thereby showed that a function defined by a power series is continuous on any domain strictly inside its circle of convergence, because the convergence is necessarily uniform there. This result had been previously stated by Abel but not given an adequate proof. But few if any could know of Weierstrass's discovery other than by going to the University of Berlin to hear him developed his ideas during his many lecture courses.

[9]See Heine (1872, 172).

This practise drew a large number of students to Berlin, but it was ultimately an unsuccessful way to promote his ideas.

Casorati's Notes

The Italian mathematician Felice Casorati was so irked by this secrecy that he tried to visit Weierstrass in Berlin and corresponded intensely with Weierstrass's former student Schwarz to find out about Weierstrass's most recent discoveries; he hoped to present these in the planned second volume of his treatise, but in the end it never appeared. However, the historian is very fortunate in having Casorati's notes of his trip to Germany.[10]

Casorati had decided to travel to Berlin, in order to understand what was going on in the town that was becoming the centre of mathematical world. He made notes of his conversations with the Berlin mathematicians in a scientific diary, and they provide an intriguing glimpse of some of the problems that were at the forefront of mathematical research at that time. He met Kronecker and Weierstrass separately, and occasionally together, and at times, students and other mathematicians attended the meetings. Kronecker, for example, remarked that the concept of continuity "was still a confused idea", and the Dirichlet principle and the concept of analytic continuation also came up—topics that were to continue to interest Weierstrass for years to come.

Kronecker even surprised Casorati by claiming to have examples of "functions that do not admit differential coefficients, that cannot represent lines etc." Casorati's few notes unfortunately do not allow us to decide whether Kronecker actually discovered such 'pathological' functions, or simply reported what Weierstrass had done on his own. Schering, who was at this particular meeting, mentioned Riemann's still unpublished but quite well-known example of an integrable function discontinuous at every rational point.

Casorati noted that already the Berlin group were becoming more and more critical of Riemann's way of working: "Riemann's things are creating difficulties in Berlin", he recorded. For example, Kronecker criticised Riemann and his followers were for assuming that the singular points of a complex function are always such that any two non-singular points can always be joined by a path that avoids the singular points. "But this is not possible", Weierstrass told Casorati, and "It was precisely while searching for the proof of the general possibility that he realised it was in general impossible". Kronecker provided Casorati with this example to illustrate the impossibility: the function

$$\theta_0(q) = 1 + 2\sum_{n>1}^{\infty} q^{n^2}, \ |q| < 1 \tag{19.1}$$

The function $\theta_0(q)$ cannot be analytically continued outside of the disk $|q| < 1$. "There is another expression for it"—Weierstrass told Casorati—"which also has meaning outside of the circle, that is, inside and outside but not on the circumference".

[10]These have been studied by Neuenschwander (1978a, b) and more recently in Bottazzini and Gray (2013, § 6.4.1).

The latter is "entirely made of points where the function is not defined, it can take any value there". In this rather cryptic way Weierstrass announced a result he was to publish only in 1880.[11] This was intimately connected with his example of a continuous nowhere differentiable function which he was to publish some years thereafter.[12]

In order to know the behaviour of the function defined in equation (19.1) outside the circumference $|q| = 1$, Kronecker observed that "it is necessary to resort to other means, and not to those of making q follow a path" joining a point within the disk with a point outside it. According to Casorati's report, Weierstrass had thought for a long time that points at which a function "ceases to be definite [...] could not form a continuum, and consequently that there is at least one point P where one can always pass from one closed portion of the plane to any other point of it" until he realised that "this was not the case". As an example of a point at which a function ceases to be definite, Weierstrass cited $x = 0$ for the function $e^{1/x}$, saying that "it can have any possible value" there.

Weierstrass and Schwarz

Casorati's notes amplify the picture of the issues that concerned Weierstrass as he set out to produce a rigorous theory of complex analysis adequate to the demands he wished to impose upon it. As the years went by, this quest took on a markedly algebraic cast, as a much-quoted passage from a letter he sent to Schwarz in 1875 shows:

> The more I think about the principles of function theory – and I do it incessantly – the more I am convinced that this must be built on the foundations of algebraic truths, and that it is consequently not correct when the 'transcendental', to express myself briefly, is taken as the basis of simple and fundamental algebraic propositions. This view seems so attractive at first sight, in that through it Riemann was able to discover so many of the most important properties of algebraic functions. (It is self-evident that, as long as he is working, the researcher must be allowed to follow every path he wishes; it is only a matter of systematic foundations) (in: Weierstrass *Werke* II, 235).

The 'transcendental' methods Weierstrass was rejecting were those involving the integral, and he rejected them even when dealing with functions of a single variable, which, as we shall see, gave his presentations a somewhat contrived feel at times. Also to be rejected, in his own work and, he implied, in that of his students, was any lasting use of geometric intuition, although it could be allowed to assist in the process of discovery.

Weierstrass, of course, did not see these strictures as limitations. It was not a whim, or an idle prejudice, that caused him to reject methods that had served Riemann so well. Rather, as he explained later in the same letter to Schwarz he had been "especially strengthened [in this belief] by his continuous study of the theory of

---

[11] In the same paper in which he first published his M-test.

[12] This series had already been used by Jacobi (*Fundamenta nova*, § 65) in his proof of Fermat's theorem on the number of representations of a number as a sum of four squares. The connections with 'pathological' functions had been made evident by Riemann in examples he presented in his *Habilitationsschrift* by considering the limit cases of certain theta-series.

analytic functions of several variables", which was required to create a theory of Abelian functions. To this end, Weierstrass's lectures on analytic function theory would conclude with the elements of a theory of functions of several variables.

19.5 Introductions to the Theory of Analytic Functions

We can trace the developments in Weierstrass's theory only intermittently in his publications, but much more steadily in his lectures courses from WS 1861/62 to WS 1884/85, of which many notes taken by his various students survive. (They were kept, one supposes, precisely because Weierstrass was developing a systematic theory that could not be found anywhere else and made almost no reference to any existing literature.) And indeed, Weierstrass seems to have been a good lecturer, at least for the better students. Fifty years later Leo Koenigsberger recalled attending Weierstrass's lectures on elliptic functions with Lazarus Fuchs, and wrote (Koenigsberger 1917) that they could never forget the "powerful impression and incomparable excitement" exerted on them by the "deep and over-arching content" of that lecture combined with the "natural, unvarnished" way of expounding it that was "free from any rhetoric".

In his lectures, Weierstrass regarded the fundamental property of analytic functions as the fact that they can be expanded as Taylor series. He therefore approached them by starting with functions defined in terms of a finite number of the simplest arithmetical operations (addition and multiplication) and proceeding to functions defined by an infinite number of such operations.

Weierstrass agreed that analytic functions in his own sense are precisely the functions $u + iv$ that Cauchy and Riemann had characterised as the solutions of the Cauchy–Riemann equations. But he said that he rejected that approach because, in his opinion, the general definition of a class of functions could be founded on an arbitrary property, whose generality cannot be established *a priori*. Moreover, the Cauchy–Riemann requires functions $u(x, y)$ and $v(x, y)$ that admit partial derivatives, and "in the present state of science" such functions form a class that cannot be exactly defined, and for this reason to the definition was rejected as being "poorly adapted for building the theory of analytic functions on it".[13]

Weierstrass then defined uniform convergence for series of functions, and proved that the sum of a uniformly convergent series of rational functions is a continuous function. Only now did he introduce power series of one variable, when he was in a position to establish their important, basic properties. He established the a power series has a disk of convergence; more precisely, if z_0 is a point inside a domain on which a complex analytic function f is defined, then the function can be expanded as a Taylor series that converges on a disc D_0 centre z_0 and reaches as far as the nearest singular point of the function to z_0 (that is to say, any larger disc would contain a singular point of f). Weierstrass also showed that the function converges uniformly within the disk. He produced the usual inequalities for the coefficients of a series

[13]See Hurwitz's notes (Weierstrass 1988, 49).

(without appealing to Cauchy integral theorem) and he remarked that "analogous propositions" hold for series of several variables. Next came the statement of the double series theorem for one and several variables, and then the proof that of the identity theorem (two series that agree on an (arbitrarily small) open neighbourhood of the origin define the same function). Finally, Weierstrass proved the usual theorems on term-by-term differentiation for series in one or several variables.

In keeping with his rejection of 'transcendental' methods, as he called them, Weierstrass seldom lectured on the integral calculus, whereas the differential calculus is covered in almost every one of Weierstrass's lecture notes on analytic function theory. The integral comes up in Killing's notes of the course in 1868, but only at the end of the course, after the treatment of series, and in the lecture course of 1886. We can read in the letter to Schwarz cited above that when Weierstrass discussed Cauchy's theorem he remarked that "nowadays, following Riemann's example the proof is usually obtained from the consideration of a double integral" (the Green's theorem approach), but that "I do not consider this to be completely methodical"; he then gave a different proof.

Analytic Continuation

The concept of analytic continuation was familiar to Riemann, who used it extensively, but it became of foundational importance in the work of Weierstrass. Let $f(z)$ be an analytic function defined on a disc D_0 centre z_0, and suppose that z_1 is a point in the disc D_0, then the function can be expanded as a Taylor series that converges on a disc D_1 centre z_1 and reaches as far as the nearest singular point of the function to z_1. It is quite possible that the disc D_1 extends beyond the disc D_0. When this happens the function f is said to have been continued analytically outside the disc D_0. The process can be repeated until the function f has been provided with Taylor series expansions on as large an overlapping set of discs as possible.

There are two points to note. If a chain of overlapping discs returns to a point already covered, it does not mean that the original value of the function there and the value obtained by analytic continuation agree. For example, the square root 'function' may be defined uniquely on a disc that does not include the origin—this corresponds to choosing a definite sign for the square root. It can be analytically continued around the origin until the domain of definition overlaps itself, but it returns to the negative of its original value.

Note also that there may be points where the function cannot be analytically continued in this fashion. They form what is called the natural boundary of the function, and we shall see what Weierstrass made of them shortly.

Chapter 20
Weierstrass's Foundational Results

20.1 Introduction

Weierstrass is remembered for many specific discoveries in complex function theory, and here we consider two of them: his insight into the distinction between poles and essential singularities and the idea of a natural boundary of a complex function and the connection to nowhere differentiable real functions. We then look more briefly at his systematic presentation of the theory of elliptic functions, and content ourselves with a mention of his representation theorem, and his final account of a theory of Abelian functions.

20.2 Poles and Essential Singularities

Weierstrass is remembered for a number of major results in complex function theory. One such striking insight is the distinction he pioneered between what are called poles or finite singularities, and essential singularities. On the face of it, a function such as $f(z) = 1/z^2$ is not defined at the origin, but its reciprocal is. Also, rather trivially, there is an integer n such that the function $z^n f(z)$ is analytic at the origin—in this case $n = 2$ and the function $z^n f(z) = 1$. But it is also true that the function $g(z) = 1/z^2 + 5/z - 9 + 2z^3$, say, is not defined at the origin, but its reciprocal is. Again there is an integer n such that the function $z^n g(z)$ is analytic at the origin: $n = 2$ and the function $z^n g(z) = 1 + 5z - 9z^2 + 2z^5$. This means that functions of this sort, although undefined at the origin, may be said to tend to infinity at the origin in a straight-forward way.

Compare the function $h(z) = e^z$. As z tends to ∞ along the positive real axis, this function tends to $+\infty$. But as z tends to ∞ along the negative real axis, this function tends to 0. If we set $z = it$ then $h(z) = e^{it} = \cos(t) + i\sin(t)$ which, interestingly, doesn't seem to tend to any limit at all as t tends to ∞. But if we restrict t to the

J. Gray, *The Real and the Complex: A History of Analysis in the 19th Century*, Springer Undergraduate Mathematics Series, DOI 10.1007/978-3-319-23715-2_20

values $2k\pi$, $k = 0, 1, \ldots$ then $h(2k\pi i) = \cos(2k\pi) + i\sin(2k\pi) = 1$. If we let z go to infinity radially along the line $y = mx$, so $x = t$, $y = mt$ we are led to consider $e^t(\cos mt + i\sin mt)$, and if we restrict t to the values $2\pi k/m$, $k = 0, 1, \ldots$ this time we get infinity as a limit. More interestingly, if we let z go to infinity vertically, along the lines $x = a$, $y = t$ we are led to consider $e^a(\cos t + i\sin t)$, and if we now restrict t to the values $2k\pi$, $k = 0, 1, \ldots$ this time we get e^a as a limit. This means that any positive real number can be obtained as a limit of a sequence of points tending to infinity. Finally, if we let z go to infinity vertically, along the lines $x = a$, $y = t$ we are led to consider $e^a(\cos t + i\sin t)$, and if we restrict t to the values $b + 2k\pi$, $k = 0, 1, \ldots$ this time we get $e^a(\cos b + i\sin b)$ as a limit—so any complex number can be obtained as a limit of a sequence of points tending to infinity.

All of this shows that the function $h(z) = e^z$ cannot be said to have meaningful limit as z tends to ∞, and it follows that the function $e^{1/z}$ cannot be said to have meaningful limit as z tends to 0 and so there is no integer n such that the function $z^n e^{1/z}$ is analytic at the origin.

That is what Weierstrass proved. This shows that there is a fundamental distinction between functions $f(z)$ for which there is an integer n such that the function $z^n g(z)$ is analytic at the origin and those for which there is not. The former kind are said to have finite poles, or a pole of order n where n is the least integer with the stated property. The latter kind was said by Weierstrass to have an essential singularity. Weierstrass was also able to show in 1864 that in a neighbourhood of an essential singularity the function gets arbitrarily close to any arbitrary value. In this he had been preceded by Casorati, and the result is therefore known, correctly, as the Casorati–Weierstrass theorem.[1]

20.3 Natural Boundaries

The natural boundary of a complex function is so called because it is a natural thing to look for. Imagine that you have a complex function f defined by a power series on some open disc in the complex plane, say the unit disc centred on the origin, and you attempt to extend the domain of definition by choosing a point $\zeta \in |z| < 1$ and obtaining the power series expansion of f in an open disc centre ζ. You are interested in the case where this new disc contains points not contained in the unit disc. What can happen? Well, obviously, you either succeed or you fail. If you always fail, then there is no point $\zeta \in |z| < 1$ that is the centre of an open disc that sticks out beyond the unit disc, and so the function f cannot be analytically continued beyond the unit disc, and the boundary of the unit disc is accordingly a natural boundary for the function.

[1] And even more accurately as the Casorati-Sokhotskii-Weierstrass theorem, after the Russian mathematician Sokhotskii who came to the same ideas at about the same time, see Bottazzini and Gray (2013, § 7.7.3).

If you succeed, if only for some $\zeta \in |z| < 1$, then you carry on trying to push outside every disc you obtain until you have obtained the maximal domain of definition for the function. The points on the boundary of the domain are the natural boundary of the function.

There might be no natural boundary. A polynomial function in z is certainly defined by a power series (indeed, a polynomial) that converges everywhere. Let's say you admit poles, and want to investigate whether the function is defined at $z = \infty$ by the method of replacing z with $1/z$. Then you find that rational functions (quotients of one polynomial by another) also have no natural boundary. In fact, by Liouville's theorem, as Riemann seems to have known, the converse is also true: if a function defined on the whole extended complex plane has no natural boundary, then it is a rational function (this includes the case of a function defined by a polynomial).

What about functions with natural boundaries? Now it gets much harder, and we shall not pursue the story very far. Mathematicians, such as Weierstrass and Hermite, quickly discovered that the unit circle can be a natural boundary, and when it is this can be because every point on the unit circle is a pole for the function, or because a dense set of points are, and an industry developed investigating what can happen. Then you can start to ask about what other point sets can be natural boundaries, and which cannot. For example, a single point can be a natural boundary, as is the case for the point at infinity and the exponential function, or as the origin is for the function $e^{1/z}$ if you prefer.

There is one topic, however, in this domain that is worth pursuing just a little further here, and that is the connection with real functions.

20.4 Disjoint Domains and Non-differentiable Functions

In 1880 Weierstrass published a paper in which he showed that a single analytic expression could define an analytic function on two or more disjoint domains. This, he was pleased to note, refuted the opinion of Riemann's that every complex function could be obtained by arithmetic operations on quantities.

Weierstrass gave this example of a uniformly convergent series of rational functions to illustrate his new result, which he said he had found and presented in his lectures for years (and indeed had discussed with Casorati and Kronecker, as we saw in (19.4) above). It was given by the series

$$F(x) = \sum_{\nu=0}^{\infty} \frac{1}{x^{\nu} + x^{-\nu}}$$

which is convergent for $|x| < 1$ and $|x| > 1$. He proved (using results Jacobi had established for the theta function—see Eq. (19.1)) that in each domain this series represents an analytic function that cannot be analytically continued across their common boundary and into the other domain (Weierstrass 1880, 211). The series therefore represents two functions, not two branches of one function. Therefore the method of series can define functions that Riemann's methods cannot.

This gave Weierstrass the opportunity to clarify an essential point connecting real and complex analysis. He had always insisted in his lectures on the importance of two theorems that "did not coincide with the standard view": the continuity of a real function does not imply its differentiability; and a complex function defined in a bounded domain cannot always be continued outside it. As he put it, the points where the function cannot be defined may be "not simply be isolated points, but may also form curves and surfaces" (Weierstrass 1880, 221).

To demonstrate this point, Weierstrass introduced a series $\Sigma A_\nu x^\nu$ that is assumed to be absolutely and uniformly convergent for $|x| \leq 1$. By restriction to $|x| = 1$ one obtains the series $\Sigma A_\nu e^{\nu t i}$, which represents a continuous function $g(t)$ of a real variable t. If the power series can be analytically continued outside the circle $|x| = 1$ then there is a power series with its centre in the disc and which converges on a disc D_1 that includes part of the unit circle (where the new disc sticks out). The new power series is absolutely and uniformly convergent inside D_1, and in particular on all points of the unit circle inside D_1, and the restriction $g(t)$ is therefore infinitely differentiable as a function of t on that part of the unit circle. This means that parts of the unit circle where the restricted function $g(t)$ is not differentiable must form part of the natural boundary of the function. So if the restricted function is never differentiable on the unit circle, the unit circle is a natural boundary for the original power series.

Bottazzini has suggested that Weierstrass's power series approach to complex analysis inevitably led Weierstrass to investigate the natural boundary of a complex function and so to the study of non-differentiable real functions on the boundary.[2] This duly led him to the discovery of continuous nowhere differentiable functions via an astute use of uniform convergence. And indeed, when referring to the French translation of his 1880 paper, Weierstrass observed to Schwarz on March 6, 1881 with a fine sense of understatement:

> My latest paper created more of a sensation among the French than it really deserves; people seem finally to realize the significance of the concept of uniform convergence.

Weierstrass's Continuous Nowhere Differentiable Function

One of the highlights of Weierstrass's work on this topic was his announcement (Weierstrass 1872) of a continuous nowhere differentiable function. However, he published neither the explicit function nor, accordingly, a proof of the correctness of his claim, and it was only revealed when Paul du Bois-Reymond got permission from Weierstrass to publish it in Du Bois-Reymond (1875, 29–31).

---

[2]Riemann too, as Dedekind reported when editing Riemann's papers, found his examples of pathological functions looking at limit case of θ-functions. But, as Klein wrote in his history of mathematics in the 19th Century (1926, I, 286), Riemann's attitude was quite different from Weierstrass's: "Riemann excluded natural boundaries from his considerations. Weierstrass, on the contrary, was led precisely by his systematic manner of thought to look closer at the behaviour of an analytic function in the neighbourhood its natural boundaries".

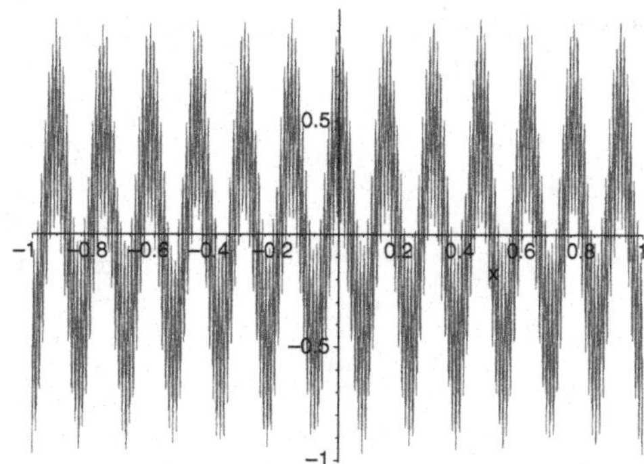

Fig. 20.1 The first few terms of a Weierstrass function

This is the function

$$f(x) = \Sigma_n b^n \cos(a^n \pi x), \quad 0 < b < 1\ ,\ ab > 1 + 3\pi/2\ ,\ a \text{ an odd integer.}$$

Continuity follows from the fact that, by Weierstrass's M-test with $M_n = b^n$, the series is uniformly convergent for all x. The large values of a^n that enter the sum very quickly make it clear that the function will be likely to oscillate a lot, which hints at non-differentiability, but the oscillations could be damped out by the shrinking powers of b. In fact, establishing non-differentiability everywhere is hard, and is nicely discussed in Bressoud (1994, 260–262), which is not that different from Weierstrass's original account.[3] The condition that the integer a be odd makes the fact that $\cos(N\pi) = \pm 1$ easier to handle. The condition that $ab > 1 + 3\pi/2$ is what makes the tail of the series oscillate so much that the function is nowhere differentiable.

We can get an impression of this by looking at graphs of the Weierstrass function for $a = 13, b = 1/2$, where Weierstrass's conditions are obeyed, and for $a = 3, b = 1/2$, where they are not.

Even the first five terms of Fig. 20.1 suggest that differentiability is failing everywhere, whereas Fig. 20.2, which shows the first 15 terms in the fake, suggests that differentiability is only failing at some points (albeit possibly a dense set).

To get an impression of how tricky these convergence arguments were, note that Riemann had claimed in his lectures in 1861 that the function

$$\Sigma_n \frac{\sin(n^2 x)}{n^2}$$

fails to be differentiable for infinitely many values of x, but it was only in his (1916) that Hardy showed that differentiability fails for a dense set of points in any interval, and only in 1970 was it shown that there are infinitely many points at which the

[3] See Du Bois-Reymond (1875) and Weierstrass (1872), the version in Weierstrass's *Mathematische Werke*, which differs from the account given by du Bois-Reymond only in the correction of an simple error.

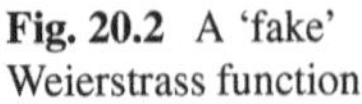

Fig. 20.2 A 'fake' Weierstrass function

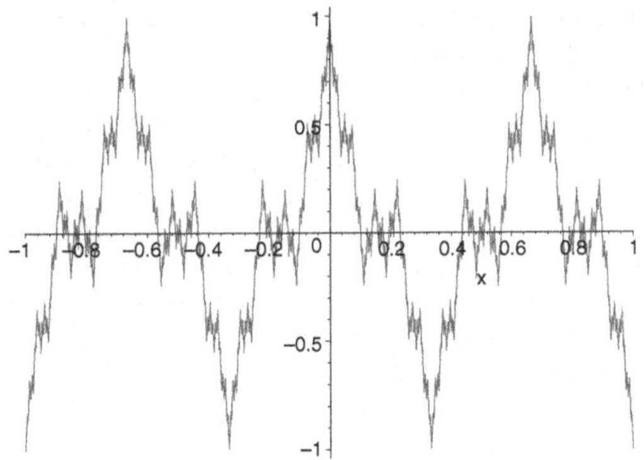

function is differentiable.[4] Weierstrass suggested in his lecture that even Cauchy, Gauss, and Dirichlet, rigorous critics as they were, had never doubted that a continuous function was other than differentiable except at isolated points where the derivative was either not defined or infinite.

In his account of Weierstrass's function, du Bois-Reymond explained carefully that a function could be continuous at a point without being continuous at any nearby point (he gave an example too complicated to be reproduced here). It was, he went on, one of the deepest recent results in mathematics that a function could be continuous at every point of an interval and differentiable nowhere, and this was Weierstrass's discovery. Mathematicians in German circles had been concerned with this issue for some years, led by those in Riemann's school, because Riemann had suggested that the function $\Sigma_n \frac{\sin(n^2 x)}{n^2}$ had this property. It was certain that the function failed to be differentiable at infinitely many points in any interval, but it seemed that no-one in that circle had put a proof down on paper. It was clear that Hankel's method of condensation of singularities allowed one to construct continuous functions that failed to be differentiable at a dense set of points, but Weierstrass's example deserved to be widely known because of its great simplicity and naturalness.

Weierstrass also asked du Bois-Reymond to send a copy of the memoir to Gaston Darboux in France, because it was apparent that he was interested in the topic but unaware of what had been done.[5]

[4] Weierstrass is our source for this information about Riemann, and interestingly he was of the opinion that differentiability fails everywhere for Riemann's function but is "somewhat difficult to demonstrate"; see Weierstrass (1872, 72).

[5] For more detail, see Schubring (2012, 570–575) and Gispert (1983, 49).

Gaston Darboux (1842–1917)

Indeed, Darboux, in his memoir (1875) also gave an example of a function that is continuous everywhere and differentiable nowhere. This is the function

$$f(x) = \sum_{n=1}^{\infty} \frac{\sin(n+1)!x}{n!}$$

It is worth pausing to note that every term in the summation is analytic and therefore infinitely differentiable, which tends to emphasise the fact that convergence and differentiability do not go easily together. Note also that Darboux's function is a trigonometric series—can it be a Fourier series?

The next two graphs show the sum of the first 4 and the first 6 terms of the series, and give a good impression of how crinkly the function is getting (Figs. 20.3 and 20.4).

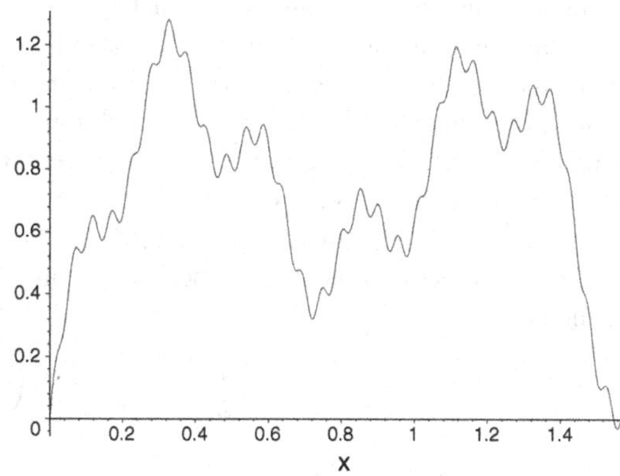

Fig. 20.3 The first 4 terms of Darboux's series

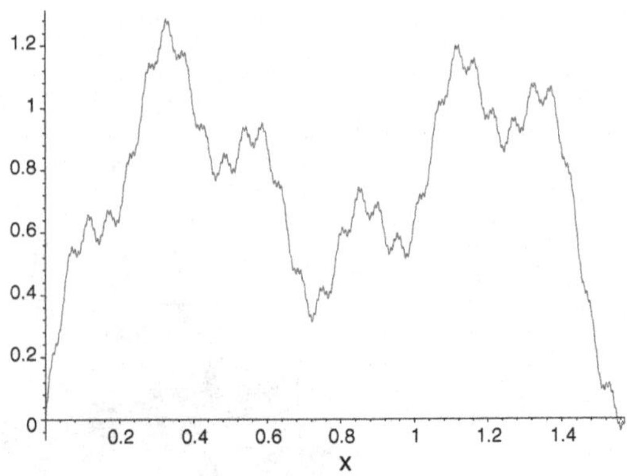

Fig. 20.4 The first 6 terms of Darboux's series

It would seem that Darboux had heard of Weierstrass's announcement of 1872, because he mentioned it in a letter to his friend Houël that year, but had come to his own example of an everywhere continuous, nowhere differentiable function.

20.5 Advanced Results

It would take us too far afield to follow Weierstrass's treatment of many topics here, but it is important to give an impression of what he accomplished that made so many mathematicians of his time regard him as the leading complex analyst, and one of the people most responsible for making complex analysis one of the most important branches of mathematics. A number of the more vital achievements will be listed.

Elliptic Functions

Weierstrass became the author of the definitive account for its day, although it is no longer regarded as the fundamental approach. His approach was to characterise the elliptic functions among the functions satisfying an algebraic addition theorem (a formula connecting the values of a function at u, v and $(u+v)$). We can notice how algebraic, and ungeometrical this is, and also that the book that eventually appeared on it was a large and successful of formulae and results needed by anyone working with elliptic functions.[6] It had no historical remarks, and no technical motivations—in this it resembled the lecture courses that were the only way Weierstrass liked for the dissemination of his ideas.

That said, in his lectures in 1874 rewrote the theory of elliptic functions in a lasting way. He denoted the fundamental periods 2ω and $2\omega'$ and introduced the function σ defined by

$$\sigma(u) = u \prod_{w \neq 0} \left(1 - \frac{u}{w}\right) \exp\left(\frac{u}{w} + \frac{1}{2}\frac{u^2}{w^2}\right), \tag{20.1}$$

[6]The *Formeln und Lehrsätze* (or, *Formulae and Results*) of 1885.

where $w = 2\mu\omega + 2\nu\omega'$, and μ, ν integers. He then defined the function $\wp(u)$, which "is one of the most important [functions] of the whole analysis", as

$$\wp(u) = -\frac{d^2}{du^2}\log\sigma(u) = \frac{1}{u^2} + \sum_{w \neq 0}\left(\frac{1}{(u-w)^2} - \frac{1}{w^2}\right), \tag{20.2}$$

where $w = 2\mu\omega + 2\nu\omega'$, as above.[7] $\wp(u)$ is doubly periodic function with exactly one infinite point of order two in the period parallelogram.

Weierstrass then deduced the crucial properties of his $\wp$ function and of elliptic functions in general. For example, "Every doubly periodic function can be represented in the form

$$\varphi(\text{u}) = \text{C}\frac{\sigma(u-u_1)\ldots\sigma(u-u_\nu)}{\sigma(u-v_1)\ldots\sigma(u-v_\nu)} \tag{20.3}$$

where the u's and v's can be chosen in such a way to build two complete systems of non-congruent magnitudes (i.e. the difference of any two of them is not a (multiple of a) period) that satisfy the condition $\sum u_v = \sum v_\nu$" (Theorem 13). Such a function is doubly periodic, and it is called an elliptic function of degree ν, and "there are only elliptic functions of degree 2 and above".

Moreover, he showed that $\wp(u+v)$ and $\wp(u-v)$ are algebraically expressible in terms of $\wp(u)$, $\wp(v)$ and their derivatives, and that $\wp(u) = s$ is a solution of the differential equation

$$\left(\frac{ds}{du}\right)^2 = 4s^3 - g_2 s - g_3, \tag{20.4}$$

where $g_2 = 60\sum(\frac{1}{2w})^4$ and $g_3 = 140\sum(\frac{1}{2w})^6$.

He then proved that if g_2 and g_3 are chosen arbitrarily a suitable pair of periods ω and ω' be found the corresponding $\wp(u)$ function satisfies equation (20.4). To do so he introduced the quantities e_1, e_2, e_3 such that

$$\wp'(u)^2 = 4(\wp(u) - e_1)(\wp(u) - e_2)(\wp(u) - e_3),$$

and established the necessary properties. This is the definitive answer that links elliptic functions and elliptic integrals and shows that there is an elliptic function for every possible set of periods. He concluded his lectures by discussing the cases $g_2^3 - 27g_3^2 \neq 0$, and $g_2^3 - 27g_3^2 = 0$.

Weierstrass's Representation Theorems

The famous representation theorem, announced in letters at the end of 1874, showed that an arbitrary point set in the plane of complex numbers can be the zero set of an analytic function if and only if the set has no accumulation point. Weierstrass's proof explicitly exhibited the required function, and the form of the function allowed

[7] The order is reversed in the *Formeln und Lehrsätze*.

mathematicians to raise, and answer other questions of this type. For example, his student Mittag–Leffler showed in 1867 that one can prescribe the set of poles of a complex function in the same way.

Abelian Functions

All of Weierstrass's work may be said to circle round the topic of abelian functions. This accounts for his algebraic preferences and for his interest in developing a theory of complex function in several variables. Despite the early successes which made his name and got him to Berlin, it is fair to say that he never got more than glimpses of how to proceed, and indeed made one or two significant mistakes in his attempts to go from one to many variables. We can only note here that this difficult topic was the promised land, not to be entered until after Weierstrass's death.

Chapter 21
Revision—and Assessment

At this point in the course we have seen a lot of mathematics emerge: the existence of harmonic functions satisfying certain boundary conditions, trigonometric functions, and complex analytic functions (which, in Riemann's hands but not those of Weierstrass, have intimate links to harmonic functions). But it is also true that very little of this was solidly established at the time: Dirichlet's principle was beginning to look too insubstantial to solve Dirichlet's problem, and with it could fall much of Riemann's complex analysis. The study of trigonometric series might open the way to new results about real functions, but intuition seemed to offer little guidance.

One way to take stock is to grapple more closely with the sources, and five seem appropriate:

1. Green on potential theory (his 'An Essay on the application of mathematical analysis to the theories of electricity and magnetism')
2. Dirichlet on the theory of Fourier series (his 'Sur la convergence des séries trigonométriques')
3. Schwarz on conformal mapping (his 'Ueber einige Abbildungsaufgaben', 1869)
4. Riemann on the Cauchy–Riemann equations (his (1851, Sects. 1 to 4)) and Maxwell on conjugate functions (his 'Theory of conjugate functions in two dimensions')
5. Stokes on uniform convergence (his 'On The Critical Values Of The Sums Of Periodic Series', 1847).

Green's essay on potential theory is reprinted in his *Mathematical Papers* (pp. 356–374), which is available in several places on the web (consult the Digital Mathematics Library and the Internet Archive). A translation of Dirichlet's paper on the theory of Fourier series is given in Appendix A.2 at the end of this book, as is this particular paper by Schwarz on conformal mapping in Appendix A.5, and Riemann's approach to the Cauchy–Riemann equations in Appendix A.3. Maxwell's *Treatise on Electricity and Magnetism* 3rd ed. Vol. I, Chap. 12, §§ 182–191, pp. 284–294, and Stokes on uniform convergence—a paper reprinted in his *Mathematical and*

J. Gray, *The Real and the Complex: A History of Analysis in the 19th Century*,
Springer Undergraduate Mathematics Series, DOI 10.1007/978-3-319-23715-2_21

physical papers, Vol. 1, 1880, 236–313 are also available in the web in the Internet Archive. Stokes's paper is also discussed below, in § 22.5.

When I taught the course, I advised the students to read all the extracts quickly over once and find (at least) one they wanted to proceed with. None of them are altogether easy to read, but for a first approach it is possible to let the obscure bits wash over you and wait for something more comprehensible to turn up. Sometimes a difficult paragraph has an easier second part, or is followed by an easier one. The aim is to form some sense of what the extract is about.

Terminology can be a problem. This is not simply a matter of old words for shared ideas, but words for ideas that may be muddled or obscure and in need of refinement. A dialogue with the text and your own understanding of the material is required. But it is perhaps more important to know what, in a short essay, to include and what to leave out. The aim is for you to walk away with a good understanding of what the authors did, so that you can state what were the key stages in their arguments, what were routine, what was difficult, what was done well and what, if anything, was left unclear. The papers by Dirichlet and Schwarz are the most rigorous, and repay careful study. At the other extreme, Green's essay is the most impressionistic, and the challenge is to see precisely how he conveyed insight when he could not provide rigour. See if you can say precisely where, and in what ways, he appealed to intuition to bolster his arguments.

Chapter 22
Uniform Convergence

22.1 Introduction

Dirichlet's papers proved rigorously and beyond dispute what Abel had already shown by an example: Fourier series can represent discontinuous functions. This established that there is an entire class of convergent series that are composed of continuous functions, but for which the sum was not continuous and that consequently contradicted Cauchy's theorem on the continuity of the sum of a series of continuous functions. But the fact that a statement has been refuted does not mean that it will be clear where the incriminating point lies. It is possible that an insight sheds some light, certainly, but leaves other parts of the problem shrouded in obscurity. It is even possible that an insight can make everything more confusing. After all, Cauchy's false 'theorem' made a clear simple statement.

In fact, it is not clear what the implications for Dirichlet's work were taken to be at the time, and Dirichlet did not comment on the contradiction that existed between the proper demonstration of the convergence of Fourier series and Cauchy's theorem. He seems to have thought that refuting Cauchy's false proof of the convergence of Fourier series was enough, and no one in the 1830s seems to have had any better idea of what is involved in establishing the continuity of a convergent series of continuous functions than Cauchy's in the *Cours d'analyse*. Not until in the 1840s did there begin to emerge, with difficulty and in differing contexts, other ways of considering the continuity and convergence of a series of functions.[1]

---

[1]Chapter closely follows *Bottazzini* § 5.4. The Uniform Convergence of Series, but draws on later historical literature, notably Bråting (2007) and Viertel (2014). A brief summary of the theorems relating to the properties of convergent sequences of integrable, continuous, or differentiable functions is provided in Appendix B.

J. Gray, *The Real and the Complex: A History of Analysis in the 19th Century*,
Springer Undergraduate Mathematics Series, DOI 10.1007/978-3-319-23715-2_22

22.2 Seidel

The first person to explicitly denounce the patent contradiction in which the theory of series found itself was Philipp Ludwig von Seidel, a former student of Dirichlet at Berlin who later studied under Jacobi and Franz Neumann at Königsberg. In 1847 he published an article devoted to series that represent discontinuous functions, and there he wrote:

> One finds in Cauchy's *Cours d'analyse algébrique* ... a theorem which states that the sum of a convergent series whose individual members are functions of a quantity x and continuous in the vicinity of a particular value of x, is likewise always a continuous function of the same quantity in this neighbourhood. It follows from this that series of this kind are not adapted to represent discontinuous functions in the vicinity of points where their values jump.[2]

After briefly outlining the path of Cauchy's demonstration, Seidel continued:

> Nevertheless, the theorem stands in contradiction to what Dirichlet has shown, that, for example, Fourier series also always converge if one forces them to represent discontinuous functions – in fact, the discontinuity will frequently be found in the form of those series whose individual members are still continuous functions. [...] When one begins from the certainty thus obtained that the proposition cannot be generally valid, then its proof must basically lie in some still hidden supposition. When this is subjected to a precise analysis, then it is not difficult to discover the hidden hypothesis. One can then reason backwards that this cannot occur with series that represent discontinuous functions.[3]

The theorem that Seidel found to characterize these series was the following[4]:

If one has a convergent series which represents a discontinuous function of a quantity x, whose individual members are continuous functions, then one must be able to give values of x in the immediate neighbourhood of the point where the function jumps for which the series converges *arbitrarily slowly*.

An outline of Seidel's demonstration can be sketched as follows. If $s(x)$ is the sum of the series and $s_n(x)$ and $r_n(x)$ are the sums of the first n terms of the series and the remainder respectively, then

$$s(x) = s_n(x) + r_n(x)$$

When an increment of δ is added to x, we have

$$s(x + \delta) - s(x) = [s_n(x + \delta) - s_n(x)] + [r_n(x + \delta) - r_n(x)].$$

This equation will decide the continuity or discontinuity of $s(x)$ in the neighbourhood of x. Since, in order for $s(x)$ to be continuous, $|s_n(x + \delta) - s_n(x)|$ must be $< \tau$, the entire question is reduced to studying the behaviour of $|r_n(x + \delta) - r_n(x)|$. Now, in order for the series to converge, both $r_n(x + \delta)$ and $r_n(x)$ must be smaller than an

[2]See Seidel, (1847, 35), quoted in Bottazzini, *The* Higher Calculus, 202.

[3]See Seidel (1847, 36–37), quoted in Bottazzini, *The* Higher Calculus, 202–203.

[4]See Seidel (1847, 37), quoted in Bottazzini, *The* Higher Calculus, 203.

arbitrary small ρ^1 and ρ respectively, when n is suitably large. If the function $s(x)$ is continuous, then for positive ε smaller than ρ^1 and ρ, one has $|r_n(x+\delta) - r_n(x)| < \varepsilon$ for $n > n_0(\varepsilon)$. And if, as δ tends to zero, there exists an integer N that is the largest of the successively determined n_0 such that, for $n > N$

$$|s(x+\delta) - s(x)| < \tau + \rho + \rho^1 \tag{22.1}$$

then this inequality says precisely that the limit function $s(x)$ is continuous.

If, however, the number n_0 increases beyond every finite value when δ, beginning from an initial value η, tends to zero, then Eq. (22.1) ceases to be true and the convergence of the series in the neighbourhood of x becomes arbitrarily slow. This led Seidel to ask whether the continuity of the sum of the series by itself implies uniform convergence. It was thirty years before the question was answered in the negative, in the form of suitable counterexamples constructed by Darboux (1875), Du Bois-Reymond (1876), and Cantor (1880). These will be discussed in the next chapter.

Viertel (2012, 110) has made the interesting observation that Seidel mentions neither Abel's nor Dirksen's doubts about Cauchy's proof; it seems that it may well have been a question that Dirichlet thought it worthwhile for his former student to attack.

22.3 Björling

The Swedish mathematician Emanuel Gabriel Björling published a paper in Latin in 1846 on the convergence of series that he then reworked in 1853 in Swedish, when he took the opportunity to claim that his result was equivalent to Cauchy's in his (1853) and was therefore correct. It was argued in Grattan-Guinness (1986) that Cauchy must have read the Latin version by 1852 at the latest, which gives Björling's paper an extra level of interest, and so we briefly consider it here, following (Bråting 2007), which should be consulted for the details, including an English translation of Björling's Swedish article.[5]

Björling's theorem says[6]

> If a series of real-valued terms $f_1(x)$, $f_2(x)$, $f_3(x)$, ... is convergent for every value of x from x_0 up to X, and in addition its particular terms are continuous functions of x between the given limits; then the sum $f_1(x) + f_2(x) + f_3(x) + \cdots$ necessarily has to be a continuous function of x between the given limits.

---

[5]Grattan-Guinness argued on the basis of similarities between the two papers including the citations and the fact that Björling had also published some results on convergence teats in Liouville's *Journal* that Cauchy would surely have seen. It is no argument against this view that Cauchy did not mention Björling by name in his (1853).

[6]See Björling (1846, 21) quoted in Bråting (2007, 520).

The obvious problem with this is that Abel's exception, the Fourier series $\sum_k (-1)^{k+1} \frac{1}{k} \sin(kx)$ would still seem to be counter-example, so everything hangs on how Björling interpreted the statement when he came to prove it. If we fix a value of x and ask that the limit function be continuous at x the statement is false. If we ask that the series converge for all x then the corresponding statement in epsilons and deltas can be made to look like a claim of uniform convergence. Bråting's view is that Björling was alert to the difficulties but insufficiently precise. She argues that Björling intended that the difference between a condition holding "for every value of x" and "for every given value of x" is that the former expression holds for x-values indefinitely close to a particular point and that the latter expression does not. In other words, a condition that holds for all x means that it holds on some interval around any given x.

There has been an extensive discussion of whether Björling possessed the concept of uniform convergence, which can be traced through Bråting's paper. Her view is that this imposes modern conceptions on his work and that an ultimately unclear sense of what it is for a function to be continuous might have been at play that was more typical of the period, and we can note that this contributes to the argument that the $\varepsilon - \delta - N$ formalism was needed in delicate matters like this one.

22.4 Gudermann and Weierstrass

By the time Seidel's paper appeared, other mathematicians were already picking out the key issue of what is today called the uniform convergence of a series of functions. Among them was Weierstrass, who was probably inspired by the article Gudermann published in the 1838 issue of Crelle's *Journal*. In it, Gudermann hinted the uniform convergence of certain infinite series that give the expansion of elliptic functions for the first time. As he put it, "It is a fact worth noting that …the series just found have all the same convergence rate".[7] Weierstrass was more precise[8]:

> Because the power series under consideration …converges uniformly, given an arbitrary positive quantity δ, a finite number of terms of the series can be discarded so that the sum of all the remaining terms is, for every value in the specified domain …in absolute value $< \delta$'.

In his lectures Weierstrass gave the now-classic definition of uniform convergence in an interval:

Definition 1 (U_1) The series $\Sigma u_n(x)$ is said to be uniformly convergent in an interval $[a, b]$ when, for every arbitrarily small positive ε, there exists an $n_0(\varepsilon)$ such that $|r_n(x)| < \varepsilon$ for $n > n_0$ and for every $x, a \leq x \leq b$.

---

[7] See Gudermann (1838, 251–252). However, as Viertel has noted (2012, 142–144), Gudermann seems to have used this and other terms about the nature of convergence impressionistically and without defining them; other terms refer to the rate of convergence, which may well have meant the number of terms n that must be considered before a sequence is within a given ε of its limit.

[8] See Weierstrass (1841b, 68–69).

As Bottazzini has remarked, "this is not the sense in which Seidel had introduced his notion. Seidel had not been interested in the 'global' properties characterizing the convergence of a series in an entire interval, but rather in the small but predetermined neighbourhood of a point." That concept is captured by this definition:

Definition 2 (U_2) The series $\Sigma u_n(x)$ is uniformly convergent in the neighbourhood of a point ζ of the interval $[a, b]$ if there exists a $\delta(\zeta)$ such that $|r_n(x)| < \varepsilon$ for every arbitrarily small positive ε, for $n > n_0(\zeta, \varepsilon)$ and for $|x - \zeta| < \delta$.

To quote Bottazzini again: "The uniform convergence of a series in an interval as defined by (U_1) naturally implies uniform convergence in the neighbourhood of every point in the interval. The converse is also true, but the demonstration is not easy and was given by Weierstrass for the first time in 1880.[9]

22.5 Stokes

A contemporary of Seidel, but one as ignorant of his work as he was of Weierstrass's, was a young mathematical physicist who was investigating the question of the mode of convergence of series, the Irishman George Stokes. Stokes was one of the first to establish Cambridge as a leading centre in the world for applied mathematics, a topic the utilitarian Victorians were always much more willing to support than pure research.[10] He was familiar with the work of French mathematicians such as Cauchy and Poisson, but was almost completely ignorant of what Dirichlet had done, so his account of the convergence of Fourier series followed Poisson's account instead. He therefore approached the study of series using Cauchy's false 'theorem' on the continuity of the sum of a series of continuous functions, which he took not from the *Cours d'analyse* but rather from Moigno's *Leçons de calcul différentiel et de calcul intégral* of 1840–1844, which in turn was based on Cauchy's lectures. However, Stokes was cautious where Cauchy had been rash. He used Cauchy's definition of the continuity of a function and adopted his distinction between 'convergent' and 'divergent' series, but introduced an additional distinction between 'essentially' and 'accidentally' convergent series. This corresponds to the modern distinction between absolute and conditional convergence, also introduced at the same time by Dirichlet.

Stokes set out to prove the convergence of the Fourier sine series of a class of functions, although what class it could be exactly Stokes never made clear. He first assumed that the Fourier series of a function f converges absolutely (what he calls essential convergence). He then used the method of Abel summation to show that the series converges to the function it is supposed to represent. This involved him in finding a second series that depends on a parameter g and converges for $0 \leq g < 1$,

[9]See Weierstrass, 'Zur Functionenlehre', (1880). Today the result is obtained by applying the Heine–Borel theorem.

[10]He was also a descendant of the tradition begun by Peacock and Babbage, as can be seen by his use of the first difference operator, Δ.

with the further property that if this second series converges as g tends to 1 from below then the first series will converge correctly.

His argument is delicate and worth describing. To get the sum of the Fourier series, he used a standard trigonometric identity, $2\sin A \sin B = \cos(A-B) - \cos(A+B)$. To show that the summed g-series agrees in the limit as g tends to 1 with the function where the function is continuous, and takes the appropriate average value at the points where the function has a finite jump discontinuity, he argued as follows. Fix attention on a point x in the interval of integration $[0, a]$. In any range of the dummy variable x' not including x, the limiting value of the integral is 0. When the range of integration, however small, includes x the limit as g tends to 1 of the integral is improper and a subtle analysis yields the conclusion. Here Stokes, without saying so, made further assumptions about f that shows he knew what continuity at a point means (his verbal definition of continuity earlier is not clear)—or, if you are less generous, that he did not know what a merely continuous function could look like, and supposed it to be locally monotone.

The next paragraph shows that the original Fourier series does in fact converge, so the Abel limit as g tends to 1 is equal to this limit, and therefore the original series converges as the g-series does: pointwise to the function except at jump discontinuities.

The rest of the paper is less exciting, but it does at least show that Stokes knew that term by term differentiation of a convergent Fourier series is not generally valid.

The most original point in Stokes's work lies in his study of Fourier series in the neighbourhood of a point of discontinuity of the function. It is the same problem as Seidel's but Stokes followed a completely different approach. In his words, he dealt with the following question:

> Let
> $$u_1 + u_2 + \cdots + u_n + \cdots \tag{22.2}$$
> be a convergent infinite series having U for its sum. Let
> $$v_1 + v_2 + \cdots + v_n + \cdots \tag{22.3}$$
> be another infinite series of which the general term v_n is a function of the positive variable h, and becomes equal to u_n when h vanishes. Suppose that for a sufficiently small value of h and all inferior values the series (22.3) is convergent, and has V for its sum. It might at first sight be supposed that the limit of V for $h = 0$ was necessarily equal to U. This however is not true.[11]

This is the problem that Abel had treated for the special case of power series—a double limiting process. As Stokes went on to show, the question becomes that of knowing when we can exchange the order of two limits, and indeed we cannot always do this.

After a general discussion of the issue, Stokes stated and proved the following theorem:

[11] See Stokes (1849, 279), quoted in Bottazzini, *The* Higher Calculus, 205.

> The limit of V can never differ from U unless the convergency of the series (22.3) becomes infinitely slow when h vanishes. The convergency of the series is here said to become infinitely slow when, if n be the number of terms which must be taken in order to render the sum of the neglected terms numerically less than a given quantity e which may be as small as we please, n increases beyond all limit as h decreases beyond all limit.[12]

The English mathematician who successfully brought modern analysis to the country (as late as the early 20th Century!) was G.H. Hardy and he shrewdly observed that there is a clear distinction between what Stokes meant by an "infinitely slow" convergence and Seidel's "arbitrarily slow" convergence.

> Stokes is considering an inequality satisfied for a special value of n, or at most an infinite sequence of values of n, and *not* necessarily for all values of n from a certain point onwards.[13]

In modern terminology, Stokes introduced quasi-uniform convergence that is defined as follows: A series $\Sigma u_n(x)$ is quasi-uniformly convergent in the neighbourhood of a point ζ in an interval $[a, b]$ if for every arbitrarily small positive ε, there exists a $\delta(\zeta) > 0$ such that $|r_n(x)| < \delta$ for every N and n_0.

In relation to Stokes research, it is also important to introduce the definition of quasi-uniform convergence at a point $x = \zeta$. This is written like the definition in the neighbourhood of a point, with the essential difference that ε depends on the preceding choice of ζ, ε, and N. In his (1878, 107–108) the Italian mathematician Ulisse Dini demonstrated that a necessary and sufficient condition for $s(x)$ to be continuous at $x = \zeta$, is that the series be quasi-uniformly convergent at the point ζ.

22.6 Summary

It is interesting that it took more than 20 years for Cauchy's false theorem in his *Cours d'analyse* to begin to be clarified. Abel's counterexample did not stir a debate—it did not even stir Cauchy—and Dirichlet was silent on the subject. Only in the 1840s did four mathematicians, all unaware of each other's work and acting with different motives and objectives, produce new arguments, and they were all different. Björling's position remains ambiguous. Weierstrass introduced the idea of uniform convergence in an interval; Seidel that of uniform convergence in the neighbourhood of a point; Stokes that of quasi-uniform convergence in the neighbourhood of a point. Common to all four is the desire to surmount the idea of simple convergence as defined by Cauchy in order to ensure that the (new style) convergence of a series of continuous functions guarantees that the limit is itself continuous.

As we noted before, Cauchy returned to his own theorem only in a note in the *Comptes rendus* for 1853, when he observed simply that it is verifiable for a power series, but that it required restrictions "for other series", and gave the example that Abel had earlier given. But Cauchy mentioned neither the Norwegian mathematician

---

[12]See Stokes (1849, 281), quoted in Bottazzini, *The* Higher Calculus, 206.

[13]See Hardy (1918, 155), quoted in Bottazzini, *The* Higher Calculus, 206.

nor the more recent work of Seidel and Stokes. Instead, as we saw in § 4.3, he tried to play down the problem (and with it, the size of his mistake), saying (Cauchy 1853, 31–32) "it is easy to see how one can modify the statement of the theorem so that it will no longer have any exception. This is what I am going to explain in a few words". His new theorem (Cauchy 1853, 33) said, correctly:

> Theorem (of uniform convergence). If the different terms of the series $\Sigma u_n(x)$ are functions of a real variable x, continuous with respect to this variable within the given limits; and if, in addition, the sum $u_n + u_{n+1} + \cdots + u_{n'}$ always becomes infinitely small for infinitely large values of the whole numbers n and $n' > n$, then the series will be convergent and the sum of the series will be, within the given limits, a continuous function of the variable x.

The language of infinites and infinitesimals that Cauchy used here seemed ever more inadequate to treat the sophisticated and complex questions then being posed by analysis. Furthermore, when Cauchy wrote this note in 1853, France was even less capable of preserving its leadership in research mathematics, as Lamé pointed out that very year in an alarming report to the Académie des sciences. He exhorted mathematicians to do 'pure' research, lamenting that the almost exclusive interest in applied questions impeded the development of mathematics. He thereby denounced the limitations inherent in the polytechnic tradition.

The problems posed by the study of nature, such as those Fourier had faced, now reappeared everywhere in the most delicate questions of 'pure' analysis and necessarily led to the elaboration of techniques of inquiry considerably more refined than those that had served French mathematicians at the beginning of the century. Infinitesimals were to disappear from mathematical practice in the face of Weierstrass's $\varepsilon - \delta$ notation, and even the analysis of functions of complex variables, which Cauchy had initiated, was to be reformulated on the basis of the works of Riemann and the lectures of Weierstrass.

Chapter 23
Integration and Trigonometric Series

23.1 Introduction

Riemann's previously unpublished Habilitation essay on the integration of trigonometric series was published for the first time in 1867, in an issue of the *Göttingen Nachrichten* that came out shortly after his death and carried several of his papers that he had left in almost a fit state to print. By then his reputation as a remarkable, but difficult, mathematician had spread, and what he wrote was read. This chapter considers some of the responses to his ideas about real functions and looks at the changes wrought in the relationship between continuity and differentiability, and between the implications of convergence and uniform convergence. This will lead us to look at the world of functions defined by series that do not converge uniformly, and were often considered in the late 19th century to be the most general kind of function, which is why they were often called 'assumptionless functions'.

23.2 Heine, Cantor, and Nowhere Dense Sets

Among the first to respond was Eduard Heine. His paper (1870) on trigonometric series was animated by the following question: Are the coefficients a_n and b_n in a Fourier series unique if the series is not uniformly convergent, and so term by term integration is illegitimate? Heine noted that Dirichlet had proved that the Fourier series of a piecewise monotonic continuous function converges uniformly. The convergence is necessarily non-uniform at jump discontinuities where $f(x+) \neq f(x-)$ (a series of continuous functions that converges uniformly converges to a continuous function). Heine now observed that the Fourier series converges uniformly in general, as he called it, meaning that it converges uniformly on any subinterval omitting a specific finite set of points.

J. Gray, *The Real and the Complex: A History of Analysis in the 19th Century*,
Springer Undergraduate Mathematics Series, DOI 10.1007/978-3-319-23715-2_23

Inspired by Georg Cantor, a colleague of his at the University of Halle, Heine claimed that the Fourier series representation of a function is unique if the convergence of the Fourier series is uniform in general. To prove this, he argued as follows. If the Fourier coefficients are not unique, consider two Fourier series representing the same function. The difference of these two series represents the zero function, so the theorem will be proved if it can be shown that all the coefficients of the Fourier series formed by taking this difference are 0. In this, he succeeded; that is, Heine established that if a function is represented by a Fourier series that is uniform in general with respect to a finite set then the Fourier series is unique, and its coefficients are determined by the function.

Cantor (see Cantor 1872) then developed an approach that dropped the uniform in general condition, and aimed at showing uniqueness for any set of discontinuities. First Cantor showed that given a trigonometric series $\sum_{j=0}^{\infty} a_j \sin jx + b_j \cos jx$ that is convergent on an interval, the coefficients a_j and b_j tends to zero as $j \to \infty$. This enabled him to show that if a function defined on an interval is represented by two trigonometric series, $\sum_{j=0}^{\infty} a_j \sin jx + b_j \cos jx$ and $\sum_{j=0}^{\infty} a'_j \sin jx + b'_j \cos jx$, that converge and agree everywhere on the interval, then the coefficients of the two trigonometric series are the same: $a_j = a'_j$ and $b_j = b'_j$. Put another way, the only everywhere convergent trigonometric series that represents the zero function on an interval is the one with all coefficients zero.

Cantor then investigated what could be said if a function was represented by two trigonometric series, $\sum a_j \sin jx + b_j \cos jx$ and $\sum a'_j \sin jx + b'_j \cos jx$, that converge and agree everywhere on an interval I except at a set P where either convergence fails for at least one of them or the two trigonometric series take different values. He began with sets P that have only finitely many limit points, or, calling the set of limit points of a set P the derived set of P and denoting it P', sets whose derived set is finite. He called these sets of type 1. Sets of higher type he defined inductively. A set of type 2 is one whose derived set is of type 1. If P is a set, then P' is its derived set, and the derived set of P' is denoted $P^{(2)}$. So P is of type 2 if $P^{(2)}$ is finite. A set of nth type is one for which $P^{(n)}$ is finite. Interestingly enough, Cantor had almost no examples of the sets he was talking about, and noted only that the set $1/n : n = 1, 2, \ldots$ is of type 1, and that the set of all the rational numbers in $[0, 1]$ is of no type.

Cantor showed that the representation is unique if convergence holds except on a set of type n. His proof was a modification of Heine's, extended by induction to any integer n.

Now that we know why these sets matter, it will be convenient to notice some things about them. A set of type 1 has a finite derived set, that is, a finite set of points p' each with the property that every neighbourhood of p' contains infinitely many points of P. Note that the points p' need not belong to P. However, the points of P somehow bunch up or cluster around the points in P'. Let me introduce a term that Cantor was only going to introduce later: a set will be said to be nowhere dense if its closure has an empty interior. It follows at once that a set of type 1 is nowhere dense, and indeed that a set of type n is nowhere dense.

23.3 Assumptionless Functions

In the 1870s a number of mostly German mathematicians started to pick up on the idea that one of the things Riemann had done with his paper on trigonometric series was point towards a class of functions with no specific properties at all. They were not, for example, analytic, or differentiable, or even, necessarily, continuous. They were simply functions, in the sense that given a point in their domain (as it might be, the unit interval) they took a value at that point. As matters stood, there was nothing that could be said about them.

That being the case, most mathematicians tried to delineate a class that could be discussed, marginalising, at least for the moment, those that defied analysis. For example, series that were uniformly convergent could be discussed, series that were not lay beyond the pale. Functions that were integrable were admitted, functions that were not were excluded. This, of course, invited discussion of what it is for a function to be integrable, and the feeling of du Bois-Reymond and many other mathematicians was that Riemann's idea was the best possible: if a function was not Riemann-integrable it was not going to be integrable on any other conception of the integral worth having.

Hermann Hankel

In 1870 the mathematician Hermann Hankel published his dissertation at he University of Tübingen. After his early death in 1873 at the age of 34 it was republished in 1882in volume 20 of the *Mathematische Annalen*, because it was recognised as an important contribution to the study of assumptionless functions. In it Hankel presented an important technique for constructing functions that are far from being continuous, an analysis of at least some of the ways such functions can behave, and a theorem about the integrability of some, but not all, of these functions. That his main theorem turned out to be incorrect is further evidence, for us, of just how difficult the fundamental ideas of real analysis were proving to be.

Hankel had been a student of Riemann's in 1860 before going on to Berlin in 1861 to study under Weierstrass and Kronecker. His analysis was motivated by a desire to go, as Riemann had done, beyond the safe class of functions studied by Dirichlet. That meant dispensing with one or both of the requirements that the function by piecewise continuous and piecewise monotonic.

First, discontinuity. If a function f defined on an interval I is not continuous at the point $a \in I$ then it must be the case that for some $\varepsilon > 0$ there is no $\delta > 0$ such that $|x - a| < \delta \Rightarrow |f(x) - f(a)| < \varepsilon$, and that means there is at least one x in any interval containing x_0 with the property that $|f(x) - f(a)| < \varepsilon$. Such an ε is called a jump of the function f at x_0, and the least upper bound, denoted σ of all the jumps of f at x_0 Hankel called the *jump* of f at x_0. I shall write it $J_f(x_0)$. It will also be convenient to define $S_f(\sigma) = \{a \in I : J_f(a) > \sigma\}$. Hankel did not define these symbols, which may indicate how far mathematicians still were from thinking set-theoretically. To get beyond Dirichlet's class of functions, Hankel had to contemplate functions discontinuous at infinitely many points.

Second, functions that are not piecewise monotonic on an interval. These are functions that are not monotonic on a finite number of intervals. Hankel referred to them as functions with infinitely many maxima and minima, which is a reasonable way to think of them, and he gave the examples of the function which is $\sin 1/x$ when $x \neq 0$ and some value, say 0, when $x = 0$, and the function $x \sin 1/x$.

Because Hankel's main theorem is incorrect it will be convenient to follow it slowly and carefully, but it is helpful to know where he was going. His aim was to divide highly discontinuous functions into two classes and show that one class was integrable and the other was not. He used Riemann's approach to the integral, according to which the crucial property of an integrable function is not continuity but the wildness of its oscillation. So his division was into functions that do not oscillate too much, and those that do, and that led him to consider the set of points at which a function oscillated a lot: was this set, so to speak, small (my term) in which case the function would be integrable, or large (my term again) in which case the function would not be integrable. Such, at least, was his hope.

What Hankel meant by small and large sets is captured by his terminology. He said that a set A was *scattered* in an interval I if for all pairs $a_1, a_2 \in A, a_1 < a_2$ there is a subinterval J of I such that $J \subset (a_1, a_2)$ and $J \cap A = \emptyset$. Such sets are small. This corresponds to the modern concept of a nowhere dense set (the interior of the closure of the set is empty). He said that a set A filled an interval I if every subinterval $J \subseteq I$ meets A, so $J \cap A \neq \emptyset$. Such sets are large. This corresponds to the modern concept of a dense set. Hankel seems to have thought that a subset of an interval was either nowhere dense or dense—this is incorrect, but it is evidence of just how hard these concepts were to master when they were being introduced.

Hankel now looked at the class of what he called linear discontinuous functions (linear here means defined on an interval). These are functions defined on an interval I with the property that they are discontinuous at infinitely many points. He divided this class into two subclasses. The first, which he called pointwise discontinuous, consists of the functions f for which the set $S_f(\sigma)$ is scattered in I for all $\sigma > 0$. The second class, which he called the totally discontinuous functions, consists of functions for which there is a $\sigma > 0$ such that $S_f(\sigma)$ fills at least a subinterval of I.

Hankel introduced a further concept, which I shall call the total length of a covering. Let $|I|$ denote the length of an interval so if $I = (a, b)$ then $|I| = b - a$. If a set A is covered by a finite set $\{I_k\}$ of intervals, then the total length of the covering is $L_I = \sum_k |I_k|$. The crucial quantity is the least upper bound of all finite coverings as the number of intervals in the covering increases arbitrarily. Hankel was most interested in the sets for which this quantity is zero; I shall call them *zero coverable*. There modern name is (temporarily) withheld to make the history stand out more clearly.

We can now indicate the thrust of Hankel's argument. He believed that a linear discontinuous function was either pointwise or totally discontinuous and (Riemann) integrable in the first case and not (Riemann) integrable) in the second. This is very tidy, and it is in fact correct that a function totally discontinuous on an interval is not (Riemann) integrable. But the other way round is false: it is not true that a pointwise discontinuous function is (Riemann) integrable. This amounts to saying that a nowhere dense set is zero coverable, and that, as we shall see, is not true. The most likely reason for his mistake is an impoverished idea of what nowhere dense sets can be like, and historians have noted that every nowhere dense set in his paper was like a typical one of his examples of such a set, the set $\{\frac{1}{2^n} : n = 1, 2, \ldots\}$.

Du Bois-Reymond

Du Bois-Reymond was another mathematician who thought that the class of Riemann-integrable functions was the largest subclass of the 'assumptionless' functions worth discussing, and that Riemann's definition of integrable was as broad as possible. He is worth mentioning here for another attractive family of sets he introduced in a paper of 1875. These are the zero sets of $\sin(1/x), \sin(1/\sin(1/x)), \ldots$ are they indeed sets of type $1, 2, \ldots, n, \ldots$. Unfortunately, du Bois-Reymond seems to have believed that these sets, which are nowhere dense in any interval, are the only possible sets of type n for some appropriate n. If so, he would have believed that all nowhere dense sets are zero coverable. Certainly, nowhere in his work did he find fault with Hankel's account.

Henry Smith

It seems that the first mathematician to sort this out was the Oxford mathematician Henry Smith. He produced two (or, as you will see, perhaps we should say three) interesting examples of families of sets for the purposes of integration theory. The first family was constructed this way. The set P_1 consists of all the rational numbers of the form $1/n_1$ for positive integers n_1. The set P_2 consists of all the rational numbers of the form $1/n_1 + 1/n_2$ for positive integers n_1 and n_2, and so on, so the set P_k consists of all the rational numbers of the form $1/n_1 + 1/n_2 + \ldots + 1/n_k$ for positive integers $n_1, n_2, \ldots n_k$. It is a nice exercise to confirm Smith's result that the jth derived set of P_k, $k > j$ is $P_k^{(j)} = \{0\} \cup P_1 \ldots \cup P_{k-j}$. Smith showed that a function which is discontinuous on a set P_k is integrable, because these sets are zero coverable.

The second family you may recognise under another name. Smith's term for nowhere dense sets was sets 'in loose order'.[1]

> Let m be any given [integer] greater than 2. Divide the interval from 0 to 1 into m equal parts; and exempt the last segment from any subsequent division. Divide each of the remaining $m-1$ segments into m equal parts; and exempt the last segment of each from any subsequent division. If this operation be continued ad infinitum, we shall obtain an infinite number of points of division P upon the line from 0 to 1. These points lie in loose order ....

To understand this, look at what Smith defined that produced a nested family of sets. At stage zero, so to speak, we have the unit interval $[0, 1]$. At stage one, we have $m - 1$ equal subintervals that together form the interval $[0, \frac{m-1}{m}]$; the half-open interval $(\frac{m-1}{m}, 1]$ at the end has been excluded ("exempted", as Smith put it). At stage two, we have $(m-1)^2$ intervals, and so on.

We get a good account if we let $m = 3$ and label points in the interval $[0, 1]$ by ternary decimals. A ternary decimal is an expression of the form

[1] See Smith (1875, 147), quoted in Hawkins (1975, 38).

0.10221002011102..., where the entries are either 0, 1, or 2, and the corresponding real number is written in powers of (1/3). So the above ternary decimal is equal to

$$1 \cdot (1/3) + 0 \cdot (1/3)^2 + 2 \cdot (1/3)^3 + 2 \cdot (1/3)^4 + 1 \cdot (1/3)^5 + \cdots.$$

We can now see that Smith's process with $m = 3$ first of all drops all the ternary decimals that begin $0 \cdot 2\ldots$, because they occupy the end interval, then it drops all the ternary decimals that begin $0 \cdot 02\ldots$ or $0 \cdot 12\ldots$, because these ternary decimals occupy the final thirds of the intervals [0, 1/3] and [1/3, 2/3] respectively.

The set P that Smith is interested in is the intersection of the sets obtained at each stage, and we can see that it will consist of all ternary decimals that never contain a 2.

The set thus obtained is the first 'Cantor' set, S. It is nowhere dense, its complement at the nth stage is of length $(1 - 1/m)^n$, which tends to 1, so the set is zero-coverable and a bounded function discontinuous at precisely the points of S is integrable.

Smith then modified this family (or produced a new family, if you prefer) which is still nowhere dense but not zero-coverable, so a function discontinuous on this set is not integrable, and therefore nowhere dense does not imply zero coverable. This he did as follows. Again, let m be any integer greater than 2, divide the interval from 0 to 1 into m equal parts, and exempt the last segment from any subsequent division. But now divide each of the remaining $m - 1$ segments into m^2 equal parts; and exempt (i.e. exclude) the last segment of each from any subsequent division. If this operation be continued ad infinitum, the points that have never been excluded form an infinite number of "points of division" Q in the interval [0, 1]. This set Q is nowhere dense. But now at stage n the complement of the set is length $(1 - 1/m)(1 - 1/m^2)\ldots(1 - 1/m^n)$ which does not tend to zero as n tends to infinity but instead tends to the positive limit $\Pi_n^\infty(1 - 1/m^n)$. For example, if $m = 3$ this limit is over one-half (at about 0.560126). So the set Q is not zero coverable.

The name 'Cantor set' is derived from the very similar set that Cantor described in his (1883). This set is obtained by following Smith's construction with $m = 3$ but excluding the middle third each time. In terms of ternary decimals, the resulting 'Cantor set' consists of all ternary decimals that never contain a 1.

This illustrates the sad truth that Smith's works were not much read on the Continent. So important is the set Q, however, that it really deserves a name, and because the name 'Cantor set' is surely permanent in the literature, the set Q will here be called a Smith–Cantor set. Its merit is that it is a nowhere dense set in the unit interval that is not zero coverable.

So we see that there was a confusion, and not just in Hankel's work, between a set being nowhere dense and its being zero coverable. The confusion may have come from the fact that Cantor's sets of type n are nowhere dense and are zero coverable. This may well have suggested that the right way to single out point sets that are negligible for the purposes of integration—the zero coverable sets—is to select the nowhere dense sets.[2] This entirely topological criterion was not, however, the right way to go, as the Smith–Cantor set indicated.

Some side remarks with which to conclude. Smith, in this paper, also disproved Hankel's theorem that if a function f is Riemann-integrable then all its one-sided limits $f(x-)$ and $f(x+)$ exist. This was also disproved by Johannes Thomae in 1875, who gave this example: $f(x) = \sin(1/\sin(\pi/x))$, which is integrable, but the discontinuities (which are at $x = 1/m$) are such that $f(x+)$ and $f(x-)$ do not exist. Thomae also failed to find fault with Hankel's identification of piecewise discontinuous functions with the Riemann-integrable ones.

Here's a digression that is somehow relevant: The Taylor series of $P(x) = (1 - x)(1 - x^2)\dots(1 - x^n)\dots$ begins

$$1 - x - x^2 + x^5 + x^7 - x^{12} - x^{15} + x^{22} + x^{26} - x^{35} - x^{40} + x^{51} + x^{57} + O(x^{60}).$$

The exponents are the pentagonal numbers $\frac{1}{2}n(3n+1)$, $n = 0, -1, +1, -2, +2, \dots$. Verify Euler's result that

$$\prod_{j=1}^{\infty}(1 - x^j) = \sum_{j=-\infty}^{\infty}(-1)^j x^{n(3n+1)/2}.$$

Euler discovered this result in 1741 and published it in 1751, in E 158, and also in his *Introductio* (1748) Chapter XVI, § 323; see also Weil (1984, 279). The series is intimately involved in the study of the number of ways a given number can be written as the sum of the numbers $1, 2, 3, \dots m$. You may check that, for example, 7 may be written as the sum of numbers less than or equal to 7 in 15 ways, as Euler observed in his *Introductio* § 324.

23.4 Beyond Uniformity

Once the concept of uniform continuity was elucidated, it was clear both that the uniform convergence of a sequence or series of continuous functions ensured that the limit function was continuous, and that the uniform convergence of a sequence or series of integrable functions ensured that the limit function was integrable and also that term by term integration was valid.

We have already seen that the non-uniform convergence of a sequence or series of continuous functions is not enough to ensure that the limit function is continuous. We therefore turn to considering if a non-uniform limit of integrable functions has any useful properties.

Let us fix some notation and some ideas. Suppose that (f_k), $k = 0, 1, \dots$ is a sequence of integrable functions on an interval $[a, b]$, and define $F_n(x) =$

[2] I can now withdraw the term 'zero coverable' in favour of the later term 'measure zero'.

$\sum_{k=0}^{n} f_k(x)$. Let the series F_n tend pointwise to a limit function $F(x)$. We can now do two things.

The first is to ask if this limit function $F(x)$ is integrable. If it is, we naturally write the integral as $\int_a^x F(t)dt$. We mean by this $\int_a^x (\lim_{n\to\infty} F_n(t))\,dt$.

Another thing we can do is consider the integrals $\int_a^x F_n(t)dt$ (certainly the function F_n is integrable, being a finite sum of integrable functions). This is term-by-term integration of the series. Then we consider the limit as n increases of these integrals, if it exists. If it does, we write it as $\lim_{n\to\infty} \left(\int_a^x F_n(t)dt\right)$.

Now we can ask if these limits are the same. In symbols, does $\int_a^x (\lim_{n\to\infty} F_n(t))\,dt$ equal $\lim_{n\to\infty} \left(\int_a^x F_n(t)dt\right)$?

The answer to all these questions is yes when the convergence is uniform, so the question is: what happens when it is not?

A Non-integrable Limit of Integrable Functions

First of all, it is entirely possible that a limit of Riemann-integrable functions is not Riemann-integrable. To obtain a sequence of (Riemann)-integrable functions tending to a non-integrable limit, take an ordering on the rationals between 0 and 1, and for each integer n enclose the first n rationals, $r_1, \ldots, r_k, \ldots r_n$, in the open intervals $I_{k,n} := (r_k - n^{-3}/2, r_k + n^{-3}/2), k = 1 \ldots n$ of length $\frac{1}{n^3}$. Denote the union of these intervals by S_n, so $S_n := \bigcup_{k=1}^{n} I_{k,n}$. Let χ_n be the characteristic function on the complement of S_n in $[0, 1]$. Trivially, χ_n is integrable for each n.

Let $S := \lim_{n\to\infty} S_n$. Note that this is not $\bigcap_{n=1}^{\infty} S_n$, because a new set is acquired at each stage. In fact $S = \{x \mid \exists m \ : \ x \in S_n \forall\, n > m\}$.

In fact, $S = \mathbb{Q}$. Certainly $r_n \in S_n$ and indeed $r_n \in S_m$ for all $m > n$, so $\mathbb{Q} \subset S$. Let $\zeta \in [0, 1] - \mathbb{Q}$. Then $\zeta \in S$ implies that there is an m such that $\zeta \in S_n$ for all $n > m$, and so $\zeta \in \bigcup_{k=1}^{n} I_{k,n}$ for all $n > m$. If this holds for a constant value of k then the intersection is contained in any arbitrarily small interval around r_k, which is to say the intersection is r_k, but this is a contradiction because ζ is irrational. If on the other hand $\zeta \in \bigcup_{k=1}^{n} I_{k,n}$ for all $n > m$ for at least two distinct values of k then $\zeta \in I_{k,n} \cap I_{l,n}$ for all $n > m$, but this is impossible because that intersection is empty. So $S = \mathbb{Q}$, as claimed.

It follows that the sequence of characteristic functions χ_n is a sequence of integrable functions tending to the characteristic function on the set $[0, 1] - \mathbb{Q}$, and this is obviously not Riemann-integrable.

Darboux's Example of the Failure of Term by Term Integration

Darboux, in his fascinating *Mémoire* of 1875, showed that a uniformly convergent series of integrable functions is integrable and can be integrated term by term. He also gave this example of a series that converges, but not uniformly, to an integrable (indeed, a continuous) function, for which the result of term by term integration of the series does not agree with the integral of the function defined by the series.

This is the series $s(x) = \sum_n u_n$ whose nth term is

$$u_n = -2n^2 x e^{(-n^2x^2)} + 2(n+1)^2 x e^{-(n+1)^2x^2}.$$

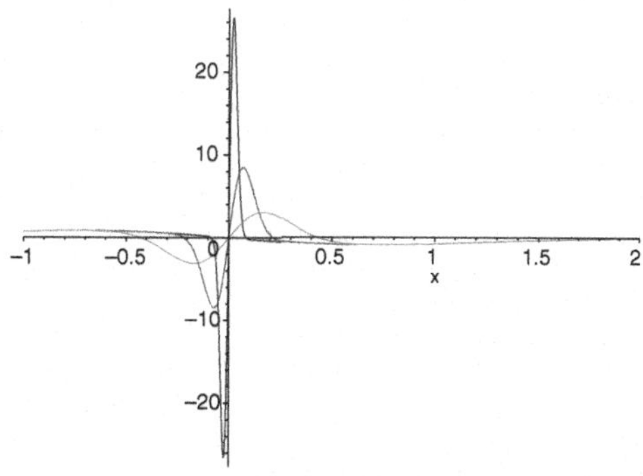

Fig. 23.1 The functions u_3, u_9, u_{30} in Darboux's sequence

If we write $f_n(x) = -2n^2xe^{(-n^2x^2)}$ then $u_n(x) = f_n(x) - f_{n+1}(x)$ and so

$$s_N(x) = \sum_{n=1}^{N} u_n(x) = f_1(x) - f_{n+1}(x).$$

Figure 23.1 shows the 3rd, 9th, and 30th terms of the sequence of functions involved (they get narrower and higher). The sharp, moving peak suggests that the convergence will not be uniform.

Figure 23.2 shows

$$s_{500}(x) - s_{100}(x) = f_{101}(x) - f_{501}(x),$$

the difference between the sum of the first 500 and the sum of the first 100 terms, so it gives an impression of the tail after the first 100 terms that define the series. Notice how narrow it is, and how big.

To prove Darboux's claim, we note that the series $s(x)$ converges to the function $f_1(x) = -2xe^{-x^2}$, which can be integrated from 0 to x and the integral is $e^{-x^2} - 1$. Term by term integration of the series $s(x)$ over the same interval, however, gives the sequence defined by

$$v_n(x) = (e^{-n^2x^2} - 1) - (e^{-(n+1)^2x^2} - 1) = e^{-n^2x^2} - e^{-(n+1)^2x^2},$$

which has partial sums

$$e^{-x^2} - e^{-(N+1)^2x^2}$$

and so the sum of the term by term integrals in the limit is e^{-x^2}. This is because the convergence is non-uniform, as Darboux explicitly noted, the tail $\sum_N^\infty u_n(N)$ being $2N/e$.

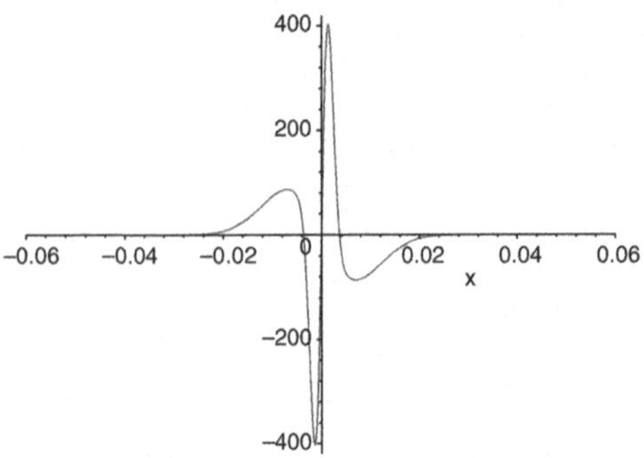

Fig. 23.2 The tail in Darboux's limit function

23.5 Weierstrass

Shortly after becoming a professor of mathematics in Berlin, Weierstrass taught a course on the theory and applications of trigonometric series and definite integrals, related to the problem of representing arbitrary single-valued functions of one real variable in Fourier series and integrals. Many years later, in a letter to Schwarz on March 14th, 1885, he wrote that "the lack of rigour" he found in all the relevant works he had at hand, and his "at that time unfruitful efforts to supply this lack" convinced him not to teach that course again.

Weierstrass was among the few to be dissatisfied with Riemann's definition of the definite integral. He expressed his concern about that, as well as his attempts at extending Riemann's definition, in letters to his friends and students. Writing to Paul du Bois-Reymond he lamented he had always complained that Riemann's definition had a limitation which he had to work around because he did not know how to put right.

> It has always seemed to me that Riemann's definition of the definite integral is subject to a disadvantage which I had to ignore since till now I did not know how to remove it. It is this: in determining the greatest oscillation that the value of the function to be integrated makes in an interval (therefore in the statement of the integrability condition), those values it has at the points of discontinuity are considered while, however, the value of the integral, when it exists, depends solely on the values which the function takes at the points of continuity. (1923, 215).[3]

On April 20th, 1885 he wrote that the definition in question should be modified "more energetically than he himself had done". On 16th May 1885 he wrote Sophie Kovalevskaya that "Riemann's definition, which is believed to be the most general one could think of, is insufficient".[4] Less than ten days later, on May 25th he repeated this statement word for word to Schwarz, adding that "it is neither general enough nor, above all, admissible. Perhaps it must be replaced by another, very different one, for

[3]The letter was reprinted in *Acta Mathematica*, vol. 39 (1923), 199–225, quote from 215, and is taken from Hawkins (1975, 67).

[4]In *Acta Mathematica*, vol. 39 (1923), p. 215.

whose foundations Cantor's newest researches offered to me essential services". This was connected with the problem of the representation of functions by Fourier series, and with the concluding remarks of Dirichlet's 1829 paper in particular. Weierstrass had begun to study this subject thoroughly in early 1885, as he announced in a letter to Schwarz on March 14th, 1885. He made it the object of a paper (Weierstrass 1885), and eventually he expounded it in his summer 1886 lectures in Berlin. There he presented his approximation theorem and his critical remarks on Riemann's definition (Weierstrass 1886, 110–112). In spite of all his efforts, however, Weierstrass was unable to succeed in finding the sought-for extension.

Chapter 24
The Fundamental Theorem of the Calculus

24.1 Introduction

It seems that our intuition, or at least the intuitions of 19th Century mathematicians, was that a function is naturally differentiable. It is differentiable, they felt, unless it is obviously not, perhaps because it has a sudden change of direction or because it is not even continuous, but such events would only happen at isolated points.

Insofar as this was their belief they were totally mistaken, and once they had sufficiently precise definitions they were gradually able to discover their error, but this was a process of discovery that had to be made many times, and led on the way to interesting discoveries of suggestive compromises. What, for example, of the fundamental theorem of the calculus? Could it be that if a function has a derivative then the derivative is necessarily integrable (and its integral differed from the original function by only a constant)? Or might there be differentiable functions with non-integrable derivatives?

24.2 Bernhard Bolzano

The first person to see clearly that continuity and differentiability had very little to do with each other (beyond the simple fact that differentiability implies integrability) was the obscure figure of Bernhard Bolzano. He became much better known in the 20th Century than he ever was in his own day (1781–1848). For most of his adult life he was banned from teaching—he was an ordained priest of the Catholic Church and a professor of theology at the University of Prague—on the grounds of heresy. This meant that his voluminous philosophical writings were by and large not published. Some of his mathematical work, which interests us here, was published, but only obscurely and it only created a stir when it was re-read after its salient novelties had been rediscovered independently by others. Much remains to be done in rediscovering

J. Gray, *The Real and the Complex: A History of Analysis in the 19th Century*,
Springer Undergraduate Mathematics Series, DOI 10.1007/978-3-319-23715-2_24

Bolzano's significance for the history of mathematics, but an auspicious start has been made in the 21st Century with the publication of *The Mathematical Works of Bernhard Bolzano* (Russ 2004).

Bolzano is a fascinating figure, driven, it would seem by a philosopher's desire for clarity to scrutinise the foundations of the calculus and, more importantly, to rework them. It was not original in the 1820s to find fault with those foundations, and in particular with the way that intuitive geometric considerations were smuggled into proofs about infinitesimals and limits where one might suppose that intuition was blind. It was highly original to find a way to resolve this problem.

Bolzano was adamant that geometric considerations should not enter proofs of theorems in analysis, most notably the fundamental theorem of algebra. This is evident both in the title of his crucial work, the *Rein analytischer Beweis, etc.* (or, A purely analytic proof, etc.)—and in the methods Bolzano used. Indeed one of the things Bolzano meant by analytic was non-geometric. Purity of method, as present-day philosophers of mathematics call the insistence on the use of only the most appropriate methods, was a deep concern of Bolzano when he was younger. But, as Russ shows, he seems to have felt that it would earn him few readers and he largely gave up the activity for a number of years. Historians of mathematics have had a tendency to talk him up as a precursor of those who rigorised the calculus by means of a sophisticated limit concept and an emphasis on the arithmetic character of the key definitions; Russ argues that this is overdone. Instead he suggests that Bolzano's proof of the binomial theorem may be better appreciated for its way of dealing with finite variable quantities and not for its account of something like the modern limit concept. Even more importantly, as Russ shows, Bolzano's definition of what it is for a function to be continuous at a point in the *Rein analytischer Beweis, etc.* seems to be the modern one, and therefore clearer than Cauchy's definition a little later, who only considered continuity on an interval.

Cauchy, as we have seen, was clear about limits but not clear by later standards about the real numbers. Bolzano certainly had and used the concept of what are called Cauchy sequences today. He claimed that if a sequence of quantities u_n, say, has the property that the absolute value of the difference $u_n - u_{n+r}$ can be made arbitrarily small for all values of r by taking n large enough, then the sequence converges to a limit. Did he too fail to see that this required a precise theory of the real numbers? Russ suggests that the insight that sequences with this 'bunching up' property are defining a number is a profound insight into the nature of number that Bolzano was perhaps the first to express. This marks, in fact, his contribution to the Bolzano–Weierstrass theorem.

What excited mathematicians in the 1930s (!) about Bolzano's work was his description of a continuous, nowhere differentiable function. What a remarkable thing to look for, and what a delicate microscope is needed to exhibit it. His example rested on the observation that it is always possible to put a crinkle into the graph in any interval, no matter how small. Start with, say, the function $f(x) = 1-2|x-1/2|$ on the interval $[0, 1]$. Replace each sloping segment with a zig-zag in three parts, the first and last of which are steeper than the segment they replace and which are joined by a short segment of the opposite slope and passing through the midpoint of the segment they are replacing. Do this indefinitely, and there is every chance that the graph you end up with is that of a function which is continuous in the interval [0, 1] but nowhere differentiable—but that requires some care to establish!

24.3 Non-differentiable Functions

The recognition by mainstream mathematicians that continuity does not imply differentiability seems to have occurred to Riemann and Weierstrass separately by 1861. Hankel took Riemann's strange function $r(x)$ and defined

$$F(x) := \int_0^x \sum_{n=1}^{\infty} \frac{r(nt)}{n^2} dt.$$

He showed that this function is continuous but fails to have a derivative at a dense set of points.

Hankel

Riemann and Weierstrass confined their examples to lectures, but Hankel went into print. He observed that the technique for constructing a singular point of a function at one point can be made to yield functions singular at infinitely many. This he called the method of condensation of singularities. Once you see the trick, it's very easy. In his example, he supposed he had a function, $\varphi(x)$, whose values were bounded on the interval $[-1, 1]$ by ± 1 and which was infinitely differentiable on the interval $[-1, 1]$ except at the origin—you might take the example $g(x) = x \sin(1/x)$. The origin is called a singularity because the function is not differentiable there. Hankel then constructed a function in which the single singularity of $\varphi(x)$ so impregnated a function $f(x)$ that it had infinitely many singular points in every interval (Hankel 1882, 78). This was the function

$$f(x) = \sum_{n=1}^{\infty} \frac{\varphi(\sin(n\pi x))}{n^s},$$

where $s \geq 3$—the factor n^{-s} is just there to ensure convergence. As Hankel checked, this function is singular at every rational point in the interval $[-1, 1]$.

The next graph, Fig. 24.1, is of the function $g(x) = x \sin\left(\frac{1}{x}\right)$.

More generally, suppose the function $\varphi(x)$ has a singularity at the origin (you may continue to think of $g(x) = x \sin(1/x)$). Let $\{a_n\}$ be your favourite infinite discrete set of points, and consider the sum $\sum_n \varphi(x-a_n)$. If this sum makes sense the function it defines has singular points at every a_n. How could the sum fail to make sense? Only by being infinite, so let us insist that the original function $\varphi(x)$ is never greater than 1 in absolute value, and consider this sum instead: $\sum_n \frac{g(x-a_n)}{2^n}$—you may use any other fudge factor that ensures convergence. This ensures pointwise convergence,

Fig. 24.1 $g(x) = x \sin\left(\frac{1}{x}\right)$

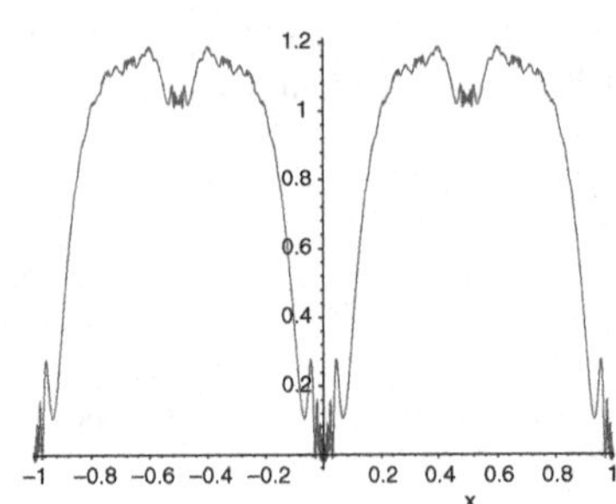

Fig. 24.2 Hankel condensation on $[-1, 1]$

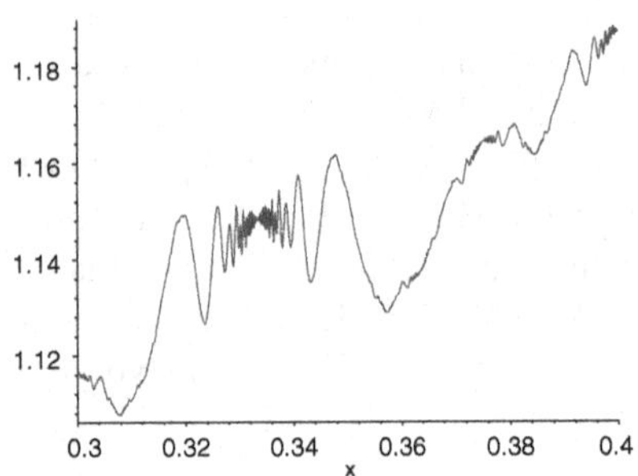

Fig. 24.3 Hankel condensation closer up

and it remains now to check what is surely plausible that the new summation has not somehow smoothed out the singularities.

If you take for $\varphi(x)$ the singular function $g(x) = x\sin(1/x)$ then the function $\sum \frac{1}{j^2} g(\sin j\pi x)$ should have lots of singularities. This is what the next two graphs suggest. Figure 24.2 is a graph of the function on $[-1, 1]$, Fig. 24.3 of the function on $[0.3, 0.4]$.

Condensing Singularities

The next two graphs show the effect of taking the function $g(x) = x \sin\left(\frac{1}{x}\right)$ and applying the method of condensing singularities to it at the points $\frac{1}{n}$. Figure 24.4 shows the first 4 and Fig. 24.5 the first 10 terms of the sum

$$\sum_{1}^{\infty} \left(x - \frac{1}{k}\right) \sin\left(x - \frac{1}{k}\right).$$

Fig. 24.4 The first 4 points

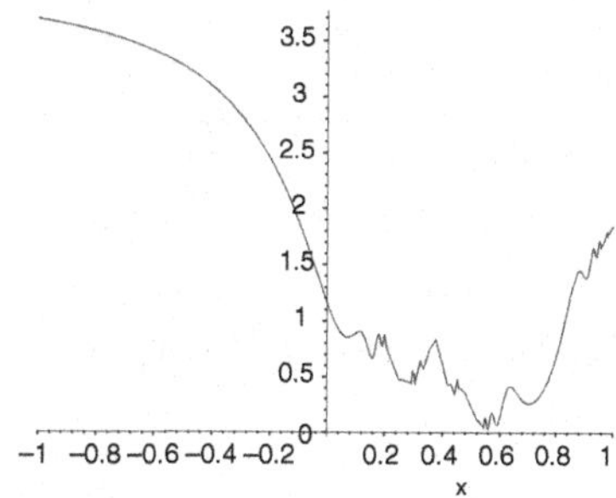

Fig. 24.5 The first 10 points

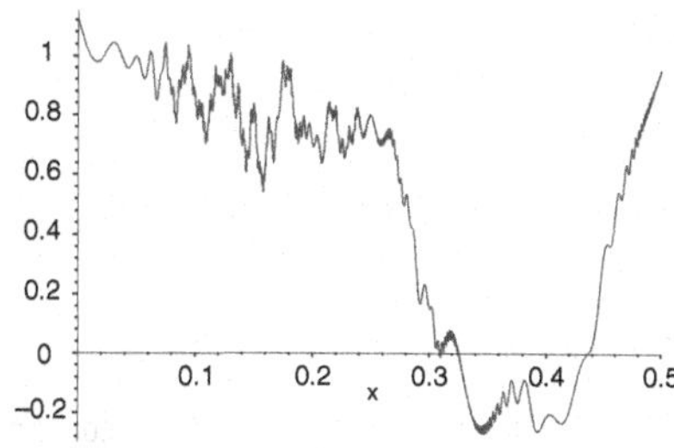

Hankel's Example

Hankel gave this example:

$$f(x) := \sum_{n=1}^{\infty} g(\sin n\pi x)/n^s, \quad s > 2,$$

where the function g is continuous on $[-1, 1]$ but not differentiable at $x = 0$, and he claimed that the function $f(x)$ is continuous but not differentiable at all the rational points.

What follows are some graphs of Hankel's function with $g(x) = x \sin\left(\frac{1}{x}\right)$. The first, Fig. 24.6, just hints at how the singularities appear.

Evidently something non-differentiable is happening at $x = 0$ and $x = 1/2$. Figures 24.7 and 24.8 show this in closer detail.

But weird things are also going on elsewhere, near $x = 3/5$ and near $x = 3/4$, as Figs. 24.9 and 24.10 show in closer detail.

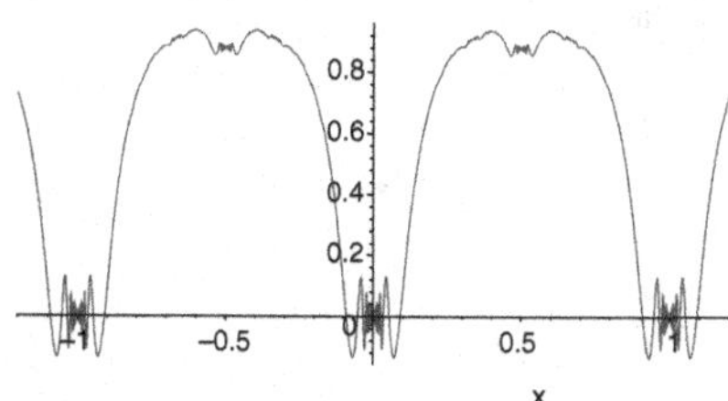

Fig. 24.6 A Hankel function

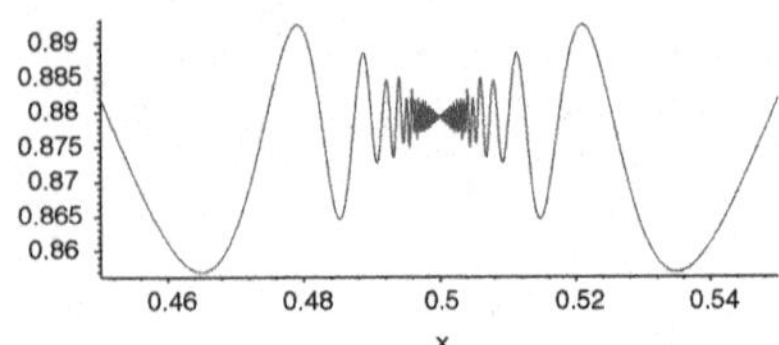

Fig. 24.7 A Hankel function near $x = 1/2$

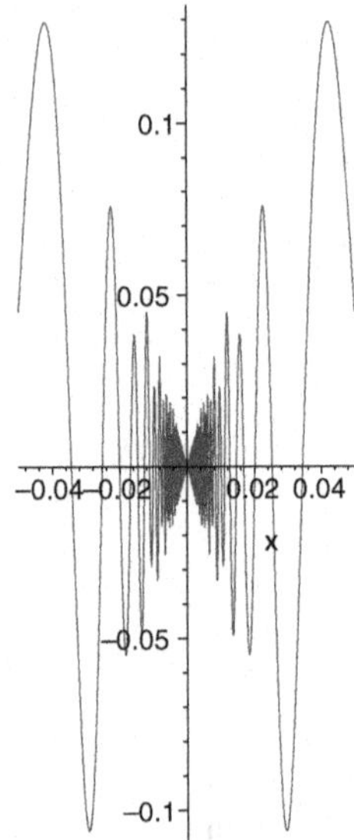

Fig. 24.8 A Hankel function near $x = 0$

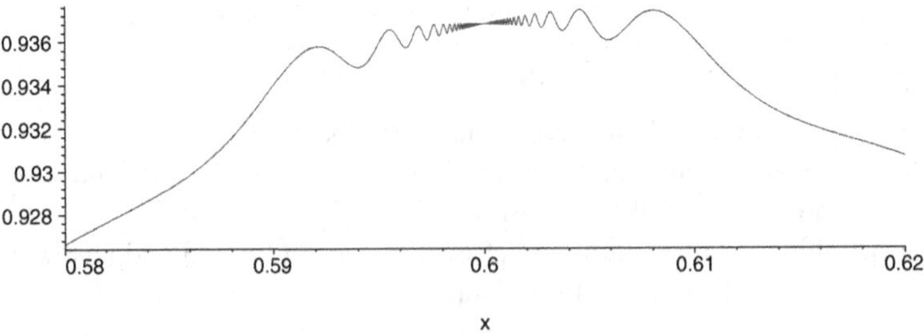

Fig. 24.9 A Hankel function near $x = 3/5$

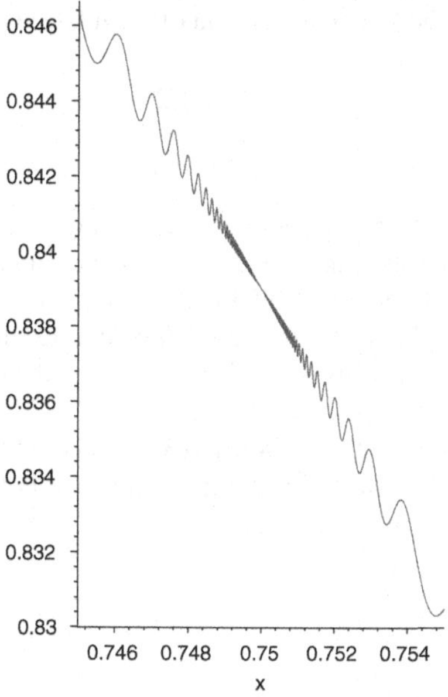

Fig. 24.10 A Hankel function near $x = 3/4$

Schwarz

Hermann Amandus Schwarz (1843-1921)

Hankel's argument's were not always correct, and they were criticised, but Schwarz then came forward with a solid example (Schwarz 1878). He obtained it from the function $g(x) = \lfloor x \rfloor + (x - \lfloor x \rfloor)^{1/2}$, where $\lfloor x \rfloor$ is the floor function (as we say today)[1] by using the method of condensation of singularities to define the function

$$f(x) := \sum_{n=0}^{\infty} \frac{g(2^n x)}{2^{2n}}$$

which fails to be differentiable at the dense set of points $m/2^n$. What is striking about this example is that the function is strictly increasing. It is possible to think that functions which fail to be differentiable do so at points where they oscillate up and down too much, but this is not the case. (It resembles the view that a function might fail to be differentiable if its derivative 'should' be 0, but not otherwise, which is implausible as soon as you say it.)

The next three graphs show the 1st, 4th and 30th stages in Schwarz's construction of function. To my eyes the nature of the limit function is very hard to see (Figs. 24.11, 24.12 and 24.13).

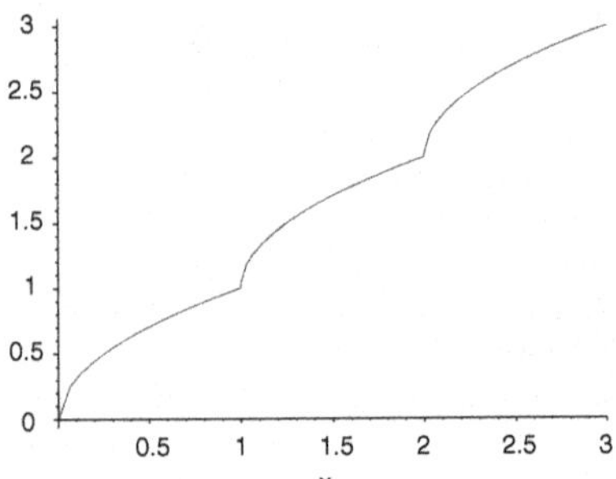

Fig. 24.11 The 1st Schwarz function

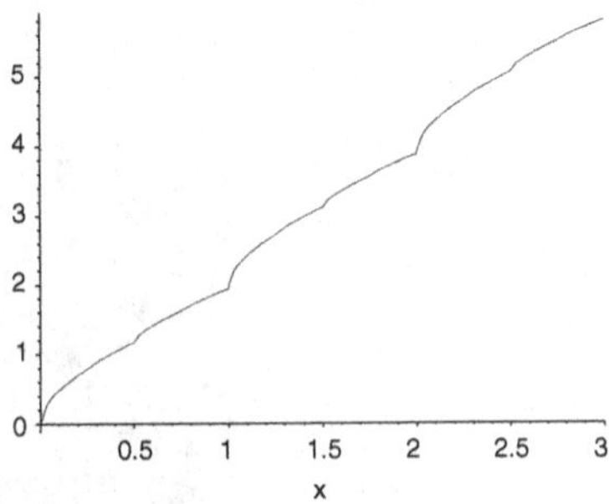

Fig. 24.12 The 4th Schwarz function

[1] The value of the floor function at x is the greatest integer less than or equal to x.

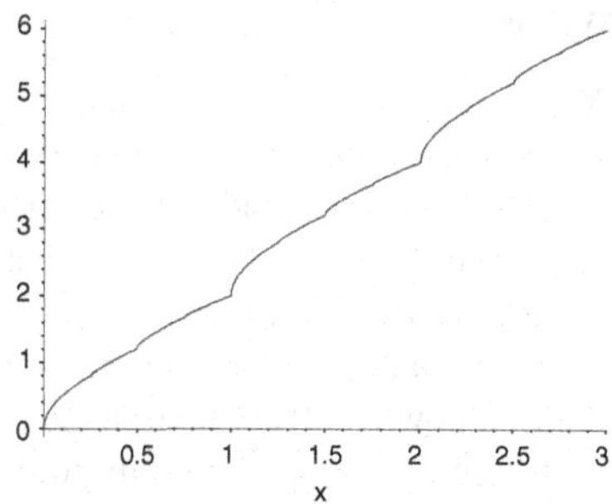

Fig. 24.13 The 30th Schwarz function

A Non-integrable Function with a Bounded Derivative

Brodén's Function

Another good example is obtained from the function $g(x) = x^{1/3}$ which is monotonic increasing and has a graph that is vertical at the origin. In 1896 Torsten Brodén (following in the steps of both Cantor and Weierstrass) asked his readers to let $\{w_r\}$ be a dense set of points in $[-1, 1]$, and $\{c_n\}$ a set of positive numbers, and to define the function

$$f(x) = \Sigma_n c_n g(x - w_n).$$

This function is strictly increasing, so it has an inverse h which is also strictly increasing even though its derivative h' vanishes on a dense set of points (the values of $f(w_n)$). So the function h' is a bounded function that is not Riemann integrable on its domain (which is $f([-1, 1])$). This shows that the fundamental theorem of the calculus cannot always be true.

An impression of the function f is given by Fig. 24.14, which fits just seven vertical tangents into the interval $[-1, 1]$, but it gives an impression of what the finished object would be like with its infinitely many vertical tangents densely packed.

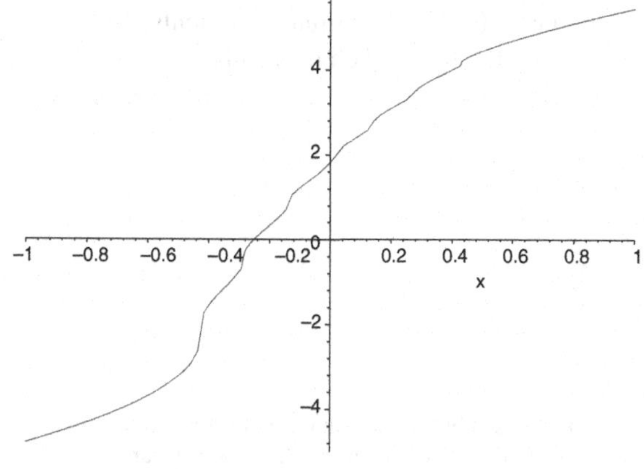

Fig. 24.14 A first approximation to Brodén's function

Weierstrass

Weierstrass enters the story here in a number of ways. As we noted above, he had heard of Riemann's ideas about continuous nowhere differentiable functions through the grapevine, and had deduced from the existence of complex functions with the unit circle as a natural boundary that there are real-valued functions of a real variable that are continuous everywhere and differentiable nowhere.

He also had a student, Leo Koenigsberger, who was a good enough mathematician to have known better than to suggest (as he did in his (1874, 13)) that a function would be differentiable if it had only finitely many maxima and minima and did not have jump discontinuities on a dense set of points (this last condition being intended to rule out Riemann's example). Evidently he did not know of Schwarz's example, which shows that this claim must be wrong.

Koenigsberger said a copy of his book to his former professor, and Weierstrass replied, offering his own counter-example.[2]

Weierstrass first distinguished carefully between what it is for a function to have some property at every point on its domain, and what is for a function to have that property at some point of any interval in its domain. Generally, it is easy to find functions of the second kind, but it can be much harder to find functions of the first kind. He then took the function $g(x) = (1/2)x\sin(\log(x))$, which fails to be differentiable only at the origin. He then chose a $k < 1$ and two sequences of real numbers $\{a_n\}$ and $\{b_n\}$, $n = 0, 1, \ldots$, the first of which is arbitrary and the second of which is such that $\sum_n b_n$ is finite. Then he defined

$$f(x) = b_0 g(x) + \sum_{n=1}^{\infty} b_n(g(x - a_n) + g(x + a_n)).$$

This function, he indicated, is continuous everywhere, monotonic (so it has a finite number of maxima and minima—zero in each case) but it fails to be differentiable at the set of points $\{a_n\}$. Weierstrass observed that the set $\{a_n\}$ can be taken to be the dense set of all algebraic numbers, which Cantor had recently proved was countable. He also speculated on whether there could be a nowhere differentiable monotonic function with bounded difference quotients, $\frac{f(x+h)-f(x)}{h}$ and thought that they would be were hard to find. Much later Lebesgue showed that in fact they cannot exist.

[2]The letter was published in volume 39 of *Acta Mathematica*, which Mittag-Leffler devoted to the life and work of Weierstrass in 1923, and which carried a number of historically valuable sources.

Ulisse Dini

Ulisse Dini (1845-1918)

The Italian mathematician Ulisse Dini studied mathematics in Pisa before going to Paris for the year 1865. On his return he began to doubt the truth of some of the propositions that he had heard concerning continuity and differentiability, which the publications of Hankel, Schwarz, Darboux and Weierstrass confirmed. He undertook further investigations of his own, which he published in his widely read *Fondamenti per la teorica delle funzioni di variabili reali* (or, *Foundations of the theory of functions of real variables*) (1878).

His study rested on four useful and elementary terms, defined as follows. He considered the fraction $\frac{f(x+h)-f(x)}{h}$ with $0 < h < b - x$, and let G_x and L_x be respectively the greatest lower bound and the least upper bounds of this fraction in the interval stated; they are functions of x and b. Then Dini defined

$$D_{+}f(x) := \lim_{b\downarrow x} G_x = \liminf_{h\downarrow 0}\left(\frac{f(x+h)-f(x)}{h}\right)$$

$$D^{+}f(x) := \lim_{b\downarrow x} L_x = \limsup_{h\downarrow 0}\left(\frac{f(x+h)-f(x)}{h}\right)$$

$$D_{-}f(x) := \lim_{b\downarrow x} G_x = \liminf_{h\downarrow 0}\left(\frac{f(x-h)-f(x)}{-h}\right)$$

$$D^{-}f(x) := \lim_{b\downarrow x} L_x = \limsup_{h\downarrow 0}\left(\frac{f(x-h)-f(x)}{-h}\right)$$

where the limits are from above ($h > 0$). Note that lim inf and lim sup are later ideas, Dini gave only the first half of these definitions.

Dini applied these ideas to the fundamental theorem of the calculus. Cauchy had shown
(1) that for a continuous function f on an interval,

$$\left(\int_a^x f\right)' = f(x).$$

and also
(2) that if F is a function with a continuous derivative then

$$\int_a^b F' = F(b) - F(a).$$

However, Hankel had shown that the first of these claims fails for Riemann's strange function on a dense set of points.

Dini was able to show that if f is integrable and $F(x) = \int_a^x f$ then

$$F(b) - F(a) = \int_a^b DF,$$

where DF is any of the four of Dini's derivatives. So (1) is true in the sense that $F(x)$ has four derivatives, and they differ from one another by functions whose integral over any interval vanished. This does not mean that they differ by a nothing, merely by a function whose integral is negligible. The set of points at which such functions could be non-zero was still unclear in Dini's time.

As for (2), Dini was able to show that if F is a continuous function and one of its four Dini derivatives is integrable (denote it DF) then

$$\int_a^b DF = F(b) - F(a).$$

In short, the familiar fundamental theorem of the calculus holds for a differentiable function F if its four Dini derivatives agree, and if they do not something can still often be said.

24.4 Exercises

1. How would you answer somebody (in the 1840s or 1850s) who said that it makes no sense to speak of a function being continuous at a point and that continuity has to mean something like continuous on an arbitrarily small interval?
2. How different do you take differentiable and merely continuous functions to be?
3. In the light of the differences between differentiable and merely continuous functions, and of suitable examples drawn from his work, what do you suppose Cauchy's concept of continuity might have been? (Be prepared to offer several tentative and different opinions).

Chapter 25
The Construction of the Real Numbers

25.1 Introduction

Much of early 19th-century analysis rested uneasily on an intuitive notion of quantity that embraced all the measurable objects for which the natural numbers were inadequate, in particular lengths. Real analysis was the study of real varying quantities, and at least informally it was compatible with infinite and infinitesimal magnitudes. Cauchy had redefined these infinite and infinitesimal magnitudes in terms of limits, but it was still unclear what the domain of real quantities comprised, and it struck several German mathematicians that complications in real analysis might be alleviated by providing a good definition of what the real numbers might be. Here we look at Weierstrass's not entirely successful account, then at Dedekind's more elegant version, and finally and very briefly at the theories of Cantor and Heine.

25.2 Weierstrass's Ideas

One concept that disappeared from analysis under Weierstrass's influence was that of the infinitesimal. It was shut out by his creation in the 1860s of a concept of real number, and the identification of ordinary magnitudes with real numbers. This meant that the concept of a magnitude could be replaced by that of a real number, and the concept of a variable magnitude could be replaced by that of a real variable. In the event, Weierstrass's way of doing this was not adopted, and other ways of achieving this goal were preferred, but Weierstrass did identify the main features of the problem and its solution. He was clear that a system of 'numbers' must be created that exactly captured the intuitive properties of finite magnitudes, and that they must be created out of rational numbers, whose existence and properties were taken to be more or less

J. Gray, *The Real and the Complex: A History of Analysis in the 19th Century*,
Springer Undergraduate Mathematics Series, DOI 10.1007/978-3-319-23715-2_25

unproblematic. That is to say, positive whole numbers were collections of identical thought-objects, positive rational numbers were finite collections of units and aliquot parts ($1/n$ where n is a integer). Irrational numbers were made up as infinite sums of these rational numbers, or rather, as appropriate equivalence classes of them.

The Bolzano–Weierstrass Theorem

The details of this construction need not concern us. Of more interest is this major achievement in Weierstrass's creation of a theory of real numbers: his proof that every bounded infinite set of real numbers has a limit point. This result (called today the Bolzano–Weierstrass Theorem) encapsulates a key difference between Weierstrass's ideas and those of Cauchy. Cauchy was vague on this point, out of a belief that nature provided such a limit point. Its existence was somehow built in to the concept of magnitude. Weierstrass, who had constructed the real numbers out of more primitive concepts, knew that he had to prove that his creation had this fundamental property, and that is what he did. As Cantor (1882, 566) and Jourdain (1915, 14) pointed out, Weierstrass was the first to avoid assuming that the limits exist that are taken to define real numbers.

One may ask if Weierstrass was creating a class of objects where previously there had been none (and only a comforting illusion) or if he thought of himself as providing an improved definition of what was present in the world. On the first account he would say there had simply not been a foundation for the calculus. On the second account there had been foundations which he was now making more precise and more amenable to the new level of mathematical precision. It seems that Weierstrass's answer would have been the second, with the proviso that it was only the new, mathematically precise system of numbers that could enter into proofs. Mathematics was no longer connected to intuition at the level of continuous magnitudes, but at the more basic level of the rational numbers, or even at the level below that, of the positive integers.

Weierstrass's ideas provoked quite some debate in Berlin. As early as 1870 Kronecker was moving towards a position critical of all talk of real numbers. Schwarz wrote to Cantor that Kronecker regarded the Bolzano–Weierstrass principle as an "obvious sophism" and expected to be able to produce functions "that were so unreasonable that, despite satisfying all of Weierstrass's assumptions, they would have no upper bounds". Schwarz and Cantor, then at the start of their careers, defended their mentor Weierstrass's views as indispensable for analysis, and indeed if Kronecker ever did seek such functions he failed to find any.[1]

[1] See Ferreirós (1999, 37).

25.3 Dedekind's Construction of the Real Numbers

Weierstrass's clumsy construction was soon replaced by two others, one due to Dedekind, the other to Cantor. Dedekind came to his in 1858, and wrote about it in a letter to Durège, a friend of his from his time in Göttingen, but only published it in 1872 when he learned that he would otherwise lose priority in the matter. He had been sent a memoir by Heine on the same subject and felt that although he agreed with the substance of the memoir his own presentation seemed to be simpler. And even as he wrote his own memoir up, he was sent a paper by Cantor that also captured what he, Dedekind, saw as the essence of continuity, although he was less persuaded of the higher realms of abstract numbers that Cantor conjured up.

Dedekind's construction is his famous idea of what are today called Dedekind cuts. He had already noted that, given an arbitrary unit of length, every rational number can be associated with a unique point on a line, but the converse is false: there are lengths that are not measured by any rational multiple of the unit length. A cut, as he called it, is a division of the rational numbers into two disjoint sets such that every rational number in one set is less than every rational number in the other set. It is convenient to call two such sets L for 'left' and R for 'right', with every l in L less than any r in R. Dedekind showed how to consider cuts as numbers, calling the necessary work time-consuming but not difficult.

He distinguished between cuts for which there is a rational number q, say, in L such that $l \leq q < r$ for all l in L and all r in R (or, similarly, for a q in R) and those for which there is no such rational number. The former kind of cut he identified with the rational number q, while irrational numbers he defined as cuts of the second kind. The fact that there are cuts of the second kind was the reason, he said, for the incompleteness or discontinuity of the rational numbers. For example $\sqrt{2}$ is not a rational number, but it is defined by the cut which assigns the rational number q to L if and only if either q is negative or $q^2 < 2$ and assigns the rational number q to R if and only if q is non-negative and $q^2 > 2$. This cut is of the second kind.

Each cut is taken to define a real number, and Dedekind then showed that if one attempts to define cuts in the set of real numbers, the cuts are always of the first kind, so no new numbers can be obtained by iterating the construction.[2]

> Beside these properties, however, the domain $\mathbb{R}$ possesses also *continuity*; i.e., the following theorem is true:
>
> IV. If the system $\mathbb{R}$ of all real numbers breaks up into two classes $\mathbb{A}_1$, $\mathbb{A}_2$ such that every number α_1 of the class $\mathbb{A}_1$ is less than every number α_2 of the class $\mathbb{A}_2$ then there exists one and only one number α by which this separation is produced.
>
> *Proof.* By the separation or the cut of $\mathbb{R}$ into $\mathbb{A}_1$ and $\mathbb{A}_2$ we obtain at the same time a cut (A_1, A_2) of the system R of all rational numbers which is defined by this: that A_1 contains all rational numbers of the class $\mathbb{A}_1$ and A_2 all other rational numbers, i.e., all rational numbers of the class $\mathbb{A}_2$. Let α be the perfectly definite number which produces this cut (A_1, A_2). If β is any number different from α, there are always infinitely many rational numbers c lying between α and β. If $\beta < \alpha$, then $c < \alpha$; hence c belongs to the class A_1 and consequently

[2] See Dedekind *Continuity and irrational numbers*, 20–21.

also to the class $\mathbb{A}_1$, and since at the same time $\beta < c$ then β also belongs to the same class $\mathbb{A}_1$ because every number in $\mathbb{A}_2$ is greater than every number c in $\mathbb{A}_1$. But if $\beta > \alpha$, then is $c > \alpha$; hence c belongs to the class A_2 and consequently also to the class $\mathbb{A}_2$, and since at the same time $\beta > c$, then β also belongs to the same class $\mathbb{A}_2$ because every number in $\mathbb{A}_1$ is less than every number c in $\mathbb{A}_2$. Hence every number β different from α belongs to the class $\mathbb{A}_1$ or to the class $\mathbb{A}_2$ according as $\beta < \alpha$ or $\beta < \alpha$; consequently α itself is either the greatest number in $\mathbb{A}_1$ or the least number in $\mathbb{A}_2$, i.e. α is one and obviously the only number by which the separation of R into the classes $\mathbb{A}_1, \mathbb{A}_2$ is produced. Which was to be proved.

Dedekind insisted that the real numbers he had defined were creations of the human intellect, and he went on to insist that the same was true of the rational numbers and even the integers. On this basis he spelled out, more clearly than Weierstrass had done, what the relationship was between the real numbers he had defined and magnitudes along a line. Dedekind saw that if real numbers and magnitudes were in a one-to-one correspondence, then it was necessary that whenever a line was divided into exactly two sets, such that every member of one set lay to the left of every member of the second set, there was exactly one point that marked the separation of the sets.

Dedekind did remarkably little with his cuts. He showed how to add them, and how they could be used to make limiting arguments rigorous, but the rest of the work (showing how to subtract, multiply and divide them) which is not entirely trivial when the cuts define negative numbers, he never wrote down. Maybe it was trivial for him.

25.4 The Ideas of Heine and Cantor

What of the accounts of the real numbers that Heine and Cantor published, which Dedekind thought were inferior to his own? Heine observed that many of Weierstrass's as yet unpublished results would remain dubious until the nature of the real numbers was clarified. Heine then gave his account, which began with a purely formal account of the integers and the rational numbers, and then defined real numbers as (equivalence classes of) certain sequences of rational numbers. He then showed that repeating this construction did not lead to yet more types of number. Then he used his newly-defined real numbers to establish that a continuous function on a closed bounded set is uniformly continuous.

Because they became better known, let us look in a little more detail at Cantor's ideas. They grew out of his study of the possible bad point sets of a Fourier or trigonometric series. Once one was aware that these series can misbehave on some strange looking sets indeed, it becomes natural to want to know about the mother set, the set of all real numbers. So Cantor gave his own definition of the real numbers in a paper on trigonometric series of 1872.

He considered what we (but not he) would call Cauchy sequences, sequences of rational numbers $\{a_n\}$ such that the difference $|a_{n+m} - a_n|$ becomes, as he put it, infinitely small with increasing n. He said that such a sequence had a definite limit, and he then set out the arithmetic of such limits, after noting that another sequence a'_n may be said to have the same limit if the sequence $\{a_n - a'_n\}$ has the limit 0, when he said the sequence became infinitely small. So Cantor studied limits where today we might study equivalence classes of Cauchy sequences, and it is these limits that are to be the real numbers.

First he defined an order relation on limits, and showed that if a and b are two limits than exactly one of the following is true:

$$b = a, \quad b > a, \quad b < a$$

Then he defined the operations of additions, subtraction, multiplication and division on limits in terms of the sequences defining the limits, noting rather cryptically that division will be mildly troublesome for sequences a'_n for which some $a'_n = 0$.

Then he considered what would happen if, from the set of all limits, which he denoted B, one repeated the construction and formed the set C consisting of, to use our language, equivalence classes of Cauchy sequences of elements of B. Indeed he then considered repeating the construction indefinitely often. He noted that the new sets would be formally distinct, but showed that there was an order-preserving 1-1 correspondence between the sets B and C, and so that in that sense the chain of constructions stopped after the first stage.

25.5 Other Theories

Cantor's theory almost identical with his colleague Heine's in Halle, as indeed was that of the French mathematician Charles Méray, which, unknown to the German mathematicians, he had published in 1869. The Franco-Prussian war may well have played some part in this lack of communication. Their approach may be contrasted with that offered by two other mathematicians, Hermann Hankel and Johannes Thomae. Hankel gave a purely formal account of the integers and the rational numbers (which, indeed, Heine took over) but he despaired of giving a reductive account of irrational numbers, because he felt there were simply too many ways in which they could be created. Instead, he was content to leave the concept of a irrational number rooted in the intuition of magnitude.

Thomae, another student of Weierstrass's and also at Halle in the 1870s, published a book on the elementary theory of complex functions in 1880, the year after he arrived in Jena. Here, and again at greater length in the book's second edition, he married Weierstrass's theory of rational numbers with the Cantor-Heine theory of the real numbers. However, and perhaps without realising just what it could mean to have Gottlob Frege as a colleague, Thomae then gave a philosophical interpretation of the new, extended, number concept, and provoked one of the many furious squabbles that Frege enjoyed. That is another story, but briefly Thomae labelled all numbers other than the natural numbers signs, and regarded them as being entirely free of content. Their conditions of existence were simply that they had been obtained by abstraction from the rules of arithmetic. He spelled out this position at greater length in the second edition of his book, published in 1898. Arithmetic, he said was a game played with empty signs according to certain rules. The same, he observed, was true of chess, but whereas the rules of chess are arbitrary, the rules of arithmetic are such that numbers can be described by simple axioms in an intuitive way which permits their use in the study of nature. The formal point of view, he went on to claim, dissolves all metaphysical difficulties. Frege—of course—disagreed. Here, at all events, is one of the moments when the idea that mathematics might be formally very like a game of chess entered the discussion.

Chapter 26
Implicit Functions

26.1 Introduction

The short history of the implicit function theorem displays a number of characteristic features of the history of mathematics in the period we are considering. For many years the result was taken for granted, and attention paid to the task of finding explicitly functions defined only implicitly. Then it was realised that there was something to prove, but the task was assigned to complex analysis—which says something about the state of real variable theory. Only much later was the theorem proved in the context of real analysis.

26.2 Power Series and Implicit Functions

Almost from the time Descartes published his *La Géométrie* in 1637 mathematicians felt able to pass from functions defined implicitly in the form $f(x, y) = 0$ to functions defined explicitly in the form $y = g(x)$, although they allowed that the explicit form might be correct for a limited range of the independent variable x and that there might be a few values of x where this cannot be done. Their confidence, if it was examined at all, rested on the appearance of the graphs of implicitly defined functions.

The simple case of the circle $x^2 + y^2 = 1$ makes the case. At all points other than the extremities on the x-axis there are two branches of the inverse function $y = \pm\sqrt{1 - x^2}$ and it is easy to choose one, but this breaks down when the circle curves back on itself, which it does at the points where the tangent is vertical. This strongly suggests that in general an implicit function $f(x, y) = 0$ admits an expression as an explicit function $y = g(x)$ in any range of points x where there points (x, y) such that $f(x, y) = 0$ and $\frac{\partial f}{\partial y} \neq 0$, because

$$0 = df = \frac{\partial f}{\partial x}dx + \frac{\partial f}{\partial y}dy,$$

J. Gray, *The Real and the Complex: A History of Analysis in the 19th Century*, Springer Undergraduate Mathematics Series, DOI 10.1007/978-3-319-23715-2_26

and so

$$\frac{dy}{dx} = -\frac{\partial f}{\partial x} / \frac{\partial f}{\partial y},$$

and the tangent is vertical when $\frac{\partial f}{\partial y}$ is zero.

As so often in the 18th century, mathematicians posed the problem not of the very existence of an explicit function but of finding its specific expression so that it could be put to use. This was also the case with inverting functions: given a function in the form $y = g(x)$ when is there a function $x = h(y)$? In this case, the obvious condition to insist upon is that the graph of the function does not have a horizontal tangent. And because the most general expression for a function was taken to be an infinite series

$$y = g_0 + g_1 x + g_2 x^2 + \cdots,$$

the question was phrased as finding the inverse function

$$x = h_0 + h_1 y + h_2 y^2 + \cdots,$$

where the coefficients h_j will depend in some way on the coefficients g_k. In this setting the problem was sometimes called the 'reversion' of series. Thus in his (1770) Lagrange was concerned with expanding the roots of a general polynomial equation as power series. He reduced the question to that of expressing p, a root of the equation

$$\alpha - x + \varphi(x) = 0,$$

as a function of x, and he gave his answer as a power series expansion for p in terms of x that involved the derivatives of φ and the subsequent setting of α to zero. This much was formal algebra; his deeper concern was the convergence of the infinite series.

Some months later he tackled the same question in the context of celestial mechanics. Kepler's equation (Lagrange 1771),

$$t = x - c \sin x,$$

involves two known quantities t and c, and the solution is to be given as x as a power series in t with c as a parameter. Lagrange solved the problem geometrically, contrary to his usual practice, but in 1780 Laplace solved it algebraically, and Lagrange then used the result himself. After the formal work, the issue was again one of convergence of the series for the inverse function, for which, Lagrange observed, c had to be small. The problem was a difficult one, and Laplace returned to it again in 1825, which eventually caught the attention of Cauchy.

As we saw in Chap. 6 the late 1820s were a period when Cauchy was interested in complex variables and complex functions, and so he cast the problem of the reversion of series in that form and applied his residue calculus to the discussion of the convergence of the series of the inverse function. Once complex function theory was established, a full generation later, it was clear that the convergence of a power series in a real variable is determined by what happens on its maximal circle of convergence, which may well be determined by what happens for complex values of the variable (as is the case with $(1+x^2)^{-1}$). But that was not Cauchy's idea, and he used his residue calculus only to handle the formal side of estimating convergence.

Indeed, the story of the implicit and inverse function question so far illustrates the widespread belief among mathematicians that formal methods, that accommodate both real and complex values for variables, would validly take you a long way, although of course some further understanding of the convergence of series would be required, especially in applications. This approach had a lot to recommend it, but it was an obstacle to the clear understanding of what complex differentiability means and its implications for the representation of functions, and it meant that crucial distinctions between the real and the complex implicit function theorem were obscured.

26.3 Cauchy

When, then, was the question of the existence of an implicit function first addressed? Unsurprisingly, the answer is by Cauchy, not in these papers but, if only obliquely, in the lectures on differential equations that he gave at the École Polytechnique in 1820. Here he investigated when a differential equation of the form

$$\frac{dy}{dx} = f(x, y) \tag{26.1}$$

has a solution that took a particular value y_0 at a particular point x_0. After dealing with many examples, in Lectures 7 and 8 he showed that if the function f is continuous and bounded as a function of x and y on the interval $x_0 \leq x \leq X$ then there is a function y of x that satisfies the differential equation on some interval of the form $(x_0, x_0 + h)$ and for which $y(x_0) = y_0$.[1]

His argument was that the solution to the Eq. (26.1) is very close to the collection of points (x_j, y_j) given by

$$y_j - y_{j-1} = (x_j - x_{j-1}) f(x_{j-1}, y_{j-1})$$

when the points x_j are close together. This is what you would expect, because when $x_j - x_{j-1}$ is small the quotient

$$\frac{y_j - y_{j-1}}{x_j - x_{j-1}}$$

should be very close to the value of $\frac{dy}{dx}$ and so the chord joining (x_{j-1}, y_{j-1}) to (x_j, y_j) is very close to the graph of the sought-for solution of the differential equation

[1] To control what happens when the choice of initial value y_0 is varied Cauchy added the condition that the partial derivative $\frac{\partial f}{\partial y}$ is also continuous and bounded as a function of x and y on the interval $x_0 \leq x \leq X$, but we need not pursue that here.

at the point (x_{j-1}, y_{j-1}). Then, as the points x_j get closer and closer together, the conditions on the function $f(x, y)$ are sufficient to ensure convergence, the function y is represented by the integral of $f(x, y)dx$, and the theorem is proved.

Why is this connected to the idea of implicit and explicit functions? Consider the implicit function defined by

$$g(x, y) = x^2 + y^2 - 1 = 0,$$

for which

$$\frac{\partial g}{\partial x} = 2x \quad \text{and} \quad \frac{\partial g}{\partial y} = 2y.$$

It is elementary that the differential equation

$$\frac{dy}{dx} = -\frac{x}{y}$$

has $g(x, y)$ as the unique solution that passes through the point $(x, y) = (0, 0)$. But this gives the solution in implicit form; what Cauchy did, however, was to show that this equation has a solution in the form $y = h(x)$, thus exhibiting the solution as an explicit function of x.

Ten years later, Cauchy was the first person to tackle the question directly of when an implicitly defined function $f(x, y) = 0$ can be written locally as an explicit function $y = g(x)$, which he did in the memoir he wrote in Turin in 1831.[2]

In what follows we consider an equation $f(x, y) = 0$ defined on a suitable domain that contains the point (a, b) for which $f(a, b) = 0$, and look for an explicit function $y = g(x)$ such that $f(x, g(x)) = 0$ and $g(a) = b$. We shall also show that under suitable conditions the explicit function $g(x)$ is an analytic function of x on its domain.

Cauchy's proof is far from clear, and it will be helpful to suppose that $(a, b) = (0, 0)$. On the assumption that $y = 0$ is a root of $f(0, y) = 0$ of multiplicity 1, the absolute values of all other roots of the equation $f(0, y) = 0$ are greater than a certain amount, which can be denoted Z. Cauchy aimed to show that under these conditions and for each fixed value of x in a certain interval, the equations $f(x, y) = 0$ and $f(0, y) = 0$ have the same number of roots—one in each case—and so y is a single-valued function of x on that interval.

Here is a free translation of what he wrote. I have imposed the simplification $b = 0$.

Cauchy's (1831)

Now let y be an implicit function of x determined by an equation of the form $f(x, y) = 0$, and $b = 0$ be a value of y that corresponds to a particular value

[2]This account follows the version that he published in 1841. See also Bottazzini and Gray (2013, 150–155).

of x. If one sets $y = b + z = z$, we have $f(x, z) = 0$. This done, denote by $f_y(x, y)$ the [partial] derivative of f with respect to y. Suppose moreover that

- the equation $f(0, z) = 0$ has a unique real or imaginary root z such that $|z| < Z$ for some Z,
- the function $f(x, z)$ has a unique and definite value for every real or complex value of z such that $|z| < Z$,
- for such a value of z the function $F(y) = F(z)$ never becomes discontinuous or infinite.

Then, on writing

$$\tilde{z} = Ze^{qi}$$

for a real arc q one will have the equation

$$F(y) = \frac{1}{2\pi}\int_{-\pi}^{\pi} \tilde{z}\frac{f_y(x, \tilde{z})}{f(x, \tilde{z})}F(\tilde{z})dq.$$

So it remains to deduce this result from [the Cauchy integral theorem, established at the start of the memoir], which one does as follows.

When $z = \tilde{z}$ both terms of the binomial

$$\frac{f_y(x, \tilde{z})}{f(x, \tilde{z})}F(\tilde{z}) - \frac{F(z)}{(\tilde{z} - z)}$$

become infinite; but the binomial itself generally has a finite value, to wit

$$F_z(z) + \frac{1}{2}\frac{F(z)}{f_y(x, z) f_{xz}(x, z)}.$$

Therefore, if the modulus Z of z is chosen in such a way that the above statements are satisfied, namely

1. the equation $f(x, z) = 0$ has a unique root z such that $|z| < Z$,
2. $F(y) = F(z)$ never becomes discontinuous or infinite for any z such that $|z| \leq Z$,

then the product

$$\tilde{z}\left(\frac{f_y(x, \tilde{z})}{f(x, \tilde{z})}F(\tilde{z}) - \frac{F(z)}{(\tilde{z} - z)}\right)$$

will be a continuous function of $\tilde{z}$ for the stated value of Z or any smaller value.[3]

Now, by the earlier result

$$\int_{-\pi}^{\pi} \tilde{z}\frac{f_y(x, \tilde{z})}{f(x, \tilde{z})}F(\tilde{z})dq = F(z)\int_{-\pi}^{\pi} \frac{zdq}{\tilde{z} - z}$$

[3]The opaque notation is Cauchy's.

$$= 2\pi F(z) = 2\pi F(y),$$

which is the formula we want. If one sets $F(y) = 1$ this gives

$$\frac{1}{2\pi}\int_{-\pi}^{\pi} \tilde{z}\frac{f_y(x,\tilde{z})}{f(x,\tilde{z})}dq = 1. \tag{26.2}$$

If, on the contrary, one sets $F(y) = y$ one concludes that

$$y = \frac{1}{2\pi}\int_{-\pi}^{\pi} \tilde{z}(\tilde{z})\frac{f_y(x,\tilde{z})}{f(x,\tilde{z})}dq,$$

that is,

$$y = \frac{1}{2\pi}\int_{-\pi}^{\pi} \tilde{z}^2\frac{f_y(x,\tilde{z})}{f(x,\tilde{z})}dq. \tag{26.3}$$

[End of the free translation.]

Cauchy then proved two theorems that allowed him to infer that the equations $f(x,z) = 0$ and $f(0,z) = 0$ have the same number of zeros for all x less than a certain amount. In the case when this number is 1 it follows that for each x in a suitable interval around $x = 0$ there is a unique z for each x, and so the equation $f(x,z) = 0$ defines an implicit function $z = g(x)$.

Here is the second of Cauchy's theorems in the case when we are dealing with a unique root.

Theorem IV If the equation $f(0,z) = 0$ has a unique root of modulus less than Z and moreover

1. for all x such that $|x| < X$ [for some non-zero real X] and z such that $|z| < Z$ the function $f(x,z)$ always has a unique and definite value,
2. for x such that $|x| < Z$ the function $f(x,\tilde{z})/f(0,\tilde{z})$ is developable as a convergent series in ascending powers of x for all x such that $|x| < X$ [i.e. the function is complex analytic, as Cauchy would later realise].

Then for such an x the equation $f(x,z) = 0$ has a unique root of modulus less than Z.

It will help to write

$$\frac{f_y(x,\tilde{z})}{f(x,\tilde{z})} = h(x,\tilde{z}).$$

Cauchy's insight was to expand $h(x,\tilde{z})$ as a power series in X:

$$h(x,\tilde{z}) = h(0,\tilde{z}) + x\frac{du_1}{d\tilde{z}} + x^2\frac{du_2}{d\tilde{z}} + \cdots,$$

where the seemingly artificial expressions for the coefficients of the powers of x are obtained are the derivatives of $h(x, \tilde{z})$ at $x = 0$ and so are expressions in the derivatives of $f(x, \tilde{z})$. But because u_1 and u_2 vanish, the result is that

$$\int_{-\pi}^{\pi} h(x, \tilde{z}) = \int_{-\pi}^{\pi} h(0, \tilde{z}),$$

But the expression on the left counts the number of zeros of $h(x, \tilde{z})$ inside $|x| < X$, so it is is equal the number on the left, and by Eq. (26.3) this is 1.

A further use of the Cauchy representation theorem, using the special cases above $(F(y) = 1,\ F(y) = y)$, allowed Cauchy to deduce that this root is expressible as a convergent power series in x. This uses nothing more than the fact that the function $h(x, \tilde{z})$ is analytic, which follows from the fact that $f(x, y)$ is, in ways Cauchy did not specify, well-behaved (say, in later terms, analytic in x and y).

Cauchy went on to prove the theorem that is usually called Rouché's theorem in complex analysis today, which says that if the analytic functions f and g are defined on an open disc and at all points z on the boundary of the disc we have

$$|f(z) - g(z)| < |f(z)|,$$

then the number of zeros and f and g in the open disc are the same when counted with multiplicities. Rouché gave this result in his (1862), noting as he did so that his presentation was more precise than anything that could be found in Cauchy's papers. Cauchy's argument was indeed messily presented, but the theorem is nonetheless present in Cauchy (1831), see his Theorem VII, p. 93.

But what had Cauchy proved? One entirely correct modern answer (see e.g. (Krantz and Parks 2002, 27–31)) is that he had proved the implicit function theorem for holomorphic functions—although not entirely rigorously or convincingly. But it is noticeable here, as it is throughout his work in the 1830s, that he had not appreciated the difference between real and complex differentiability. We can make his theorem precise by tying it to the holomorphic domain, but at the time Cauchy had no idea what the extent of that domain was.

The first person to be clear about the implicit function in the holomorphic context from start to finish was Weierstrass, in his lectures in Berlin in 1861. They survive in unpublished notes by Schwarz[4] and other writers, such as Pincherle, who was an enthusiast for Weierstrassian methods. He remarked[5] (Pincherle 1922, 209) that the theorem

> was first proved by Cauchy, then a little later by Weierstrass with more precision and without recourse to curvilinear integrals. The method I am about to give, that proceeds more fluently and efficiently, is a mixed method.

Pincherle considered a function $F(x, y) = 0$ that was regular and analytic in both variables when x is in a domain A and y in a domain B, and for which $F_y(x, y)$

---

[4]Mentioned without further discussion in Bottazzini and Gray (2013, 378).

[5]He used almost the same words in the edition of his lecture notes at Bologna in 1901.

does not vanish. He let (a, b) be a point with $a \in A$ and $b \in B$ and sought to define a function $y = \alpha(x)$ such that $\alpha(a) = b$ and $F(x, \alpha(x) = 0$ and α is regular and analytic for all $x \in A$.

He first observed that it is enough to prove the theorem when $a = 0 = b$. He expanded the function $F(x, y)$ as a power series

$$F(x, y) = y\varphi(y) + f(x, y),$$

where $\varphi(y)$ is given by a power series that does not vanish when $y = 0$:

$$\varphi(y) = c_0 + c_1 y + c_2 y^2 + \cdots .$$

Clearly, $f(0, 0) = 0$. He took logarithms of both sides and obtained

$$\log F(x, y) = \log y + \log \varphi(y) + \log \left(1 + \frac{f(x, y)}{y\varphi(y)}\right)$$

which he wrote as

$$\log F(x, y) = \log y + \log \left(1 + \frac{c_1}{c_0} y + \frac{c_2}{c_1} y^2 + \cdots \right) + \log \left(1 + \frac{f(x, y)}{y\varphi(y)}\right). \quad (26.4)$$

Pincherle now chose a number ρ_1 such that $|y| < \rho_1$ implies $|\varphi(y)| > A = c_0/2$, and another number ρ_2 such that $|y| < \rho_2$ implies $\left|\frac{c_1}{c_0} y + \frac{c_2}{c_1} y^2 + \cdots \right| < q < 1$, and a value for ρ' strictly less than ρ, the minimum of ρ_1 and ρ_2. Then, in the annulus $\rho' < |y| < \rho$ one has $|y\varphi(y)| > \rho' A$.

Then he chose a number r such that $|x| < r$ implies that $|f(x, y)| < \rho' A\mu$, for all y in the annulus and for some $\mu < 1$. Under these conditions ($|x| < r, \rho' < |y| < \rho$) the term $\log \left(1 + \frac{f(x,y)}{y\varphi(y)}\right)$ can be written as a uniformly convergent Laurent series in negative and positive powers of y in which the coefficients are analytic functions of x regular in the disc $|x| < r$.

The term $\log \left(1 + \frac{c_1}{c_0} y + \frac{c_2}{c_1} y^2 + \cdots \right)$ is regular and analytic for $|y| < \rho$, and so Eq. (26.4) can be written in the form

$$\log F(x, y) = \log y + \log c_o + \sum_{j=-\infty}^{\infty} a_j(x) y^j .$$

Pincherle differentiated it with respect to y and obtained

$$\frac{F_y'(x, y)}{F(x, y)} = \frac{1}{y} + \sum_{j=-\infty}^{\infty} j a_j(x) y^{j-1} .$$

He then remarked that because the second term on the right hand side is the derivative of a Laurent series it cannot contain a term in y^{-1} and so the residue of the logarithmic derivative of $F(x, y)$ is 1. But, by standard results in complex function theory that he had proved earlier, it follows that $F(x, y)$ has a unique root of first order $\tilde{y}$, for each $\tilde{X}$ such that $|\tilde{x}| < r$ given by

$$\frac{1}{2\pi i}\int \frac{yF'_y(x, y)}{F(x, y)}dy,$$

around any circle centre the origin of radius less than r. Moreover, this root, which is the coefficient of y^{-2} in the term on the right hand side, can be written as $\tilde{y} = -a_{-1}(\tilde{x})$ and is a regular analytic function on $|x| < r$. This proves the theorem.

The improvement that Pincherle imported was the use of Laurent series, which Weierstrass avoided because the most natural approach to them relied on the theory of the Cauchy integral and Weierstrass always endeavoured in his lectures in Berlin in using that body of ideas.

26.4 Dini

Clarity about the distinction between the real and complex cases of the implicit function theorem only came with the work of Dini in the 1870s, specifically his *Lezioni di analisi infinitesimale* vol. 1 1877 (Ch XIII, pp. 197 et seq.). He also gave the first rigorous extension of the theorem to several variables. In fact, and as a result, the implicit function theorem is known in Italy as Dini's theorem.

We shall follow him only in the simplest case. Dini supposed there was a function $f(x, y) = 0$ defined when x and y were in some region C—we can suppose x and y lie in two intervals. Furthermore, x_0 and y_0 in the appropriate intervals are such that $f(x_0, y_0) = 0$. The function $f(x, y)$ is finite and continuous on this region, it has both first partial derivatives on the region, and $\frac{\partial f}{\partial y}$ is non-zero. He then set about showing that for any value of $x_0 + h$ sufficiently close to x_0 there is a value y_h of y such that $f(x_0 + h, y_h) = 0$. The function $y = y(h)$ will therefore be the required explicit function.

He supposed that there were positive numbers h_0 and k_0 such that $|h| < h_0$, $|k| < k_o$ implies that the point $(x_0 + h, y_0 + k)$ is in C. He wrote

$$f(x_0+h, y_0+k) = f(x_0+h, y_0+k) - f(x_0, y_0+k) + f(x_0, y_0+k) - f(x_0, y_0),$$

using the fact that $f(x_0, y_0) = 0$, and so

$$f(x_0 + h, y_0 + k) - f(x_0, y_0 + k) = hf_x(x_0 + \theta h, y_0 + k)$$

and

$$f(x_0, y_0 + k) - f(x_0, y_0) = kf_y(x_0, y_0) + \alpha,$$

where $0 < \theta < 1$ and α is a number that, when k_0 is sufficiently small but not zero, can be made arbitrarily small. Therefore

$$f(x_0 + h, y_0 + k) = hf_x(x_0 + \theta h, y_0 + k) + k(f_y(x_0, y_0) + \alpha).$$

He then observed that because $f_y(x_0, y_0)$ is non-zero, and for k_0 and α sufficiently small, one can say that for all $k < k_0$ the quantity $f_y(x_0, y_0)+\alpha$ will always have the same sign as $f_y(x_0, y_0)$ and is greater than zero by at least an amount d. Therefore $kf_y(x_0, y_0) + \alpha$ takes values greater in absolute value than $k_0 d$ as k goes from $-k_0$ to $+k_0$.

He then defined λ as the lim sup of all the absolute values of $f_x(x, y)$ for x and y in the region considered, and noted that

$$|f(x_0 + h, y_0 + k)| < h_0\lambda.$$

Because $h_0\lambda < k_0 d$, if need be by choosing a smaller value of h_0, it follows that $f(x_0 + h, y_0 + k)$ goes from negative to positive as k goes from $-k_0$ to k_0, and so, because it is continuous, there must be at least one value of k in that range for which $f(x_0 + h, y_0 + k) = 0$.

From this it follows that for each value of h such that $-h_0 < h < h_0$ that there is a value of k such that $f(x_0 + h, y_0 + k) = 0$. Therefore, for each allowed value of x there is one or more function y that satisfies the equation $f(x, y) = 0$ and for which when $x = x_0$ we have $y = y_0$. So, provided $f_y(x, y)$ is non-zero in some subregion of C containing (xc_0, y) there is exactly one value of k and therefore one function y.

He concluded that

> In the interval $(x_0 - h_0, x_0 + h_0)$ there is a well-determined, single-valued, real function $y = y_0 + k$ that can be regarded as a function implicitly defined by the given equation $f(x, y) = 0$.

26.5 Conclusions

The implicit function theorem has been little studied by historians of mathematics; even the richly informative book (Krantz and Parks 2002) does not make it entirely clear how mathematicians made their way to it in the first instance. But it illustrates rather tidily some of the themes of any history of analysis. There was a prolonged phase when the possibility was taken for granted and the emphasis was placed on doing it in an effective way. This came to involve power series and questions of convergence, and this shows just how confused people were about real and complex differentiability, even when, as Cauchy was, they were clear that there was something to prove. Only quite late, after a proof in the complex domain, was a proof given of an implicit function in the real domain for functions satisfying a modest set of natural conditions.

Of course, the theory did not stop with Dini's work. There are versions of the theorem for several variables in the 19th century, and many developments in the 20th, as Krantz and Parks describe—but these must be themes for later historians.

Chapter 27
Towards Lebesgue's Theory of Integration

27.1 Introduction

This chapter looks briefly at the Lebesgue integral, to see what questions it answered and what it is good for. The first half takes up two problems in analysis before 1900, the second half looks at how they, and the problem of the fundamental theorem of the calculus, were resolved using Lebesgue's theory.

This chapter and the next two one are offered as pointers to some of the ways analysis was to develop in the 20th century, and to branch into new domains, abstract set theory and topology.

27.2 Problems

The Length of Curves

Consider the question of the length of a curve. A curve that has a length is said to be rectifiable, meaning that it can be straightened out without stretching it and made to measure its length along a line. The question became: what curves have lengths? Indeed, does every curve have a length?

The second question can be disposed of simply. The possibly surprising answer is that there are even simple closed curves for which the answer is 'No'—for example, the von Koch snowflake curve. The curve was introduced by Helge von Koch in 1904, and published more accessibly in his (1906), because he wanted to illustrate Weierstrass's discovery of a continuous, nowhere differentiable, function with a geometrical, rather than an analytic, example.

To obtain the curve, start with an equilateral triangle of side 1. The rule for generating each successive approximation is this: regard each line segment of length a as being in three equal parts; place an equilateral triangle of side $a/3$ on the interval

J. Gray, *The Real and the Complex: A History of Analysis in the 19th Century*, Springer Undergraduate Mathematics Series, DOI 10.1007/978-3-319-23715-2_27

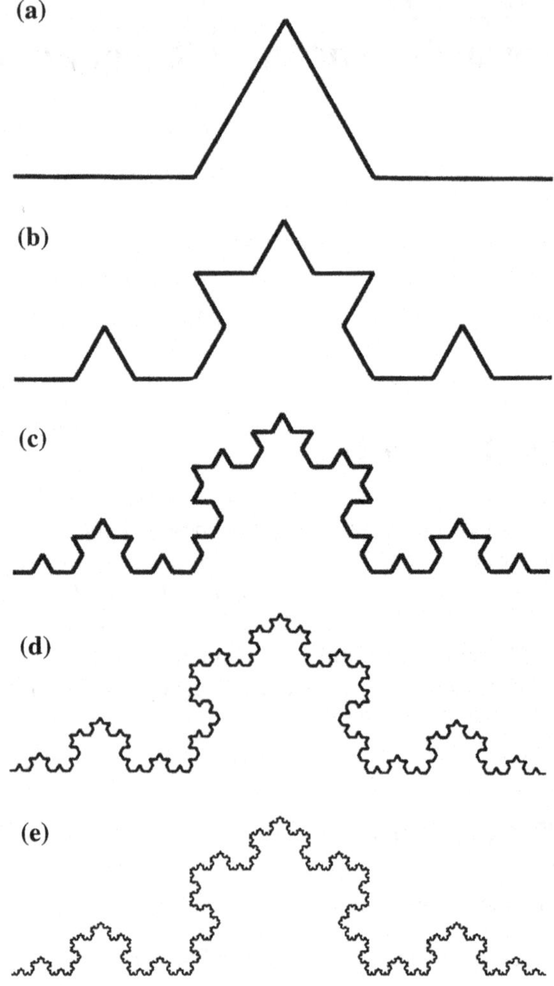

Fig. 27.1 Growing an edge of the von Koch snowflake curve

in the middle and then remove the middle interval. The first few stages are shown in Fig. 27.1 as they affect just one side.

And here is an impression of the curve (Fig. 27.2). Now you can see why it has the name it does:

Now, the original curve (the equilateral triangle) has a length of 3. At the first stage, each edge is replaced by a piece of the new curve of length 4/3, so the next figure after the triangle has length $3 \times 4/3 = 4$. Indeed, at every stage the length of the curve goes up by 4/3, so the length of the von Kochcurve can only be infinite, which is another way of saying it has a length.

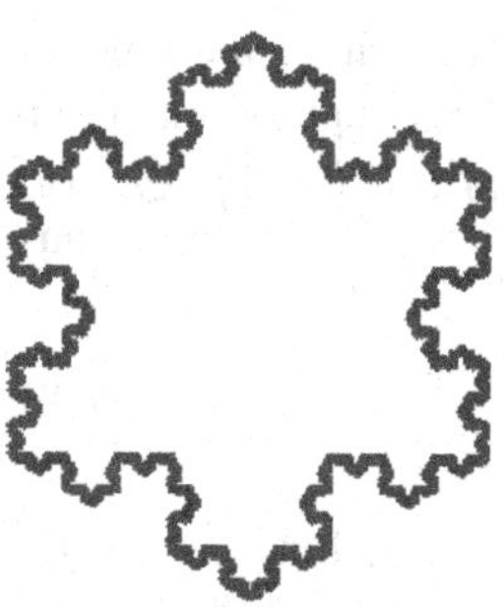

Fig. 27.2 The von Koch snowflake

But what is the right way to define the length of a curve? Suppose a curve is defined as $y = f(x),\ x \in [a, b]$. Elementary calculus says that the length is given by the integral $\int_a^b \left(1 + (f(x)')^2\right)^{1/2} dx$. This is defended by considering a sequence of points $(t_i, f(t_i))$ on the curve and arguing that when $t_{i+1} - t_i = dx_i$ and therefore $f(t_{i+1}) - f(t_i) = dy_i$ the length of the interval joining successive points on the curve is a good approximation to the length of the curve between those two points, and so the length of the curve between a and b is well approximated by the sum of these quantities, and then proceeding to the limit as i increases indefinitely. Since each interval has length $\left(dx_i^2 + dy_i^2\right)^{1/2} = \left(1 + \frac{df}{dx}(x_i)^2\right)^{1/2} dx$ the familiar calculus expression follows.

Now it might be argued that the geometry is good, and the length of the curve is given as the limit of a sum of lengths of intervals, but that the introduction of the derivative f' is spurious, and that a curve might have a length even though it is not differentiable. After some tentative explorations by du Bois-Reymond and Otto Stolz, Ludwig Scheefer took up the question. In his (1884) he claimed that there are indeed curves which have a length, but for which the derivative oscillates in every interval and so the integral has no meaning as a Riemann integral.

Scheefer's Function

Scheefer's example went as follows. Let (s_r) and (w_r) be two sequences of real numbers and suppose that each $s_r > 0$ and Σs_r converges. Define a function f as follows:

$$f(x) = \sum_{-\infty}^{x} s_r$$

where the summation is taken over all s_r such that $w_r < x$. The function is monotonic increasing, since

$$f(w_r+) - f(w_r-) = f(w_r+) - f(w_r) = s_r.$$

At each w_r the function f has a jump discontinuity of s_r as you can easily check. Now suppose that the w_r form a dense subset of (a, b). The image set $f([a, b])$ is $[f(a), f(b)] - \bigcup_{r=1}^{\infty}(f(w_r), f(w_{r+}))$.

Scheefer defined an 'inverse' function $g = f^{-1}$ on $[f(a), f(b)]$ as follows:

$$g(y) = \begin{cases} f^{-1}(y) & \text{if } y \in [f(a), f(b)] \\ w_r & \text{if } y \in (f(w_r), f(w_r+)] \end{cases}$$

The function g is flat on $(f(w_r), f(w_{r+1}))$, so $g' = 0$ in each of these intervals. The sum of the length of these intervals is $\Sigma_{r=1}^{\infty} s_r = f(b) - f(a)$ so $f([a, b])$ is a discrete subset of $[f(a), f(b)]$, and so g' is zero except on a discrete set. On the other hand, the length of the curve $y = g(t)$ plainly is $\Sigma_{r=1}^{\infty} s_r = f(b) - f(a)$, so this is an example of a curve with a definite length that is not to be evaluated by the familiar integral.

This is reminiscent of a function that defines what is sometimes called the 'devil's staircase' and is described at the end of this chapter.

A breakthrough in this subject, although it was not entirely appreciated by its author, had earlier been made by Camille Jordan in 1881 when he introduced the concept of functions of bounded variation. We say that a function $f : [a, b] \to \mathbb{R}$ is of bounded variation if for every partition of $[a, x]$, say $a = t_0 < t_1 < \cdots < t_{n-1} < t_n < x$, the sums $\Sigma'(f(t_i) - f(t_{i-1})$ and $-\Sigma''(f(t_i) - f(t_{i-1}))$, taken respectively over the increasing and decreasing segments, have finite least upper bounds for all values of x in $(a, b]$. Let these bounds be $P(x)$ and $N(x)$ respectively. Then

$$f(x) - f(a) = \Sigma(f(t_i) - f(t_{i-1})) = \Sigma'(f(t_i) - f(t_{i-1}) - \left(-\Sigma''(f(t_i) - f(t_{i-1})\right)$$

and so, in the limit,

$$f(x) - f(a) = P(x) - N(x)$$

which shows that the function f is the difference of two nondecreasing functions. The converse is trivial, so Jordan had shown that a function is of bounded variation if and only if it is the difference of two nondecreasing functions.

Jordan published this work in a note attached to the third volume of the first edition of his *Cours d'analyse*, (1887), and again in the first volume of the second edition, (1893), so it was widely read. He drew the corollary that a curve $y = f(x)$ is rectifiable if and only if the function f is of bounded variation, and in 1893 he also noted that if a curve $(x(t), y(t))$ is a rectifiable curve for which x' and y' exist and are continuous, and $s(t)$ denotes the length of the curve from, say, t_0 to t, then $s'(t) = \left(x'(t)^2 + y'(t)^2\right)^{1/2}$. But curiously, he did not deduce that the classical integral formula therefore applied in this case. Plainly he thought it had been shown to be irrelevant to the study of lengths of curves.

27.3 The Area of Surfaces

The area of a surface is seemingly evaluated by the calculus in a perfectly routine way. Suppose that the surface is given in vector form as $\mathbf{r}(u, v)$. Any elementary textbook in differential geometry will tell you that the area is obtained by integrating $\left(EG - F^2\right)^{1/2}$ where

$$E := \mathbf{r}_u.\mathbf{r}_u, \quad F := \mathbf{r}_u.\mathbf{r}_v, \quad G := \mathbf{r}_v.\mathbf{r}_v.$$

But is this the best definition? Why, indeed, is it true?

The rationale for this definition is that the expression $\left(EG - F^2\right)^{1/2}$ is the area of the image of an infinitesimal square in the (u, v) domain, and of course the area of the surface is found by adding up (i.e. integrating) these infinitesimal distorted squares over the surface in question. Now, the analogy with the question of the lengths of curves is fairly clear. On a curve one wants to attach rulers as successive points, approximate the length that way, and then steadily refine the approximation so that the limit is a good definition of the length of a curve. So, on a surface, why not attach lots of triangles and proceed in the same way? This would mean that at each stage the area of the surface is approximated by the sum of the areas of the triangles.

Schwarz's Paradox

It was Schwarz who found a most disturbing paradox by looking at this naïve idea carefully. It had been given in a textbook by the French mathematician Serret, so Schwarz communicated his observation to Hermite, who promptly put in the notes for his *Cours d'analyse* for 1881. Schwarz (1881–1882) considered a cylinder (you can almost hear the magician saying "Nothing tricky here, just an ordinary cylinder, nothing special.") and considered it as the image of a map from the rectangle $[0, 2\pi] \times [0, h]$ that sends (u, v) to $(r \cos u, r \sin u, v)$, where r and h are constants (the radius and height of the cylinder). This cylinder has, of course, the area $2\pi rh$. He then inscribed $4mn$ triangular faces on the cylinder as follows: the vertices are the images of these points on the rectangle (Fig. 27.3):

$$\left(\frac{2\mu\pi}{m}, \frac{\nu h}{n}\right), \left(\frac{2(\mu+1)\pi}{m}, \frac{\nu h}{n}\right), \left(\frac{(2\mu+1)\pi}{m}, \frac{(2\nu+1)h}{n}\right);$$

or

$$\left(\frac{(2\mu-1)\pi}{m}, \frac{(2\nu+1)h}{n}\right), \left(\frac{(2\mu+1)\pi}{m}, \frac{(2\nu+1)h}{n}\right), \left(\frac{(2\mu+1)\pi}{m}, \frac{2\nu h}{n}\right).$$

All these triangles are isosceles and congruent, and they each have area

$$r \sin\left(\frac{\pi}{m}\right) \cdot \sqrt{r^2\left(1 - \cos\left(\frac{\pi}{m}\right)\right)^2 + \left(\frac{h}{2n}\right)^2}.$$

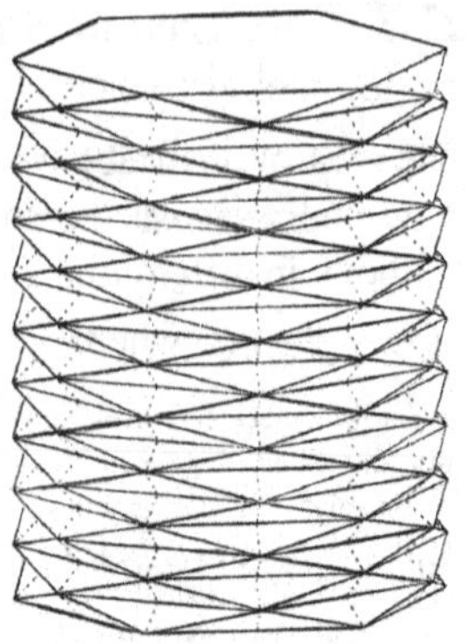

Fig. 27.3 Triangles on a cylinder

So the area of the collection of triangles is

$$4mnr\sin\left(\frac{\pi}{m}\right)\cdot\sqrt{r^2\left(1-\cos\left(\frac{\pi}{m}\right)\right)^2+\left(\frac{h}{2n}\right)^2}$$
$$=4mnr\sin\left(\frac{\pi}{m}\right)\sqrt{4r^2\sin^4\left(\frac{\pi}{2m}\right)+\left(\frac{h}{2n}\right)^2}. \qquad (27.1)$$

It is now desired to take the limit as the number of triangles increases, which happens as m and n tend to infinity. However, Schwarz observed, if one lets $m = an$ this limit is $2\pi rh$, which is indeed the expected value (the dominant term in the square root is the $\frac{h}{2am}$). But if one sets $n = am^2$ then this limit is the limit of

$$2\pi r\sqrt{a^2r^2\pi^2+h^2}$$

which depends on a and can therefore be arbitrarily large. Likewise, if $n = am^3$ the area is the limit of

$$8r^2m^4\sin\left(\frac{\pi}{m}\right)\sin^2\left(\frac{\pi}{2m}\right)$$

and again the limit can be arbitrarily large.

The naïve definition of area is therefore erroneous, and Schwarz had no suggestion as to how to put it right. Although Peano came along a few years later, as did Hermite, with suggestions for what could be done, their ideas were conspicuously patches and the damage remained done. It looked as if there was no satisfactory definition of the area of even the simplest curved surfaces.[1]

[1] For a thorough account of the contributions of Peano and Lebesgue, see Gandon and Perrin (2009).

27.4 Lebesgue's Answers

The length of a curve was defined by Henri Lebesgue as the smallest limit of the lengths of polygonal lines tending uniformly to the curve. He then showed that if the curve is given parametrically as

$$x = f(t), \quad y = \phi(t), \quad z = \psi(t),$$

then when f, ϕ, and ψ have derivatives, a necessary and sufficient condition for a curve to have a length is that the integral of $\sqrt{\left(f'(t)^2 + \phi'(t)^2 + \psi'(t)^2\right)}$ exists.

Similarly, Lebesgue defined the area of a surface L as the least limit of the areas of polyhedral surfaces that tend uniformly to L. He then showed that in the special case where the surface has continuously varying tangent planes the area is indeed given by the formula $\int\int \sqrt{EG - F^2} dudv$.

It took much longer for Lebesgue to resolve the problems with the fundamental theorem of the calculus. Recall that this can be phrased in three different ways, concerning a function $f(x)$ defined on an interval $[a, b]$, and the function $F(x) := \int_a^x f$.

- I: F is a primitive function for f, that is $F' = f$
- II: Every primitive function of f is of the form $\int_a^x f + C$
- III: (a result used by Cauchy to prove II): If G is such that $G'(x) = 0$ on $[a, b]$ then $G(x)$ is constant on $[a, b]$.

Lebesgue could now prove the fundamental theorem of the calculus (II) for functions with bounded derivatives, which had been a problem before because of the existence of bounded derivatives that are not Riemann-integrable. First he was able to show that, on his definition of measurable sets, if the function f is the limit of a sequence of measurable functions (f_n), then F is also measurable. Notice that this is not true of Riemann-integrable functions. Then he showed that if, in addition, the functions $|f_n(x)|$ are all bounded by a number B, then $\int f = \lim_{n\to\infty} \int f_n$. Finally he applied these ideas to the functions $\frac{f(x+h)-f(x)}{h}$ as follows:

$$\int_a^b f' = \lim_{h\to 0} \int_a^b \frac{f(x+h) - f(x)}{h} = \left[\lim_{h\to 0} \frac{1}{h} \int_x^{x+h} f \right]_a^b = f(b) - f(a).$$

This still left the fundamental theorem of the calculus in the form $\left(\int_a^x\right)' = f(x)$.

In 1902–1903 Lebesgue was invited to give the prestigious Cours Peccot Lectures at the Collège de France. He found the opportunity stimulating, and the published version, his *Leçons sur l'intégration et la recherche des fonctions primitives* (1904) carries a fine axiomatisation of the idea of integration. Lebesgue showed that there was essentially only one good definition of the integral, it satisfied these natural axioms, and his definitions exemplified them. The axioms are:

1.
$$\int_a^b f(x)dx = \int_{a+h}^{b+h} f(x-h)dx;$$

2.
$$\int_a^b f + \int_b^c f + \int_c^a f = 0;$$

3.
$$\int_a^b (f+g) = \int_a^b f + \int_a^b g;$$

4.
$$\text{If } f(x) \geq 0 \text{ and } b > a \text{ then } \int_a^b f \geq 0;$$

5.
$$\int_0^1 1 = 1;$$

6.
$$\text{If } f_n(x) \uparrow f(x) \text{ then } \int_a^b f_n \uparrow \int_a^b f.$$

Lebesgue could now show that if any of the four Dini derivatives Df of f are integrable then $\int_a^b Df = f(b) - f(a)$.

It is interesting to see that the Dirichlet function on [0, 1] that takes the value 1 at rational points and 0 at irrational points is Lebesgue integrable by considering the function

$$\lim_{m\to\infty}\left(\lim_{n\to\infty}(\cos(m!\pi x))^{2n}\right).$$

This example is due to Lebesgue himself.

By now numerous people were responding to Lebesgue's ideas, and finally in 1907 Lebesgue came to the triumphant result that the fundamental theorem of the calculus in the form I can be proved, indeed for any integrable (but not necessarily continuous) function f:

$$\left(\int_a^x f\right)' = f(x)$$

almost everywhere.

That is to say, equality holds except on a set of measure zero, one negligible for the purposes of education.

This is a good place to end what might seem like an extended advertisement for the Lebesgue integral.

Hausdorff's Paradox

The mathematician and philosopher Felix Hausdorff should, however, have the last word. Hausdorff was a radical in his approach to mathematics. He gave a lecture course on Time and Space in 1903–1904 in which he said that[2]

> Mathematics totally disregards the actual significance conveyed to its concepts, the actual validity that one can accord to its theorems. Its indefinable concepts are arbitrarily chosen objects of thought and its axioms are arbitrarily, albeit consistently, chosen relations among these objects. Mathematics is a science of pure thought, exactly like logic.

Hausdorff's interests in mathematics deepened after 1911. He was the first to take set theory up energetically since Cantor, and his major work, important for both set theory and topology, is his *Grundzüge der Mengenlehre* (1914). Hausdorff published what became known as his paradox in the next year, 1915.[3] It states that on any definition of measure that satisfies Lebesgue's original four axioms, there will be bounded sets that are not measurable. To derive a contradiction, Hausdorff assumed that every bounded set is measurable. He took the sphere in three-dimensional space, and divided it into four disjoint sets A, B, C, D with the property that A, B, C and $B \cup C$ are all congruent, while D has measure zero. Because the sets A, B, and C are congruent, they have equal measure, and because they cover the sphere except for the set D of measure zero, each must have measure 1/3. But the sets A and $B \cup C$ are also congruent and cover the sphere except for the set of measure zero, so they must have equal measure, which must therefore be 1/2. It follows that the set A has measure 1/2 and 1/3, which is impossible. Accordingly, the assumption that there is some definition of measure according to which every bounded set is measurable is false.

How could mathematicians respond to this unsettling and paradoxical argument? They could agree that's how it is—there will always be non-measurable sets, which is a limitation on the uses of measure theory that might prove worrying; they could reject the concept of measure and look for another concept altogether—but Lebesgue's four axioms are very natural ones with many successes to their credit; or they could scrutinise the proof and hope to find a flaw. Borel took this third way out. Hausdorff had used the axiom of choice in his proof, and Borel replied that the paradox came about not because measure was an inherently flawed concept, but because the set A was not properly defined.[4] To construct a set using the axiom of choice was for Borel no construction at all. "If one scorns precision and logic", he wrote, "one arrives at contradictions".[5] Hausdorff, of course, was not persuaded. He was happy with the axiom of choice and simply not bothered that any definition of the area of a set is inherently imperfect—even though this was a conclusion that could never have been dreamt of by researchers a generation before.

---

[2]See Corry (2006, 148).

[3]Hausdorff (1915).

[4]See the second edition of his *Leçons sur la théorie des functions* (1914).

[5]Borel (1914, 256), quoted in Moore (1982, 188), who also notes that a number of Italian mathematicians had already explicitly rejected the idea that one can make infinitely many arbitrary choices, Peano among them, see Moore (1982, 76–82).

The Staircase Function

The Cantor function of Kolmogorov and Fomin is defined as follows. Recall that the Cantor set is obtained by deleting open intervals from $I = [0, 1]$ in the following order: first delete $J_1 = (1/3, 2/3)$. Then delete $J_2 = (1/9, 2/9)$ and $J_3 = (7/9, 8/9)$. At the nth stage, delete the 2^n open intervals which, going from left to right can be called $J_{2^{n-1}}, J_{2^{n-1}+1}, \ldots, J_{2^{n-1}+2^{n-1}-1} = J_{2^n-1}$. Define a function f_k on J_k which is $\frac{2r+1}{2^{m+1}}$ where $k = 2^m + r,\ 0 \leq r < 2^m$ (so on J_{13} the function is $11/16$). Consider the function f on $\cup_k J_k$ that agrees with f_k on J_k, and define the Cantor function to be the unique continuous function on the interval I that agrees with f on $\cup_k J_k$ (which is of course, the complement of the Cantor set).

This remarkable function is continuous, monotonic increasing, bounded, and indeed differentiable on [0, 1]. It takes every value from 0 to 1, but its derivative is zero everywhere except on a countable set of measure zero!

Here's another Cantor function. Write every number in [0, 1] as a ternary decimal, with a convention about numbers that end in $\ldots 22222222\ldots$ which can also be written to end in $\ldots 100000\ldots$. The function maps a number with no 1's in its ternary expansion to itself, and a number of the form $0.a_1 a_2 \ldots a_k \ldots$, where a_k is the first 1 to $0.a_1 a_2 \ldots a_k 00000\ldots$.

Here are two graphs that give an impression of the staircase function

$$st(x) := \sum_1^{\infty} \frac{\lfloor kx \rfloor}{2^k}.$$

Figure 27.4 shows the sum of the first 30 terms.

Figure 27.5 shows the difference between the 20th and the 10th terms, showing how crinkly the additions are becoming.

The function is monotone increasing. It is continuous at x if and only if x is irrational, and if x is irrational then the value $st(x)$ is transcendental.

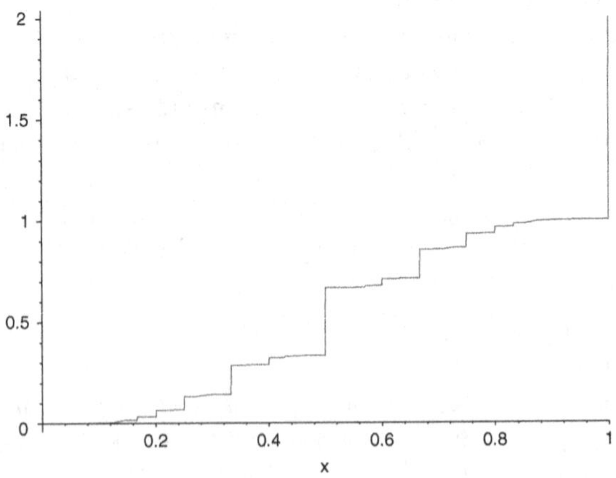

Fig. 27.4 The sum of the staircase function to 30 terms

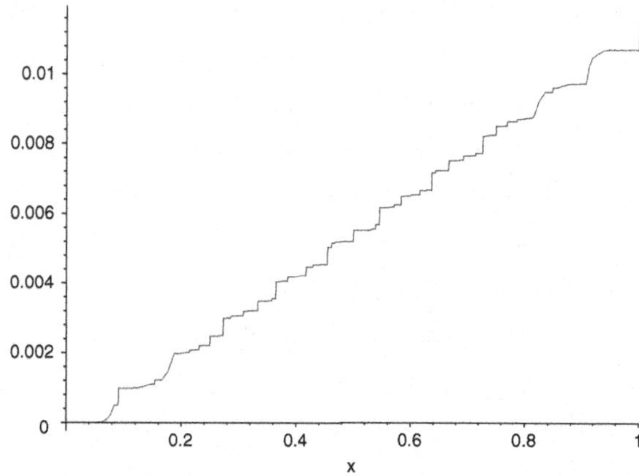

Fig. 27.5 The difference between the 20th and the 10th terms of the staircase function

27.5 Exercises

1. Do we need a theory of the real numbers if our intuitions are reliable? Are they reliable?
2. What was meant by a domain of a function of two variables? Compare the likely answers of Riemann, Harnack, and Poincaré.
3. What is good, and what is not so good, about the Riemann integral?

Chapter 28
Cantor, Set Theory, and Foundations

Georg Cantor (1845–1918)

28.1 Introduction

Dedekind had ushered in the idea that mathematical concepts might profitably be defined in terms of a naive conception of sets. His friend Cantor extended this to a theory of sets of various sizes, and came to a number of paradoxical conclusions, such as the existence of a one-to-one correspondence between a line and a square, which threatened the fundamental idea of the dimension of these and other domains. He also discovered that infinite sets could have different sizes, as for example the set of rational numbers and the set of real numbers, but he was never able to establish his belief that the smallest uncountable set can be put in a one-to-one correspondence

J. Gray, *The Real and the Complex: A History of Analysis in the 19th Century*,
Springer Undergraduate Mathematics Series, DOI 10.1007/978-3-319-23715-2_28

with the real numbers (the continuum hypothesis). Indeed, as we shall see, the whole edifice of set theory came to be threatened by a number of paradoxes that were put forward around the start of the 20th century.

28.2 The Continuum

For much of his life, Georg Cantor was interested in the continuum, a.k.a. the Reals. His problem was: how to characterise it among the collection of all sets. In this endeavour, he never succeeded, because there were substantial, indeed permanent obstacles in his way. Some problems were topological, others proved to belong in the uncharted but shifting domain of the foundations of mathematics, and while both are interesting we shall take only the foundational route here.

The first question concerns the size of the continuum, and eventually it became known as the continuum hypothesis. The first landmark is Cantor's discovery that infinite sets can actually have different sizes. Cantor came to this conclusion in 1873, when he showed that there are both countable and uncountable sets. His first proof is not so well known. It is not difficult to count the rational numbers, or even the algebraic numbers (the roots of polynomials with integer coefficients). To show that the real numbers are not countable, Cantor argued by contradiction.

Suppose that the real numbers are denumerable, and list them:

$$\omega_1, \omega_2, \ldots, \omega_n, \ldots$$

Cantor now exhibited a real number not in this list, as follows. Choose $\alpha < \beta$, and let the first two numbers in the list that lie between α and β be α' and β'. In the same way chose α'' and β'', and so on. Either you must stop after finitely many steps, or not. It you stop at, say, α_n and β_n, it is easy to find two real numbers between α_n and β_n, but they cannot both lie in the list. If you do not stop, consider $\lim \alpha_n = \alpha_\infty$ and $\lim \beta_n = \beta_\infty$. If these are equal, say to ω, then ω is not in the list. For, if it is, then $\omega = \omega_n$ for some n and then it would have been an α or a β. If, on the other hand, $\lim \alpha_n < \beta_\infty$, then any real number between $\lim \alpha_n$ and β_∞ produces a contradiction.

Cantor's second proof was the famous diagonal argument of 1891.

28.3 The Line and the Square

In 1877 Cantor wrote to Dedekind to describe a 1-1 correspondence between the points of the line and the plane. He was so struck by this that he switched from German to French to say 'I see it, but I don't believe it!' (see Cavaillès and Noether 1937, 34). His argument was that one can take two infinite decimals, the coordinates of a point in the plane, and riffle shuffle them together to produce a single decimal. Equally, one can take a single infinite decimal and take it apart, thus establishing the remarkable 1-1 correspondence. Cantor believed his discovery undermined almost everything that had been said about geometry and the dimension of figures, but Dedekind almost spoiled the party by observing that the argument was flawed. There has to be a convention about decimals ending in a run of 9s, for they are equal to decimals ending an infinite run of 0s. Pick one, say we allow all 9's and disallow all 0s. Consider now the numbers $0.a_1 1 a_2 0 \ldots a_n 0 a_{n+1} 0 \ldots$ and $0.a_1 0 a_2 9 \ldots a_n 9 a_{n+1} 9 \ldots$. They are different, but they map to $(0.a_1a_2a_3\ldots, 0.1000\ldots)$ and $(0.a_1a_2a_3\ldots, 0.0999\ldots)$ respectively, and these are the same; Cantor's map is not 1-1. Cantor agreed at once, but repaired the proof within two days, although the initial simplicity was lost in the process. Dedekind continued to observe, however, that the damage to mathematics was not as great as Cantor had feared (or hoped?)—Cantor's map was very obviously not continuous.

28.4 Transfinite Ordinals

We saw in Sect. 23.2 above that Cantor had studied nowhere dense sets on the real line. Transfinite ordinals grew out of this work, and the first transfinite symbols were indices before they acquired a new significance, and for that they had to wait a decade.

In the work on Fourier series, Cantor had the idea of forming a sequence of sets, each one the derived set of the set before:

$$P, P^{(1)}, P^{(2)}, \ldots, P^{(n)}, \ldots,$$

and continuing this process ad infinitum, thus obtaining a set he denoted $P^{(\to)}$. But of course it is possible to start with this set, and form the sequence

$$P^{(\to)}, P^{(\to+1)}, P^{(\to+2)}, \ldots, P^{(\to+n)}, \ldots,$$

and so on, to form $P^{(\to\to)}$, and beyond.

In his *Grundlagen* (1883) Cantor replaced the arrow with the Greek letter ω, and regarded the indices as numbers. He realised he had the familiar sequence of finite ordinals $1, 2, 3, \ldots$ (which are to be thought of as the first, second, third and so on rather than as one, two, three and so on) but that he had more. He had the first transfinite ordinal, ω, and a succession of ordinals beyond it:

$$\omega + 1, \omega + 2, \ldots, 2\omega, 2\omega + 1, \ldots$$

He also saw that they possessed an arithmetic, in which, however, $\omega + 1 \neq 1 + \omega$, for $\omega + 1$ is $\{2, 3, 4, 5, \ldots\}, \{1\}$ and $1 + \omega = \{1\}, \{2, 3, 4, \ldots\} = \omega$. So addition of these new numbers is defined, but it is not commutative.

Cantor now began to outline the corresponding theory of transfinite cardinals, which should measure the size of infinite sets. Here he was both remarkably successful and unsuccessful. He put his theory on the map, despite the indifference if not indeed dislike of some older mathematicians, it weathered the first storms it attracted when the set-theoretic paradoxes were discovered, and Hilbert hailed it as a Paradise from which mathematicians would never be banished. But in the absence of good definitions his ideas were at best intuitive, and very few of the important results in the subject were established by him. Yet ultimately it did become the proving ground for many discussions of the foundations of mathematics, and it remains so to this day.

The Power Set Construction

This comes from Cantor's (1890). Compare $I = (0, 1)$ and the set of all its subsets, $F = \{f : I \to \{0, 1\}\}$. Suppose there is a 1-1 correspondence from I to F that to $x \in I$ assigns $f_x \in F$, and to $f \in F$ assigns $x_f \in I$. Define $\Phi(x, y) = f_x(y)$. We now deduce there is a 1-1 correspondence between the set of maps of the form $\Phi(x, -)$ and the set F. The map is onto, because given $f \in F$ pick $x_f \in I$. Then $\Phi(x_f, y) = f_{x_f}(y) = f(y)$. The map is 1-1, for if it is not then there are $x, x' \in I$ such that $\Phi(x, y) = \Phi(x', y)$ for all $y \in I$, but this implies $f_x(y) = f_{x'}(y)$ for all $y \in I$, which implies $f_x = f_{x'}$, which implies $x = x'$. On the other hand, every map of the form $\Phi(x, -)$ satisfies the equation $\Phi(x, x) = f_x(x)$, and there is a map g such that $g(x) \neq \Phi(x, x)$—just let g take the other value.

It follows that a set is strictly smaller than its power set, and so there is a chain of larger and larger sets. By way of notation, I shall write the power set of a set S either as $\wp(S)$ or as 2^S. So one has $S \prec \wp(S) \prec \wp(\wp(S)) \prec \cdots$. This in turn opens the question of what is meant by the size of sets, which Cantor measured by what he called their 'power' and later became known as their cardinality. This is both a more elementary concept than the order type of a set and yet harder to define. We might, for example, consider the cardinality of a set to be the set of all sets with which it can be put in a 1-1 correspondence—but if we do we shall be disappointed, as we shall see.

28.5 The Continuum Hypothesis

We can now state the continuum hypothesis, as it came to be known (Cantor seems never to have called it this). The set $\mathbb{R}$ has a cardinality, call it c, which is greater than the cardinality of $\mathbb{Q}$, the rational numbers, denoted $\aleph_0$, the smallest infinite cardinal. In symbols, $\aleph_0 < c$. From the construction of the real numbers it is clear that $\mathbb{R} \subset \wp\{\mathbb{Q}\}$, where $\wp\{\mathbb{Q}\}$ is the power set of $\mathbb{Q}$. So $c \leq 2^{\aleph_0}$, and in fact these cardinals are equal. The continuum hypothesis asks for a comparison of this cardinal with the smallest cardinal larger than $\aleph_0$, namely $\aleph_1$, and it claims that these are equal. If the claim is false then $c = 2^{\aleph_0}$ is some other aleph, and the question of which one would arise. It is one of the great achievements of recent mathematics to show that on the usually accepted axiomatisation of set theory (ZF for Zermelo-Frankel), and even granting the axiom of choice (making set theory into ZFC), the continuum hypothesis is independent of the axioms and so there are set theories in which the continuum hypothesis is true and others in which it is false.

28.6 The Paradoxes of Set Theory

A set is said to be well-ordered (by us, Cantor's definition was different but equivalent) if there is an ordering on it, often denoted $<$, with the properties that:

1. given any two elements in the set, a and b, exactly one of the following holds: $a < b, a = b, b < a$;
2. every non-empty set has a least element.

Well-ordered sets have the same order type if there is a 1-1 order preserving correspondence between them. Ordinal numbers are the order types of well ordered sets.

So you might ask if every set can be well-ordered. If you do, what, for example, might be a well ordering of the real numbers? That seems very hard, if not impossible to produce. There are other difficult questions about infinite sets. The continuum hypothesis has already been mentioned. Another is the trichotomy of cardinals, the idea that given any two cardinal numbers, either one is bigger than the other, or they are the same size. Even the axiom of choice, which is the claim that given a collection of sets there is a map which picks out an element of each set and forms of set of these elements, is not obvious. In fact, as already noted, in axiomatic set theory Choice is taken as an axiom.

The Greatest Ordinal?

On the other hand, consider, with the Italian mathematician Burali-Forti in the 1890s, the collection of all ordinal numbers. Ordinal numbers form a well-ordered set, so this collection has an ordinal, which must be larger than any ordinal in the collection, but that collection contained all the ordinals. A paradox, if not indeed an antinomy—a fatal logical flaw. Or, consider the collection of all cardinals. An exactly similar argument also produces a paradox.

Now why are these paradoxes taken so seriously? One very reasonable answer is that they should not be taken any more seriously than any of the other problems just listed (the continuum hypothesis, the well-ordering principle, Choice). They show that a certain, new, topic in mathematics is not properly understood. That's interesting, challenging, and quite typical of mathematics. Burali-Forti, for example, thought that his argument just showed that trichotomy for cardinals did not hold, a result he deduced from his paradox.

Cantor was also not bothered. He had discovered these paradoxes for himself, but he had strong theological views about accessible infinities of various sizes and the truly inaccessible infinities appropriate to talk about God, so the set-theoretic paradoxes were small oddities to him.

Nor was it the case that nothing could be said. To be entirely ahistorical for a moment, it eventually became clear that starting from ZF and adding an axiom produced a great deal of unity: ZF + Choice is equivalent to ZF + well-ordering, and to ZF + trichotomy. This isn't entirely good news. ZF + Choice strikes most people as reasonable in a way ZF + well-ordering does not.

Russell's Paradox

In 1902 matters became much more serious. What are, or should be, the foundations of mathematics? Dedekind's construction of the real numbers out of sets of rational numbers is exemplary, and when he also constructed the natural numbers out of mere sets (not an easy thing to do and not to be described here) the result was to show that the most fundamental objects in mathematics can all be thought of as sets. Mathematics no longer needed to be founded in real-world objects, which was just as well because it was plainly capable of describing situations where real-world intuition was completely inadequate. How then could mathematics be true? A mixed group of philosophers such as Bertrand Russell, mathematicians turned philosophers such as Frege, and also a number of distinguished mathematicians such as Poincaré (and Hilbert, Zermelo, and Hausdorff) became interested in this question. To some, but not all of this group, the answer lay in logic. The truths of logic seem indisputable, so if mathematics could be reduced to logic theorems in mathematics would be logically true.

Set theory is not logic, so the task of reducing the one to the other is far from trivial, and by 1900 the front runner in this enterprise was Frege. As he was finishing the second volume of a work that was intended to show that the whole logicist enterprise, as this philosophy of mathematics is called, can be made to work, he received a postcard from Russell. It concerned the class of all predicates that cannot be predicated of themselves. In more usual mathematical language, let us say that a set contains itself if it is an element of itself: $S \in S$. We define $A := \{S : S \notin S\}$, and ask if A contains itself or not. By definition, $A \in A => A \notin A$, and $A \notin A => A \in A$. Trouble either way!

Russell's paradox, known already in Göttingen to Ernst Zermelo and Hilbert, can be given a much more logical, less mathematical, presentation. Frege saw at once that it destroyed his life's work, and he had the honesty to say so. Indeed, the damage cannot be put right. There is an irreducible step in any foundation of mathematics that goes beyond elementary logic. But the paradoxes, enjoyable though they are, only destroy the logicist approach to the foundations of mathematics. Other approaches, closer to naïve set theory, have other problems but more life in them. That, of course, is another story.

The only way round the paradoxes is to put limitations on what collections of objects can be sets. However, when mathematicians have attempted to characterise what sets are, say by a system of axioms guaranteeing their properties, it seems that the bounds of elementary logic are irreversibly crossed.

Chapter 29
Topology

29.1 Introduction

Here we look briefly at problems to do with curves and surfaces, lengths and areas, starting with the great question of: what is a continuous curve? In particular, does a simple closed curve on a sphere have a simply-connected interior and a simply-connected exterior? Then we return to the problem of distinguishing a the unit interval from the unit square, and we conclude by showing how curves can have areas (non-zero Lebesgue measure). In this way we shall pick up some of the roots of both point-set and algebraic or geometric topology.

29.2 The Jordan Curve Theorem

The Jordan curve theorem states that a simple closed curve divides the plane into two regions. It was refined in Schoenflies (1906) to the claim that one of these regions is topologically a disc and the other is a disc with a point removed.

Jordan first stated the claim in his *Cours d'Analyse* (1887).[1] He had in mind the question that until the theorem is proved, what sense does it make to evaluate the Cauchy integral of a meromorphic function round a curve and count the contribution of the points *inside*? This is, of course, easy for simple contours (the disc, a semicircle, a rectangle, a half-plane) but by the late 1890s very general curves were being investigated from a number of standpoints: what then?

How can the theorem be difficult? Let us start with the simplest cases. A triangle with three straight sides satisfies the theorem because we can orient the sides of the triangle in the same way (so that we may follow round the triangle in a way that

[1] See vol. 3, 587–594. It does not seem to occur in some later reprints, e.g. in 1959.

J. Gray, *The Real and the Complex: A History of Analysis in the 19th Century*, Springer Undergraduate Mathematics Series, DOI 10.1007/978-3-319-23715-2_29

respects the orientation on the sides). With respect to this orientation we may say that each side extends to a line that divides the plane into two regions, and the inside of the triangle that is on the same side of all three lines. This proof fails, however, even for quadrilaterals (precisely, for re-entrant quadrilaterals), but it highlights a relevant feature of the plane (that it is orientable). The argument is incorrect when applied to the projective plane or the Möbius band.

For a polygon with a finite number of straight sides a popular informal proof asks us to consider a point that does not lie on any of the lines obtained by continuing any of the sides. Any half-line through such a point crosses the sides a finite number of times, and if we count the intersection points (counting any such vertices of the polygon as two points) we find that the inside of the polygon may be defined as those points for which this number is odd (and the exterior as the points for which the number is even). The remaining points, other than those that lie on a side, are assigned to the inside or the outside by looking at their immediate neighbours. But we need to be sure that we have not accepted this proof because we believe the Jordan curve theorem and go "inside, outside, inside, …" as we cross each side until we escape.

But what if the polygon has infinitely many straight sides? What if the curve is a rectifiable curve? Can we proceed by approximating it sufficiently accurately by polygons with finitely many sides—but surely the inside of an approximating polygon need not always be a subset of the inside of the curve? Matters are easier if the curve is differentiable, because then one may use segments of tangents of curves as sides of a suitable polygon. One can hope that differentiable curves have convex and concave sections—but what about curves that oscillate infinitely often? And non-differentiable but rectifiable curves and non-rectifiable curves are still out of consideration.[2]

Jordan offered a proof that began with the polygonal case and addressed these difficulties. Some ten years later a second proof was offered by Oswald Veblen in 1904, and this proof seems to have become generally accepted, although several different proofs have since been given. A consensus then grew up that Veblen's was the first correct proof and that Jordan's argument was either obscure in some way or even flawed, based perhaps on Veblen's complaint that Jordan's proof "is unsatisfactory to many mathematicians. It assumes the theorem without proof in the important special case of a simple polygon and of the argument from that point on, one must admit at least that all details are not given".[3]

However, this consensus has recently been challenged by Thomas Hales, who consulted Jordan's proof and looked for the alleged error, only to find that

> In view of the heavy criticism of Jordan's proof, I was surprised when I sat down to read his proof to find nothing objectionable about it. Since then, I have contacted a number of the authors who have criticized Jordan, and each case the author has admitted to having no

[2]Non-rectifiable curves make for other problems too: What is the integral in the Cauchy integral theorem for closed curves that do not have a length, such as the von Koch snowflake curve?

[3]Quoted in Hales (2007, 45) from which the next few paragraphs are drawn.

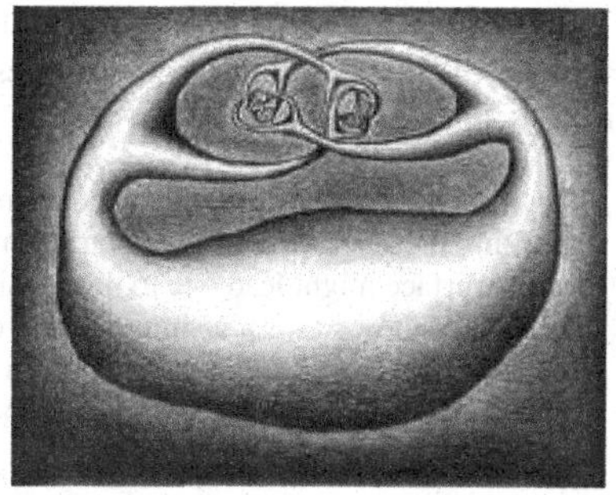

Fig. 29.1 Alexander's horned sphere

direct knowledge of an error in Jordan's proof. It seems that there is no one still alive with a direct knowledge of the error.

Hales then gives a modernised version of Jordan's argument that preserves the essentials of his argument.

The analogue of the Jordan curve theorem is also true in n dimensions, but the claim about topological simplicity of the regions is false, as the Alexander horned sphere shows. This was introduced to the world by the American topologist James Waddell Alexander II in his paper (1924) (Fig. 29.1).

The construction of the horned sphere is an iterative one. It starts with a torus. A slice is removed from it, leaving a cylinder with two edges β and γ facing each other. To each end of the cylinder a torus is attached, in such a way that the two tori are linked. The construction is now repeated on each torus ad infinitum, to produce a surface Σ. In Fig. 29.2, taken from Alexander's paper, the left and right had ends

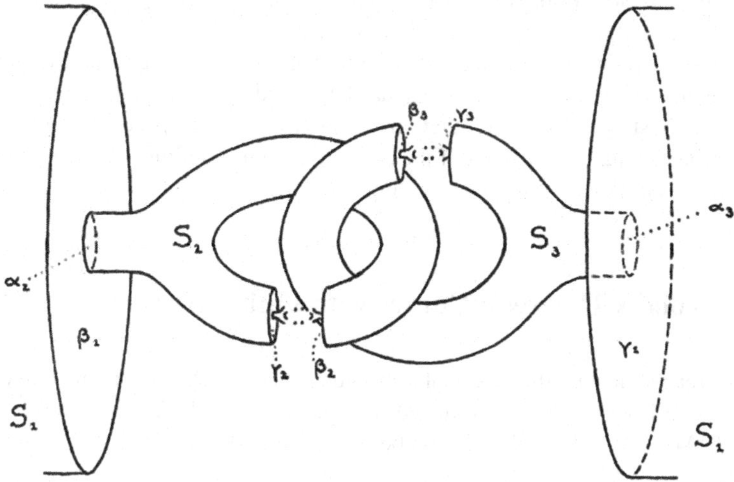

Fig. 29.2 The construction of Alexander's horned sphere

of the cylinder are labelled S_1; S_2 and S_3 are the tori attached at the ends, and a hint is given of the next stage in the process.
Alexander remarked (pp. 9, 10)

> It will be seen without difficulty that the interior of the limiting surface Σ is simply connected, and that the surface itself is of genus zero and without singularities, though a hasty glance at the surface might lead one to doubt this last statement. The exterior R of Σ is not simply connected, however, for a simple closed curve in R differing but little from the boundary of one of the cells γ_i cannot be deformed to a point within R. It is easily shown, in fact, that the group of R requires an infinite number of generators. [...]
>
> This example shows that a proof of the generalized Schönfliess theorem announced by me two years ago, but never published, is erroneous.

29.3 The Problem of Dimension

Naively, curves and surfaces are different. The Cantor map shows that some care must be used to say how they are different, but a clue might be that the map is not continuous. In 1890, the Italian mathematician Guiseppe Peano caused a great stir with his example of a continuous map of the unit interval onto the plane, the first so-called space-filling curve. Peano had a particular distrust of geometric reasoning, which distrust his novel curve was intended to promote, the more comprehensible descriptions are, however, pictorial and due to David Hilbert in his (1891), later discussed by the American mathematician E.H. Moore in his (1900).[4]

Start with the series of approximations shown in Fig. 29.3.

Observe how straight and hooked segments are replaced as each square is divided into 4. The curve Hilbert produced is the limit of the series that is produced by iterating this construction. A better picture of it is given by the sixth step, which is shown in Fig. 29.4, taken from Sagan (1994, 11).

With these examples in mind, it is worth noting the idea of a covering dimension of a space that was introduced by Lebesgue (1911) and later studied by many authors. It is the smallest number n such that any open cover of the space contains at least one point belonging to $n + 1$ sets of the covering. On this definition, a real manifold of dimensions n happily has covering dimension n.

29.4 Osgood's Theorem: Curves with Finite Area

Here by area we mean 2-dimensional Lebesgue measure! The American mathematician William Fogg Osgood introduced this curve in his (1903). It is described in Sagan (1994, 133–135), which should be consulted for the details.

[4]Moore also discussed continuous curves with no tangents in the same spirit.

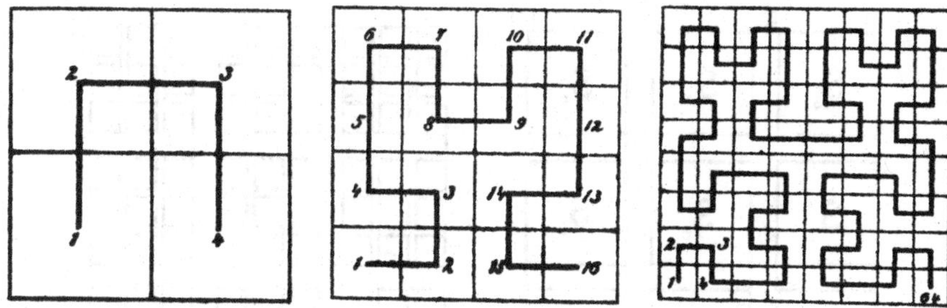

Fig. 29.3 The first three stages in the construction of Hilbert's space-filling curve

Fig. 29.4 The sixth stage in the construction of Hilbert's space-filling curve

Remove 2 bars horizontally and 2 bars vertically from a unit square, leaving 9 squares, as shown in Fig. 29.5. If the widths of the bars are w then the side of each little square s_1 is given by $3s_1 + 2w = 1$, so $s_1 = (1 - 2w)/3$ and the area of the 9 squares is $(1 - 2w)^2$. Now, from each little square, remove bars of width $w/6$, and get 81 squares. Note that at each stage in this part of the construction we select solid squares, not parts of an 'honest' curve. You should compute the total area of the 81 squares.

At each stage, join the 9 new squares as shown in Fig. 29.6. To get a curve, draw a curve that lies in each square as shown (and joins up to the next square)—note that the curve also lies in all the previous sets of squares, too.

Osgood's curve is the limiting curve as the process is repeated indefinitely. It is possible to parameterise it so to prove that it is a continuous Jordan curve. Nonetheless, by choosing w to be arbitrarily small the curve has two-dimensional Lebesgue measure arbitrarily close to 1.

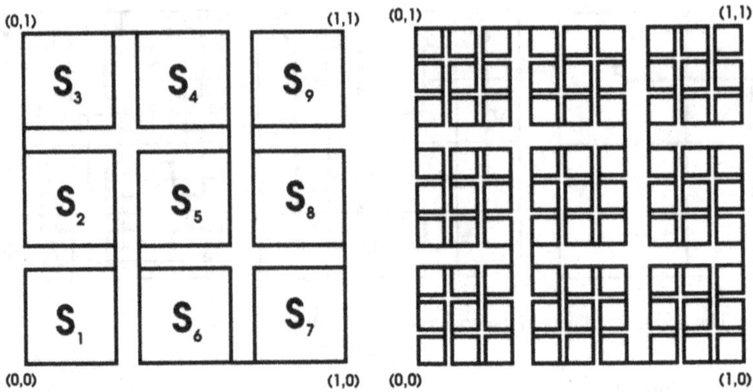

Fig. 29.5 The first two stages in the construction of Osgood's squares

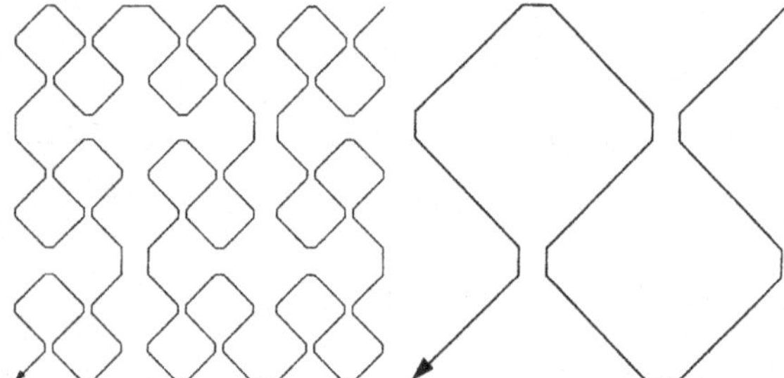

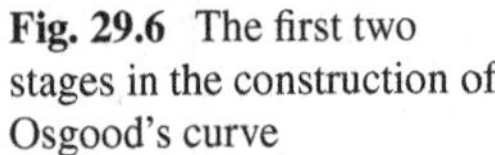

Fig. 29.6 The first two stages in the construction of Osgood's curve

29.5 Concluding Remarks

It will be clear by now that this last chapter belongs to the genre of final lectures, in which some of the ideas of the course are to run around briefly free of the constraint to flow clearly and yet rigorously from what has gone before, in the hope that they will excite the students. And indeed, these examples were chosen with that aim partly in mind. But they were also presented to show that analysis did not stop around 1900, and that the social and historical processes thorough which mathematics is done also continued.

In a period when fractal sets and Hausdorff dimensions are much discussed the liveliness of the mathematics needs no emphasis. It is also worth noting that with every passing year the mathematical analysis of the last century also needs its historians.

Chapter 30
Assessment

On the view that end-of-course assessment ought to ask students to review their course and especially the later parts of the course, I tended to set rather straightforward essay titles. A direct request for a survey of the century could only produce a rushed composite of the material, so I chose to narrow the period to the period 1850–1900 or 1860–1900 (the difference being Riemann). I therefore asked students first to describe the main developments in mathematical analysis in the years 1850 (or 1860) to 1900, making clear what was important about them, and then to consider what were the most important problems or opportunities facing mathematical analysts as the 19th century came to an end. I added that the dates are approximate and may be varied by a few years if it helps their argument, but may not be narrowed.

I also offered the following advice.

Pick three or four interesting topics for each part of the essay, and look for moments when the nature of the subject changed, or the criteria for its importance shifted.
Remember to consider who found it important, and why, and try to give their reasons, not yours (or, if you prefer, give their reasons as well as yours). Try to assess their influence and to look for dissenting voices, if there were any.
The key words people use on such occasions (novel, fundamental, profound, rigorous) are often revealing: what developments merit what words? How and to what extent were judgements of this kind grounded in the mathematics?
In the second part, do not let your knowledge of current mathematics lead you to suggest that the future could be predicted. The best answers will be grounded in what was said in the first part.

As for the writing of the essay, a wealth of mathematical detail is not necessary, but mathematical accuracy is.

Don't end up with a final paragraph full of superlatives that don't stand up to scrutiny.

Finally, all manner of conclusions are OK—if well argued!

J. Gray, *The Real and the Complex: A History of Analysis in the 19th Century*, Springer Undergraduate Mathematics Series, DOI 10.1007/978-3-319-23715-2_30

Two currents were at work: a wish that the students would gain some insight into why the subject they have spent so much time studying, and which can seem remote from practical and utilitarian concerns, has the shape and scope that it does; and a hope that they would see the growth to some extent from the inside, as a mathematician active in the 19th century might have done. If it is not seen as a succession of facts (definitions, theorems, examples, counter-examples, proofs) but as a creation of people engaged in various aspects of what is involved in making good sense of the greatest idea in modern mathematics—the calculus, broadly defined—then it can seem like a challenge. And if the course has managed to convey something of the messy, undirected, uncertain ways in which research is done and new mathematics discovered it may even be a resource for students after they have graduated and begun to work on their own.

But now it seems to me that I could have been more adventurous, and set titles like these:

- What is the importance of the concept of oscillation (in the values of a function) for the development of mathematical analysis in the period 1860–1910?
- How true is it to say that by 1900 real analysis had become counter-intuitive?
- Darboux began his address to the International Congress of Mathematicians in Rome in 1908 (Darboux 1909, 105) by saying in reference to complex analysis that "A profound change has been brought about before our eyes in the orientation of mathematical studies in the last 30 years". To what extent would you say this was true of the period 1850–1880?

Appendix A
Papers in Translation

A.1 Fourier on Fourier Series

[From Fourier, *The analytical Theory of Heat*, tr. A Freeman.]
§220. We see by this that the coefficients a, b, c, d, e, f, &c., which enter into the equation

$$\frac{1}{2}\pi\phi(x) = a \sin x + b \sin 2x + c \sin 3x + d \sin 4x + \&c,$$

and which we found formerly by way of successive eliminations, are the values of definite integrals expressed by the general term $\int \sin kx\phi(x)dx$, k being the number of the term whose coefficient is required. This remark is important, because it shews how even entirely arbitrary functions may be developed in series of sines of multiple arcs. In fact, if the function $\phi(x)$ be represented by the variable ordinate of any curve whatever whose abscissa extends from $x = 0$ to $x = \pi$, and if on the same part of the axis the known trigonometric curve, whose ordinate is $y = \sin x$, be constructed, it is easy to represent the value of any integral term. We must suppose that for each abscissa x, to which corresponds one value of $\phi(x)$, and one value of $\sin x$, we multiply the latter value by the first, and at the same point of the axis raise an ordinate equal to the product $\phi(x) \sin x$. By this continuous operation a third curve is formed, whose ordinates are those of the trigonometric curve, reduced in proportion to the ordinates of the arbitrary curve which represents $\phi(x)$. This done, the area of the reduced curve taken from $x = 0$ to $x = \pi$ gives the exact value of the coefficient of $\sin x$; and whatever the given curve may be which corresponds to $\phi(x)$, whether we can assign to it an analytical equation, or whether it depends on no regular law, it is evident that it always serves to reduce in any manner whatever the trigonometric curve; so that the area of the reduced curve has, in all possible cases, a definite value, which is the value of the coefficient of $\sin x$ in the development of the function. The same is the case with the following coefficient b, or $\int \phi(x) \sin 2x dx$.

J. Gray, *The Real and the Complex: A History of Analysis in the 19th Century*, Springer Undergraduate Mathematics Series, DOI 10.1007/978-3-319-23715-2

In general, to construct the values of the coefficients a, b, c, d, &c, we must imagine that the curves, whose equations are

$$y = \sin x,\ y = \sin 2x,\ y = \sin 3x,\ y = \sin 4x,\ \&c.$$

have been traced for the same interval on the axis of x, from $x = 0$ to $x = \pi$; and then that we have changed these curves by multiplying all their ordinates by the corresponding ordinates of a curve whose equation is $y = \phi(x)$. The equations of the reduced curves are

$$y = \sin x\, \phi(x),\ y = \sin 2x\, \phi(x),\ y = \sin 3x\, \phi(x), \&c.$$

The areas of the latter curves, taken from $x = 0$ to $x = \pi$, are the values of the coefficients a, b, c, d, &c., in the equation

$$\frac{1}{2}\pi\phi(x) = a \sin x + b \sin 2x + c \sin 3x + d \sin 4x + \&c,$$

§221. We can verify the foregoing equation (D), (Art. 220), by determining directly the quantities a_1, a_2, a_3, a_4, &c.., in the equation

$$\phi(x) = a_1 \sin x + a_2 \sin 2x + a_3 \sin 3x + a_4 \sin 4x \&c.;$$

for this purpose, we multiply each member of the latter equation by $\sin kxdx$, k being an integer, and take the integral from $x = 0$ to $x = \pi$, whence we have

$$\int \phi(x) \sin kx = a_1 \int \sin x \sin kx + a_2 \int \sin 2x \sin kx + \&c.$$
$$+ a_j \int \sin jx \sin kxdx + \cdots \&c.; \tag{A.1}$$

Now it can easily be proved, 1st, that all the integrals, which enter into the second member, have a nul value, except only the term $a_k \int \sin kx \sin kxdx$; 2nd, that the value of $\int \sin kx \sin kxdx$; $\frac{\pi}{2}$; whence we derive the value of a_j, namely

$$\frac{2}{\pi} \int \phi(x) \sin kxdx.a$$

The whole problem is reduced to considering the value of the integrals which enter into the second member, and to demonstrating the two preceding propositions. The integral

$$2 \int \sin jx \sin kxdx,$$

taken from $x = 0$ to $x = \pi$, in which k and j are integers, is

$$\frac{1}{k-j}\sin(k-j)x - \frac{1}{k-j}\sin(k+j)x + C.$$

Since the integral must begin when $x = 0$ the constant C is nothing, and the numbers k and j being integers, the value of the integral will become nothing when $x = \pi$; it follows that each of the terms, such as

$$a_1 \int \sin x \sin kx, \; a_2 \int \sin 2x \sin kx, \; a_3 \int \sin 3x \sin kx dx \&c.$$

vanishes, and that this will occur as often as the numbers k and j are different. The same is not the case when the numbers k and j are equal, for the term $\frac{1}{k-j}\sin(k-j)x$ to which the integral reduces, becomes $\frac{0}{0}$, and its value is π. Consequently we have thus we obtain, in a very brief manner, the values of $a_1, a_2, a_3, \ldots, a_j$, &c., namely,

$$a_1 = \tfrac{2}{\pi}\int \phi(x)\sin x dx, \quad a_2 = \tfrac{2}{\pi}\int \phi(x)\sin 2x dx$$
$$a_3 = \tfrac{2}{\pi}\int \phi(x)\sin 3x dx, \quad a_4 = \tfrac{2}{\pi}\int \phi(x)\sin 4x dx$$

Substituting these we have

$$\frac{1}{2}\pi\phi(x) = \sin x \int \phi(x)\sin x dx + \sin 2x \int \phi(x)\sin 2x dx + \&c.$$
$$+ \sin kx \int \phi(x)\sin kx dx + \&c. \qquad \text{(A.2)}$$

§222. The simplest case is that in which the given function has a constant value for all values of the variable x included between 0 and π; in this case the integral $\int \sin kx dx$ is equal to $\frac{2}{k}$, if the number k is odd, and equal to 0 if the number k is even.

Hence we deduce the equation

$$\frac{1}{4}\pi = \sin x + \frac{1}{3}\sin 3x + \frac{1}{5}\sin 5x + \frac{1}{7}\sin 7x + \&c.,$$

which has been found before.

It must be remarked that when a function has been developed in a series of sines of multiple arcs, the value of the series

$$a \sin x + b \sin 2x + c \sin 3x + d \sin 4x + \&c.$$

is the same as that of the function $\phi(x)$ so long as the variable x is included between 0 and π; but this equality ceases in general to hold good when the value of x exceeds the number π.

Suppose the function whose development is required to be x, we shall have, by the preceding theorem,

$$\frac{1}{2}\pi x = \sin x \int x \sin x dx + \sin 2x \int x \sin 2x dx + \&c.$$
$$+ \sin kx \int x \sin kx dx + \&c. \qquad \text{(A.3)}$$

The integral $\int_0^\pi x \sin kx dx$ is equal to $\pm\frac{\pi}{k}$; the indices 0 and π which are connected with the sign $\int$, shew the limits of the integral; the sign + must be chosen when k is odd, and the sign − when k is even. We have then the following equation,

$$\frac{1}{2}x = \sin x - \frac{1}{2}\sin 2x + \frac{1}{3}\sin 3x - \frac{1}{4}\sin 4x + \frac{1}{5}\sin 5x - \&c.$$

§223. We can develope also in a series of sines of multiple arcs functions different from those in which only odd powers of the variable enter. To instance by an example which leaves no doubt as to the possibility of this development, we select the function $\cos x$, which contains only even powers of x, and which may be developed under the following form:

$$a \sin x + b \sin 2x + c \sin 3x + d \sin 4x + e \sin 5x + \&c.,$$

although in this series only odd powers of the variable enter.

We have, in fact, by the preceding theorem,

$$\frac{1}{2}\pi \cos x = \sin x \int \cos x \sin x dx + \sin 2x \int \cos x \sin 2x dx$$
$$+ \sin 3x \int \cos x \sin 3x dx \&c. \qquad \text{(A.4)}$$

The integral $\int \cos x \sin kx dx$ is equal to zero when k is an odd number, and to $\frac{2k}{k^2-1}$ when k is an even number. Supposing successively $k = 2, 4, 6, 8$, etc., we have the always convergent series

$$\frac{1}{4}\pi \cos x = \frac{2}{1.3}\sin 2x + \frac{4}{3.5}\sin 4x + \frac{6}{5.7}\sin 6x + \&c.\ ;$$

or,

$$\cos x = \frac{2}{\pi}\left\{\left(\frac{1}{1} + \frac{1}{3}\right)\sin 2x + \left(\frac{1}{3} + \frac{1}{3}\right)\sin 4x + \left(\frac{1}{5} + \frac{1}{7}\right)\sin 6x + \&c.\right\}.$$

This result is remarkable in this respect, that it exhibits the development of the cosine in a series of functions, each one of which contains only odd powers. If in the preceding equation x be made equal to $\frac{1}{4}\pi$, we find

$$\frac{1}{4}\frac{\pi}{\sqrt{2}} = \frac{1}{2}\left(\frac{1}{1} + \frac{1}{3} - \frac{1}{5} - \frac{1}{7} + \frac{1}{9} + \frac{1}{11} - \&c.\right).$$

This series is known ([Euler] *Introd. ad analysin. infinit.* cap. x.).

[In §224 Fourier showed that "a similar analysis may be employed for the development of any function whatever in a series of cosines of multiple arcs." He concluded that "This and the preceding theorem suit all possible functions, whether their character can be expressed by known methods of analysis, or whether they correspond to curves traced arbitrarily."]

....

§230. If we apply these principles to the problem of the motion of vibrating strings, we can solve difficulties which first appeared in the researches of Daniel Bernoulli. The solution given by this geometrician assumes that any function whatever may always be developed in a series of sines or cosines of multiple arcs. Now the most complete of all the proofs of this proposition is that which consists in actually resolving a given function into such a series with determined coefficients.

In researches to which partial differential equations are applied, it is often easy to find solutions whose sum composes a more general integral; but the employment of these integrals requires us to determine their extent, and to be able to distinguish clearly the cases in which they represent the general integral from those in which they include only a part. It is necessary above all to assign the values of the constants, and the difficulty of the application consists in the discovery of the coefficients. It is remarkable that we can express by convergent series, and, as we shall see in the sequel, by definite integrals, the ordinates of lines and surfaces which are not subject to a continuous law. We see by this that we must admit into analysis functions which have equal values, whenever the variable receives any values whatever included between two given limits, even though on substituting in these two functions, instead of the variable, a number included in another interval, the results of the two substitutions are not the same. The functions which enjoy this property are represented by different lines, which coincide in a definite portion only of their course, and offer a singular species of finite osculation. These considerations arise in the calculus of partial differential equations; they throw a new light on this calculus, and serve to facilitate its employment in physical theories.

....

§235. It follows from that which has been proved in this section, concerning the development of functions in trigonometrical series, that if a function $f(x)$ be proposed, whose value in a definite interval from $x = 0$ to $x = X$ is represented by the ordinate of a curved line arbitrarily drawn; we can always develope this function in a series which contains only sines or only cosines, or the sines and cosines of multiple arcs, or the cosines only of odd multiples. [...] These trigonometric series, arranged according to cosines or sines of multiples of arcs, belong to elementary analysis, like the series whose terms contain the successive powers of the variable. The coefficients of the trigonometric series are definite areas, and those of the series of powers are functions given by differentiation, in which, moreover, we assign to the variable a definite value. We could have added several remarks concerning the use and properties of trigonometrical series; but we shall limit ourselves to enunciating briefly those which have the most direct relation to the theory with which we are concerned.

1st. The series arranged according to sines or cosines of multiple arcs are always convergent; that is to say, on giving to the variable any value whatever that is not imaginary, the sum of the terms converges more and more to a single fixed limit, which is the value of the developed function.

Fourier's Proofs

Fourier deferred the proofs of his theorems largely to the end of his book, when he availed himself of infinitely large and infinitely small quantities, writing (in §417) of one proof that

> it supposes that notion of infinite quantities which has always been admitted by geometers. It would be easy to offer the same proof under another form, examining the changes which result from the continual increase of the factor p under the symbol $\sin p(\alpha - x)$. These considerations are too well known to make it necessary to recall them.

This was the proof of equation (B), which he went on to turn into the equation (B') he referred to in the extract that follows:

$$f(x)\frac{1}{2\pi}\int_{-\infty}^{\infty} d\alpha f(\alpha)\frac{2\sin p(\alpha - x)}{\alpha - x}, \quad (p = \infty).$$

He added

> Above all, it must be remarked that the function $f(x)$, to which this proof applies, is entirely arbitrary , and not subject to any continuous law.

418. The theorem expressed by Eq. (II) Art. 234 must be considered under the same point of view. This equation serves to develope an arbitrary function $f(x)$ in a series of sines or cosines of multiple arcs. The function $f(x)$ denotes a function completely arbitrary, that is to say a succession of given values, subject or not to a common law, and answering to all the values of x included between 0 and any magnitude X.

The value of this function is expressed by the following equation,

$$f(x) = \frac{1}{2\pi}\sum\int_a^b d\alpha f(\alpha)\cos\frac{2ix}{X}(x - \alpha) \qquad (A).$$

423. The propositions expressed by Eqs. (A.) and (B'), Arts. 418 and 417, may be considered under a more general point of view. The construction indicated in Arts. 415 and 416 applies not only to the trigonometrical function

$$\frac{\sin(p\alpha - px)}{\alpha - x};$$

but suits all other functions, and supposes only that when the number p becomes infinite, we find the value of the integral with respect to α, by taking this integral between extremely near limits. Now this condition belongs not only to trigonometrical functions, but is applicable to an infinity of other functions. We thus arrive at the expression of an arbitrary function $f(x)$ under different very remarkable forms; but we make no use of these transformations in the special investigations which occupy us.

With respect to the proposition expressed by Eq. (A), Art. 418, it is equally easy to make its truth evident by constructions, and this was the theorem for which we employed them at first, It will be sufficient to indicate the course of the proof.

In Eq. (A), namely,

$$f(x) = \frac{1}{2\pi}\int_{-X}^{+X} d\alpha f(\alpha) \sum_{-\infty}^{+\infty} \cos 2ix\frac{\alpha - x}{X};$$

we can replace the sum of the terms arranged under the sign $\sum$ by its value, which is derived from known theorems. We have seen different examples of this calculation previously, Sect. III, Chap. III. It gives as the result if we suppose, in order to simplify the expression, $2\pi = X$, and denote α by r,[1]

$$\sum_{-j}^{+j} \cos jr = \cos jr + \sin jr\frac{\sin r}{\operatorname{versin} r}.$$

We must then multiply the second member of this equation by $d\alpha f(\alpha)$, suppose the number j infinite, and integrate from $\alpha = -\pi$ to $\alpha = +\pi$. The curved line, whose abscissa is α and ordinate $\cos jri$ being conjoined with the line whose abscissa is α and ordinate $f(\alpha)$, that' is to say, when the corresponding ordinates are multiplied together, it is evident that the area of the curve *produced*, taken between any limits, becomes nothing when the number j increases without limit. Thus the first term $\cos jr$ gives a nul result.

The same would be the case with the term $\sin jr$, if it were not multiplied by the factor $\frac{\sin r}{\operatorname{versin} r}$; but on comparing the three curves which have a common abscissa α, and as ordinates $\sin jr$, $\frac{\sin r}{\operatorname{versin} r}$, $f\alpha)$, we see clearly that the integral

$$\int d\alpha f(\alpha) \sin jr\frac{\sin r}{\operatorname{versin} r}$$

has no actual values except for certain intervals infinitely small, namely when the ordinate $\frac{\sin r}{\operatorname{versin} r}$ becomes infinite. This will take place if r or $\alpha - x$ is nothing; and

[1] $\operatorname{versin} r = 1 - \cos r$.

in the interval in which α differs infinitely little from x, the value of $f(\alpha)$ coincides with $f(x)$/. Hence the integral becomes

$$2f(x)\int_{)}^{\infty} dr \sin jr \frac{r}{\frac{1}{2}r^2}, \text{ or } \frac{1}{2}f(x)\int_0^{\infty} \frac{dr}{r} \sin jr \,,$$

which is equal to $2\pi f(x)$, Arts. 415 and 356. Whence we conclude the previous Eq. (A).

When the variable x is exactly equal to $-\pi$ or $+\pi$, the construction shews what is the value of the second member of the Eq. (A), $[\frac{1}{2}f(-\pi)\ or\ \frac{1}{2}f(\pi)]$.

If the limits of integrations are not $-\pi$ and $+\pi$, but other numbers a and b, each of which is included between $-\pi$ and $+\pi$, we see by the same figure what the values of x are, for which the second member of Eq. (A) is nothing.

If we imagine that between the limits of integration certain values of $f(\alpha)$ become infinite, the construction indicates in what sense the general proposition must be understood. But we do not here consider cases of this kind, since they do not belong to physical problems.

If instead of restricting the limits $-\pi$ and $+\pi$, we give greater extent to the integral, selecting more distant limits a' and b', we know from the same figure that the second member of equation (A) is formed of several terms and makes the result of integration finite, whatever the function $f(x)$ may be.

We find similar results if we write $2\pi\frac{\alpha-x}{X}$ instead of r, the limits of integration being $-X$ and $+X$.

It must now be considered that the results at which we have arrived would also hold for an infinity of different functions of $\sin jr$. It is sufficient for these functions to receive values alternately positive and negative, so that the area may become nothing when j increases without limit. We may also vary the factor $\frac{\sin r}{\operatorname{versin} r}$ as well as the limits of integration, and we may suppose the interval to become infinite. Expressions of this kind are very general, and susceptible of very different forms. We cannot delay over these developments, but it was necessary to exhibit the employment of geometrical constructions; for they solve without any doubt questions which may arise on the extreme values, and on singular values; they would not have served to discover these theor.ems, but they prove them and guide all their applications.

A.2 Dirichlet on Fourier Series

Dirichlet, On the convergence of trigonometric series which serve to represent an arbitrary function between two given limits. *Journal für die reine und angewandte Mathematik*, 1829, 4, 157–169, in *Werke*, I, 117–133.

The series of sines and cosines, by means of which one can represent an arbitrary function in a given interval, enjoy among other remarkable properties that of being convergent. This property has not escaped the illustrious geometer who has opened a new domain in the applications of analysis, in introducing there the manner of expressing arbitrary functions that is under consideration; it is found in the Memoir which contains his first researches on heat. But no one, as far as I know, has given a general demonstration. I only know one work on this topic, which is by M. Cauchy and which appeared in the Memoirs of the Paris Academy of Sciences for the year 1823. The author of this work himself admitted that his proof was defective for certain functions for which however the convergence was indubitable. A close examination of the said memoir has led me to believe that the demonstration which is expounded there is not even sufficient in the cases to which the author believes it is applicable. Before entering into this matter I am going to state in a few words the objections to which the proof of M. Cauchy appears to me to be subject. The path which this famous geometer has taken in this research requires that one considers that values of the function $\phi(x)$ that one wishes to develop, when one replaces the variable x in it by a quantity of the form $u + v\sqrt{-1}$. The consideration of these values seems foreign to the question, and in any case one does not easily see what one must understand by the result of a similar substitution when the function one is given cannot be expressed by an analytic formula. I give this objection with even more confidence because the author seems to share my opinion on this point. In fact, in several of his works he insists on the necessity of defining in a precise manner the sense one attaches to a similar substitution even when this is done to a function with a regular analytic law. Above all one finds in the Memoir which is published in volume 19 of the *Journal de l'École Polytechnique* page 567 and following, some remarks on the difficulties that imaginary quantities give rise to when placed in the signs of arbitrary functions. What ever one makes of this first observation, there is still another objection to the demonstration that M. Cauchy gives that seems to leave no doubt about its insufficiency. The consideration of imaginary quantities leads the author to a result on the decrease of terms in a series that is far from proving that the terms form a convergent sequence. The result under considerataion can be stated as follows, supposing that the interval considered extends from zero to π.

"The ratio of the term of rank n to the quantity $A\frac{\sin nx}{n}$ (A being a determinate constant, depending on the extreme values of the function) differs from unity taken positively by a quantity which diminishes indefinitely as n becomes greater."

From this result and from the fact that the series which has $A\frac{\sin nx}{n}$ as its general term is convergent, the author concludes that the general trigonometric series is too. But this conclusion is not permissible, because it is easy to arrange that two series (at least when, as happens here, the terms do not always have the same sign) can be the one convergent and the other divergent, although the ratio of the two terms of the same rank differs as little as one wishes from unity taken positively when the terms are of a very high rank.

One sees a very simple example in the two series, the one having the general terms $\frac{(-1^n}{\sqrt{n}}$ and the other $\frac{(-1^n)}{\sqrt{n}}\left(1 + \frac{(-1^n}{\sqrt{n}}\right)$. The first of these series is convergent, the second on the contrary is divergent, because on subtracting the first from it one obtains the divergent series

$$-1 - \frac{1}{2} - \frac{1}{3} - \frac{1}{4} - \frac{1}{5} - \cdots$$

and however the ratio of the corresponding two terms, which is $1 \pm \frac{1}{\sqrt{n}}$, converges to unity as n increases.

I am now going to enter into the matter, beginning with an examination of the simplest cases, to which all the others can be reduced. Let us denote by h a positive number less than or at most equal to $\pi/2$ and by $f(\beta)$ a function of β that is continuous between the limits 0 and h; I understand by that a function that has a finite and definite value for every value of β between 0 and h, and is also such that the difference $f(\beta+\varepsilon) - f(\beta)$ decreases without limit when ε becomes smaller and smaller. Suppose finally that the function is always positive between 0 and h and that it decreases constantly from 0 to h in such a way that if p and q are two numbers between 0 and h, $f(p) - f(q)$ always has the opposite sign to $p - q$. That done, let us consider the integral

$$\int_0^h \frac{\sin k\beta}{\sin\beta} f(\beta) d\beta \qquad (1)$$

in which k is a positive quantity, and let us see what this integral becomes as k increases.[2] To do this, we divide it into several others, the first taken from $\beta = 0$ to $\beta = \pi/k$, the second from $\beta = \pi/k$ to $\beta = 2\frac{\pi}{k}$, and so on, the penultimate one having the limits $\frac{(r-1)\pi}{k}$ and $r\frac{\pi}{k}$, and the last the limits $r\frac{\pi}{k}$ and h, $r\frac{\pi}{k}$ denoting the greatest multiple of π/k which is contained in h. It is easy to see that these new integrals, which are $r+1$ in number, are alternately positive and negative, the function under the integral sign evidently being positive between the first limits, negative between the limits of the second, and so on. It is no less easy to convince oneself that each one is smaller than the one before, abstraction being made of sign.[3] In fact ν denoting an integer less than r, the expressions:

$$\int_{(\nu-1)\pi/k}^{\nu\pi/k} \frac{\sin k\beta}{\sin\beta} f(\beta) d\beta, \quad \int_{\nu\pi/k}^{(\nu+1)\pi/k} \frac{\sin k\beta}{\sin\beta} f(\beta) d\beta$$

represent two consecutive integrals. In the second, let us replace β by $\pi/k + \beta$; the integral then changes into this:

$$\int_{(\nu-1)\pi/k}^{\nu\pi/k} \frac{\sin(k\beta+\pi)}{\sin(\beta+\pi/k)} f(\beta+\frac{\pi}{k}) d\beta$$

or, which comes to the same thing

$$-\int_{(\nu-1)\pi/k}^{\nu\pi/k} \frac{\sin k\beta}{\sin(\beta+\pi/k)} f(\beta+\pi/k) d\beta.$$

[2] Dirichlet wrote i for k.

[3] The 19th Century phrase for 'in absolute value'.

The two integrals which we thus wish to compare having therefore the same limits, one sees without difficulty that the second has a numerical value less than the first. For that it is sufficient to note that it follows from the supposition made about the function $f(\beta)$:

$$f\left(\frac{\pi}{k}+\beta\right) < f(\beta),$$

and that on the other hand

$$\sin\left(\frac{\pi}{k}+\beta\right) > \sin\beta,$$

the arcs β and $\frac{\pi}{k}+\beta$ each being less that $\pi/2$, from which the inequality follows that:

$$\frac{f(\beta)}{\sin\beta} > \frac{f(\beta+\frac{\pi}{k})}{\sin(\beta+\pi/k)},$$

which holding for all values of β between the limits $(\nu-1)\frac{\pi}{k}$ and $\nu\pi/k$ shows, as we have said, that each integral is bigger than the one before, abstraction being made of sign. This circumstance holds a fortiori when one compares the penultimate one with the last one, noting that the difference of the limits $r\frac{\pi}{k}$ and h of the last is less than π/k, the common difference of all the others.

Now let us examine in a little more detail the integral of rank ν, which is:

$$\int_{(\nu-1)\pi/k}^{\nu\pi/k} \frac{\sin k\beta}{\sin\beta} f(\beta)d\beta.$$

as the function of β which is found under the integral sign is a product of factors $\frac{\sin k\beta}{\sin\beta}$ and $f(\beta)$, which are both continuous functions of β between the limits of integration, and as on the other hand the first of these factors always has the same sign between these same limits, one concludes, in virtue of a known theorem, that the integral under consideration is equal to the integral of the first factor multiplied by a quantity lying between the greatest and the least value of the other factor. As the second factor decreases from the first limit to the second, the quantity under consideration lies between $f((\nu-1)\frac{\pi}{k})$ and $f(\nu\pi/k)$. Denoting this by ρ_ν, our integral will be equivalent to:

$$\rho_\nu \int_{(\nu-1)\pi/k}^{\nu\pi/k} \frac{\sin k\beta}{\sin\beta} d\beta.$$

the integral which is still contained in this expression depends both on ν and k. It is positive or negative according as $\nu-1$ is even or odd; we shall nonetheless denote it by K_ν, abstraction being made of sign. We shall however need to know the limit to which this converges when, ν remaining constant, k becomes greater and greater.

To discover this limit, let us replace β by γ/k, γ being a new variable. We shall then have:

$$\int_{(\nu-1)\pi}^{\nu\pi} \frac{\sin\gamma}{\sin(\gamma/k)} d\gamma/k.$$

In this form, it is evident that it converges to the limit

$$\int_{(\nu-1)\pi}^{\nu\pi} \frac{\sin\gamma}{\gamma} d\gamma,$$

which, to abbreviate, we denote by k_ν, abstraction being made of sign.

One knows that the integral $\int_0^\infty \frac{\sin\gamma}{\gamma}$ has a finite value and is equal to $\pi/2$. This integral can be divided into an infinity of others, the first taken from $\gamma = 0$ to $\gamma = \pi$, the second from $\gamma = \pi$ to $\gamma = 2\pi$, and so on. These new integrals are alternately positive and negative, each of them has a numerical value less than the one before, and that of rank ν is k_ν, abstraction being made of sign. The proposition we have cited reduces therefore to saying that the infinite sequence:

$$k_1 - k_2 + k_3 - k_4 + k_5 - etc. \quad (2)$$

is convergent and has a sum equal to $\frac{\pi}{2}$.

The terms of this sequence always decrease, and it follows from a known proposition that the sum of the first n terms is greater or less than $\pi/2$ according as n is odd or even, and that this sum, which one can denote by S_n, differs from $\pi/2$ by a quantity less than the following term k_{n+1}.

Let us now return to integral (1) and seek to find the limit to which it converges as k increases indefinitely. In making the number k thus increase, the integrals into which we have decomposed integral (1) change their values continually and at the same time their number will increase; it is necessary to know the result of this double change when it is continued indefinitely. For this we take an integer m (which we shall suppose is even for simplicity) and suppose that m remains fixed while k increases. The number r, which increases continually with k will eventually finish by surpassing the fixed number m, however large it is chosen.

This done, we divide the integrals into two groups whose sum is equivalent to integral (1). The first group will contain the first m of these integrals, the second will be composed of all that follow. One has, for the sum of the first group

$$K_1\rho_1 - K_2\rho_2 + K_3\rho_3 - K_4\rho_4 + \cdots - K_m\rho_m \quad (3)$$

and the second, the number of whose terms increases continually with k has for its first terms:

$$K_{m+1}\rho_{m+1} - K_{m+2}\rho_{m+2} + \cdots \qquad (4)$$

Let us consider these two groups separately. As the number k increases indefinitely, the sum (3) converges to a limit that it is easy to determine. In fact, the quantities $\rho_1, \rho_2, \ldots, \rho_m$, which are contained the first between $f(0)$ and $f(\frac{\pi}{k})$, the second between $f(\frac{\pi}{k})$ and $f(2\frac{\pi}{k})$, and the last between $f((m-1)\frac{\pi}{k})$ and $f(m\frac{\pi}{k})$ each converge to the limit $f(0)$ as, m remaining fixed, k increases without limit. On the other hand we have seen that the quantities:

$$K_1, K_2, \ldots, K_m$$

converge in the same circumstances to the respective limits:

$$k_1, k_2, \ldots, k_m.$$

Therefore the sum (3) converges to the limit:

$$(k_1 - k_2 + k_3 - \cdots - k_m)\, f(0) = S_m f(0),$$

which is to say that the difference between the sum (3) and $S_m f(0)$ will always finish, abstraction being made of sign, by being constantly less that ω, ω denoting a positive quantity as small as one wishes.

Let us now consider the sum (4) the number of terms of which increases continually. These terms being alternately positive and negative and each of them having a numerical value less than the term before, as we have seen above, and considering the integrals that the terms represent, if follows from a known principle[4] that this sum, whatever the number of terms, is positive like its first term $K_{m+1}\rho_{m+1}$and has a value less than this term. Now, this first term converging to the limit $k_{m+1} f(0)$, it follows that the sum (4) will always be less than $k_{m+1} f(0)$ increased by a positive quantity ω' as small as one wishes. In combining this result with the one we obtained for the sum (3) only a moment ago, one sees that integral (1), which is the sum of expressions (3) and (4) will always finish by differing from $S_m f(0)$ by a quantity less, abstraction being made of sign, than $\omega + \omega' + k_{m+1} f(0)$, ω and ω' being quantities of arbitrary smallness. On the other hand S_m differs from $\frac{\pi}{2}$ by a quantity numerically less than $k_m + 1$; therefore the integral will always finish by differing from $\frac{\pi}{2} f(0)$ by a quantity less than $\omega + \omega' + 2k_{m+1} f(0)$, abstraction being made of sign.

As m can be chosen so great that k_{m+1} will be less than any given quantity, it follows that integral (1) will always finish when k increases without limit, by constantly differing from $(\pi/2) f(0)$ by a quantity less, abstraction being made of sign, than any number as small as you wish. It is therefore proved that integral (1) converges to the limit $(\pi/2) f(0)$ for increasing values of k.

Let us now suppose that the function $f(\beta)$, instead of always decreasing from 0 to h, is constant and equal to unity. In this case one can determine the limit to which integral (1) converges by the same considerations as we have already used; one can

[4]The principle that we are going to apply can be stated in this way. The letters $A, A', A'', \ldots$ denoting an arbitrary number of positive quantities such that $A > A' > A''$ etc., the quantity $A - A' + A'' - A''' +$ etc. is positive and less than A. This follows immediately from the fact that the preceding quantity can always be put in one of the two following forms: $(A - A') + (A'' - A''') + \cdots$ or $A - (A' - A'') - (A''' - A^{iv}) + \cdots$.

see this at once by recalling that the preceding demonstration is based on the fact that the integrals into which we have decomposed integral (1) form a decreasing sequence. Now this decrease depends on two things, the decrease in the factor $f(\beta)$ and the increase in the divisor $\sin\beta$. If $f(\beta)$ becomes a constant number, the increase in $\sin\beta$ will always suffice to make each integral in the series smaller than the one before. One thus finds, always supposing h to be positive and at most equal to $\frac{\pi}{2}$ that the integral $\int_0^h \frac{\sin k\beta}{\sin\beta} d\beta$ converges to the limit $\frac{\pi}{2}$. It follows that the integral $\int_0^h c\frac{\sin k\beta}{\sin\beta} d\beta$, in which c is a positive or negative constant, converges to the limit $c\frac{\pi}{2}$.

We have supposed that the function $f(\beta)$ was decreasing and positive between the limits 0 and h. The first circumstance continuing to hold, that is to say the function being such that $f(p) - f(q)$ has the opposite sign to $p - q$ for values of p and q between 0 and h, let us suppose that $f(\beta)$ is not always positive. One can take a positive constant c so great that $c + f(\beta)$ always takes a positive sign from $\beta = 0$ to $\beta = h$. The integral $\int_0^h \frac{\sin k\beta}{\sin\beta} d\beta$ being equal to the difference of these:

$$\int_0^h (c + f(\beta))\frac{\sin k\beta}{\sin\beta} d\beta - \int_0^h c\frac{\sin k\beta}{\sin\beta} d\beta$$

the limit will be the difference of the limits to which these latter converge. Now these latter belong to the case previously examined ($c + f(\beta)$ being a decreasing and positive function) and converge to the limits $[c + f(0)]\frac{\pi}{2}$ and $c\frac{\pi}{2}$, whence the first converges to the limit $\frac{\pi}{2}f(0)$.

Let us now consider a function $f(\beta)$ increasing from 0 to h. In this case $-f(\beta)$ will be a decreasing function. The integral $\int_0^h -\frac{\sin k\beta}{\sin\beta} d\beta$ will therefore converge to the limit $-\frac{\pi}{2}f(0)$, and consequently the integral $\int_0^h \frac{\sin k\beta}{\sin\beta} d\beta$ to the limit $\frac{\pi}{2}f(0)$.

Putting these results together. One has this statement: Whatever be the function $f(\beta)$, provided that it remains continuous between the limits 0 and h (h being positive and at most equal to $\frac{\pi}{2}$), and whether it increases or decreases from the first of these limits to the second, the integral $\int_0^h f(\beta)\frac{\sin k\beta}{\sin\beta} d\beta$ will finish by constantly differing from $\frac{\pi}{2}f(0)$ by a quantity less than any number assignable when one lets k increase beyond any positive limit.

Let us denote by g a positive number different from 0 and less than h, and let us suppose that the function remains continuous and increases or decreases from g to h. The integral $\int_0^h f(\beta)\frac{\sin k\beta}{\sin\beta} d\beta$ will then converge to a limit that it is easy to discover. One can proceed by considerations analogous to those that we have applied to integral (1); but it is simpler to reduce this new case to those we have considered in what has preceded. The function being given only from g to h it is entirely arbitrary for values of β between 0 and g. Suppose that one understands by $f(\beta)$, for values of β between 0 and g, a continuous function that increases or decreases from 0 to g according as $f(\beta)$ increases or decreases from g to h; suppose moreover that $f(g-\varepsilon)$ differs infinitely little from $f(g+\varepsilon)$ if ε decreases without limit; having satisfied these conditions in an arbitrary way, which one can always do in infinitely many ways, the function $f(\beta)$ satisfies from 0 to h the conditions expressed in statement (5). The integrals:

$$\int_0^g f(\beta)\frac{\sin k\beta}{\sin\beta}d\beta \quad \int_0^h f(\beta)\frac{\sin k\beta}{\sin\beta}d\beta$$

will therefore converge, both of them, to the limit $\frac{\pi}{2}f(0)$. Whence one concludes that the integral $\int_g^h f(\beta)\frac{\sin k\beta}{\sin\beta}d\beta$, which is the difference of the preceding, has zero as its limit.

This new result can be put together in a single statement with the one we have obtained above. One will then have: The letter h denoting a positive quantity at most equal to $\frac{\pi}{2}$ and g being a quantity that is also positive and in addition less than h, the integral

$$\int_g^h f(\beta)\frac{\sin k\beta}{\sin\beta}d\beta$$

in which the function $f(\beta)$ is continuous between the limits of integration and is either always increasing or always decreasing from $\beta = g$ to $\beta = h$, will converge to a certain limit when the number k becomes greater and greater. This limit is equal to zero except in the case when g has the value null, in which case it has the value $\frac{\pi}{2}f(0)$.

It is evident that this result will only be lightly modified if the function $f(\beta)$ interrupts the continuity at $\beta = g$ or $\beta = h$, that is to say if $f(g)$ were different from $f(g+\varepsilon)$ and $f(h)$ from $f(h-\varepsilon)$, ε denoting an infinitely small positive quantity, provided that the values $f(g)$ and $f(h)$ do not become infinite. It is only necessary in this case to replace $f(0)$ by $f(\varepsilon)$ in the preceding statement, considering that $f(\varepsilon)$ is equal to $f(0)$.

We are now ready to prove the convergence of periodic series which express arbitrary functions between given limits. The path which we shall follow will lead us to establish the convergence and at the same time to determine their values. Let $\phi(x)$ be a function of x, having a finite and definite value for each value of x lying between $-\pi$ and π, and suppose that it is required to develop this function in a series of sines and cosines of multiple arcs of x. The series which solves this question is, as one knows:

$$\frac{1}{2\pi}\int \phi(\alpha)d\alpha$$
$$+\frac{1}{\pi}\left\{\begin{array}{l}\cos x\int\phi(\alpha)\cos\alpha d\alpha + \cos 2x\int\phi(\alpha)\cos 2\alpha d\alpha + \cdots \\ \sin x\int\phi(\alpha)\sin\alpha d\alpha \,+\, \sin 2x\int\phi(\alpha)\sin 2\alpha d\alpha + \cdots\end{array}\right\}, \quad (7)$$

the integrals which determine the constant coefficients being taken from $\alpha = -\pi$ to $\alpha = \pi$ and x denoting an arbitrary quantity lying between $-\pi$ and π. (*Théorie de la Chaleur* No. 232 ff.).

Let us consider the first $2n + 1$ terms of this series (n being an integer) and see to what limit the sum of these terms converges when n becomes greater and greater. This sum can be put in the following form:

$$\frac{1}{\pi}\int_{-\pi}^{\pi}\phi(\alpha)d\alpha\left[\frac{1}{2}+\cos(\alpha-x)+\cos 2(\alpha-x)+\cdots+\cos n(\alpha-x)\right],$$

or, on summing the sequence of cosines:

$$\frac{1}{\pi}\int_{-\pi}^{\pi}\phi(\alpha)\frac{\sin(n+\frac{1}{2})(\alpha-x)}{2\sin\frac{1}{2}(\alpha-x)}d\alpha. \quad (8)$$

All now reduces to determining the limit to which this integral continually approaches when n increases indefinitely. For this we divide it into two parts, the one taken from $-\pi$ to x, the other from x to π. If one replaces α in the first one by $x - 2\beta$, and in the second α by $x + 2\beta$, β being a new variable the two integrals change into these, abstraction being made of the factor $\frac{1}{\pi}$:

$$\int_{0}^{(\pi+x)/2}\frac{\sin(2n+1)\beta}{\sin\beta}\phi(x-2\beta)d\beta,$$

$$\int_{0}^{(\pi-x)/2}\frac{\sin(2n+1)\beta}{\sin\beta}\phi(x+2\beta)d\beta. \quad (9)$$

Let us consider the second of these two integrals. The quantity x being less than or equal to π, abstraction being made of sign, $(\pi - x)/2$ cannot fall outside the limits 0 and π. If $(\pi - x)/2 = 0$, which happens when $x = \pi$, the integral is null whatever n is; in all other cases it will converge for increasing values of n to a limit that we are going to determine. Let us suppose first of all that $(\pi - x)/2$ is less than or at most equal to $\frac{\pi}{2}$, and let us note that the function $\phi(x + 2\beta)$ can have several breaks in continuity from $\beta = 0$ to $\beta = (\pi - x)/2$, and that it can also have several maxima and minima in this same interval. Let us denote by $\ell, \ell', \ell'', \ldots, \ell^{(\nu)}$, ranged in order of size, the different values of β which arise at one or other of those circumstances, and let us decompose our integral into several others taken between the limits:

$$0 \text{ and } \ell,\ \ell \text{ and } \ell',\ \ell' \text{ and } \ell'',\ \ldots,\ \ell^{(\nu)} \text{ and } \frac{1}{2}(\pi-x).$$

All these integrals occur in the case of statement (6). They all therefore converge to the limit zero as n increases, with the exception of the first which converges to the limit $\frac{1}{2}\phi(x + \varepsilon)$, ε being an infinitely small positive number.

If $\frac{1}{2}(\pi - x)$ is greater than $\frac{\pi}{2}$, which will happen when x has a negative value, one divides the integral into two others, one taken from $\beta = 0$ to $\beta = \frac{\pi}{2}$, the other from $\beta = \frac{\pi}{2}$ to $\beta = \frac{1}{2}(\pi - x)$. The first of these two integrals is one that we have already considered, it will therefore converge to the limit $\frac{\pi}{2}\phi(x+\varepsilon)$. As for the second, it can be changed into this, on replacing β by $\pi-\gamma$, γ being a new variable:

$$\int_{(\pi+x)/2}^{\frac{\pi}{2}} \phi(x+2\pi-2\gamma)\frac{\sin(2n+1)(\pi-\gamma)}{\sin(\pi-\gamma)}d\gamma,$$

or, which comes to the same thing, n being an integer:

$$\int_{(\pi+x)/2}^{\frac{\pi}{2}} \phi(x+2\pi-2\gamma)\frac{\sin(2n+1)(\gamma)}{\sin(\gamma)}d\gamma.$$

It thus has a form analogous to the preceding, and decomposing it like the preceding into several others, one sees that it will converge to the limit zero, the sole case excepted where $\frac{1}{2}(\pi+x)$ has the value null, that is to say when $x=-\pi$; in this case it continually approaches the limit $\left[\frac{\pi}{2}\right]\phi(\pi-\varepsilon)$, ε always having the same signification.[5] In summarising all that has gone before, we find that the second integral of (9) is zero when $x=\pi$, that it converges to the limit $\frac{\pi}{2}[\phi(\pi-\varepsilon)+\phi(-\pi+\varepsilon)]$ when $x=-\pi$, and that in all other cases it continually approaches the limit $\frac{\pi}{2}\phi(x+\varepsilon)$.

The first of the integrals (9) is entirely analogous to the second; on applying similar considerations to it, one finds that it is null when $x=-\pi$, that it converges to the limit $\frac{\pi}{2}[\phi(\pi-\varepsilon)+\phi(-\pi+\varepsilon)]$ when $x=\pi$ and that in all other cases it has the limit $\frac{\pi}{2}\phi(x-\varepsilon)$. Knowing in this way the limits of each of the integrals (9), it is easy to find the limit that integral (8) continually approaches when n becomes greater and greater. It is sufficient for this to recall that each integral is equal to the sum of the integrals (9) divided by π. Now, integral (8) being equivalent to the sum of the first $2n+1$ terms of the series (7), the convergence of this series is proved, and one finds by means of the preceding results that it is equal to

$$\frac{1}{2}[\phi(x-\varepsilon)+\phi(x+\varepsilon)]$$

for every value of x between $-\pi$ and π, and that for each of the extreme values pi and π it is equal to

$$\frac{1}{2}[\phi(\pi-\varepsilon)+\phi(-\pi+\varepsilon)].$$

The preceding exposition embraces all cases; it simplifies when the value of x is not one of those that presents a break in continuity. In fact, the quantities $\phi(x+\varepsilon)$ and $\phi(x-\varepsilon)$ then each being equivalent to $\phi(x)$, one sees that the series has the value $\phi(x)$.

The preceding considerations prove in a rigorous manner that, if the function $\phi(x)$, whose values are supposed finite and definite, has only a finite number of breaks of continuity between the limits $-\pi$ and π, and if moreover it has only a definite number

[5]The factor $\left[\frac{\pi}{2}\right]$ was omitted by Dirichlet.

of maxima and minima between these same limits, then series (7), whose coefficients are definite integrals depending on the function $\phi(x)$, is convergent and generally has the value expressed by

$$\frac{1}{2}[\phi(x-\varepsilon)+\phi(x+\varepsilon)],$$

where ε denotes an infinitely small number. It remains to consider the case where the assumptions that we have made about the number of breaks in continuity and on the natural number of maxima and minima cease to hold. These singular cases can be reduced to those that we have considered. It is only necessary for series (8) to make sense when the breaks in continuity are infinite in number, that the function $\phi(x)$ satisfies the following condition.

It is then necessary that the function $\phi(x)$ be such that, if one denotes by a and b two arbitrary quantities lying between $-\pi$ and π, one can always find between a and b two quantities r and s so close that the function is continuous in the interval from r to s. One easily appreciates the necessity of this restriction on considering that the different terms of the series are definite integrals and going back to the fundamental notion of integrals. One will then see that the integral of a function only signifies something when the function satisfies the condition previously stated. One will have an example of a function that does not satisfy this condition, if one supposes that $\phi(x)$ is equal to a definite constant c when the variable x has a rational value, and is equal to a different constant d when this variable is irrational. The function thus defined has finite and definite values for each value of x, however, one cannot substitute it in the series because the different integrals which enter this series lose all signification in this case. The restriction which I have made precise, and that of not becoming infinite, are the only ones to which the function $\phi(x)$ must be subjected, and all the cases that we have not excluded can be reduced to those we have considered in the preceding. But this matter, to be done with all the clarity that one desires, requires some details related to the fundamental principles of analysis, and will be expounded in another note, where I will also consider some other quite remarkable properties of series (7).

A.3 Riemann on Elementary Complex Function Theory

G.B.F. Riemann, *Foundations for a general theory of functions of a variable complex quantity.*[6] Inauguraldissertation, Göttingen, 1851, second unaltered publication, Göttingen, 1867.

§ 1.

[6]In making this translation, I consulted the one in *A Source Book in Classical Analysis*, G. Birkhoff and U. Merzbach (eds.) 1973, pp. 48–50, and some phrases of that translation appear here.

If one thinks of z as a variable that successively takes all possible real values, then if to each value of z there corresponds a single value of w, w is called a function of z. If, when z varies continuously on an interval, w also varies continuously, this function is called continuous on this interval.

This definition evidently does not assume any fixed law describing the function since, once this function has been defined on a given interval, it may be extended arbitrarily outside it.

The dependence of w on z may be given by a mathematical law which determines from each value of z the corresponding value of w. Formerly only certain kinds of functions (functiones continuae in Euler's terminology) were considered capable of satisfying the same law of dependence for all values of z in a given interval. Recent researches have shown, however, that there exist analytic expressions by which any continuous function may be represented on a given interval. It is therefore all the same whether the function is defined arbitrarily or by a formula. Because of the theorem recalled above, the two concepts are equivalent.

However, it is otherwise when z is not restricted to real values, and complex numbers of the form $z = x + iy$ (where $i = \sqrt{-1}$) are included.

Let $x + yi$ and $x + yi + dx + dyi$ be two values of the quantity z differing infmitesimally, to which correspond the values $u + iv$ and $u + vi + du + dvi$ of the quantity w. Then if the dependence of the quantity w on z is taken to be arbitrary ... the ratio $\frac{du+dvi}{dx+dyi}$ will in general vary with dx and dy. For, if one sets $dx+dyi = \varepsilon e^{\phi i}$, then

$$\begin{aligned}\frac{du + dvi}{dx + dyi} &= \frac{1}{2}\left(\frac{\partial u}{\partial x} + \frac{\partial v}{\partial y}\right) + \frac{1}{2}\left(\frac{\partial v}{\partial x} - \frac{\partial}{\partial y}\right)\\ &\quad + \frac{1}{2}\left[\frac{\partial u}{\partial x} - \frac{\partial v}{\partial y} + \left(\frac{\partial v}{\partial x} + \frac{\partial u}{\partial y}\right)i\right]\\ \frac{dx - dyi}{dx + dyi} &= \frac{1}{2}\left(\frac{\partial u}{\partial x} + \frac{\partial v}{\partial y}\right) + \frac{1}{2}\left(\frac{\partial v}{\partial x} - \frac{\partial u}{\partial y}\right)i\\ &\quad + \frac{1}{2}\left[\frac{\partial u}{\partial x} - \frac{\partial v}{\partial y} + \left(\frac{\partial v}{\partial x} + \frac{\partial u}{\partial y}\right)i\right]e^{-2\phi i}.\end{aligned}$$

But no matter how w is determined as a function of z by tying together simple algebraic operations the value of the derivative [or differentialquotient] $\frac{dw}{dz}$ must always be independent of the particular value of dz.[7] Evidently, not every dependence of the complex quantity w on the complex quantity can be expressed in this way.

This characteristic property of all functions defined by explicit operations, will be taken as basic in what follows, where a function will be considered independent of its expression, and we, without proving the general validity and permissibility of

[7]This requirement is evidently satisfied in all cases where from the expression for w in terms of z an expression for $\frac{dw}{dz}$ in terms of z is obtained by means of the rules for differentiation; the rigorous general validity of this remains for now undiscussed.

expressing every dependence through operations on quantity, shall proceed from the following definition:

A variable complex quantity w is called a function of another variable complex quantity z when its variation is such that the value of the derivative $\frac{dw}{dz}$ is independent of the value of the differential dz.

§ 2.

Let the quantities z and w be complex be taken as variables that can take each take complex values. The presentation of their variation on a connected two-dimensional domain is made essentially easier by a connection to spatial intuition. One thinks of each value $x + iy$ of the quantity z as represented by the point O of the plane A with rectangular coordinates (x, y), and each value $u + vi$ by the point Q of the plane B with rectangular coordinates (u, v). Every dependence of the quantity w on z will then be represented by the dependence of the position of the point q on that of the point O. If to each value of z there corresponds a value of w, which varies continuously with z, in other words, if u and v are continuous functions of x and y—then to each point of the plane A will correspond a point of the plane B, every curve generally to a curve, and every connected piece of surface to a connected piece of surface. One can think of this dependence of the quantity w on the quantity z as a mapping of the plane A onto the plane B.

§ 3.

We now investigate what properties this mapping has when w is a function of the complex quantity z, that is when $\frac{dw}{dz}$ is independent of dz. We denote by o be an arbitrary point of the plane A in the neighbourhood of O, and by q its image in the plane B, also, let $x + yi + dx + dyi$ and $u + vi + du + dvi$ be the values of z and w at these points. Then dx, dy and du, dv can be regarded as rectangular coordinates of the points o and q relative to the points O and Q considered as origins, and if one writes $dx + dy = \varepsilon e^{\phi i}$ and $du + dvi = \eta e^{\psi i}$ then the quantities $\varepsilon, \phi, \eta, \psi$ will be the polar coordinates of these points for the same origins. Now if o' and o'' are any two specified locations of the point o infinitely near to O, and if one denotes the other variables by the corresponding indices, then we have

$$\frac{du' + dv'i}{dx' + dy'i} = \frac{du'' + dv''i}{dx'' + dy''i}$$

and consequently

$$\frac{du' + dv'i}{du'' + dv''i} = \frac{\eta'}{\eta''}e^{(\psi' - \psi'')i} = \frac{dx' + dy'i}{dx'' + dy''i} = \frac{\varepsilon'}{\varepsilon''}e^{(\phi' - \phi'')i},$$

whence $\frac{\eta'}{\eta''} = \frac{\varepsilon'}{\varepsilon''}$ and $\psi' - \psi'' = \phi' - \phi''$, that is to say, in the triangles $o'Oo''$ and $q'Qq''$ the angles $o'Oo''$ and $q'Qq''$ are equal and the sides enclosing them proportional.

Thus there exists equality in the smallest parts between two corresponding infinitesimal triangles and consequently between the plane A and the lane B in general. An exception occurs only in the special case when the corresponding variations in the quantities z and w does not have a finite ratio, which is here tacitly assumed.

§ 4.

If one writes the derivative $\frac{du+dvi}{dx+dyi}$ in the form

$$\frac{\left(\frac{\partial u}{\partial x} + \frac{\partial v}{\partial x}i\right)dx + \left(\frac{\partial v}{\partial y} - \frac{\partial u}{\partial y}i\right)dyi}{dx + dyi}$$

it is clear that its value will be independent of dx and dy if and only if

$$\frac{\partial u}{\partial x} = \frac{\partial v}{\partial y} \quad and \quad \frac{\partial v}{\partial x} = -\frac{\partial u}{\partial y}.$$

These conditions are therefore necessary and sufficient for $w = u + vi$ to be a function of $z = x + yi$. Consequently the individual parts of this function satisfy the following

$$\frac{\partial^2 u}{\partial x^2} + \frac{\partial^2 u}{\partial y^2} = 0, \quad \frac{\partial^2 v}{\partial x^2} = \frac{\partial^2 v}{\partial y^2},$$

which will be taken as fundamental for the study of the properties of the individual parts of such a function, considered separately.

A.4 Riemann's Definition of the Integral

From Riemann *On the representability of a function by a trigonometric series*: The concept of a definite integral and the scope of its validity.

§ 4

The indeterminancy which still prevails on a number of fundamental points of the theory of definite integrals compels us to make some preliminary remarks about the concept of a definite integral and the scope of its validity.

So first: what have we to understand by $\int_a^b f(x)dx$?

To determine this, we take a sequence of values $x_1, x_2, \ldots, x_{n-1}$ increasing in size between a and b, and for brevity denote $x_1 - a$ by δ_1, $x_2 - x_1$ by δ_2,..., $b - x_{n-1}$ by δ_n, and by ε a positive real fraction. Then the value of the sum

$$S = \delta_1 f(a + \varepsilon_1\delta_1) + \delta_2 f(x_1 + \varepsilon_2\delta_2) + \delta_3 f(x_2 + \varepsilon_3\delta_3) + \ldots + \delta_1 f(x_{n-1} + \varepsilon_n\delta_n)$$

depends on the choice of intervals δ and the quantities ε. If this now has the property that however δ and ε may be chosen, as soon as all δ's become infinitely small, it approaches infinitely close to a fixed limit A, then this value is called $\int_a^b f(x)dx$.

If it does not have this property, then $\int_a^b f(x)dx$ has no meaning.

However, in several cases, attempts have been made to attribute a meaning to this symbol, and among these expansions of the concept of a definite integral there is one that is accepted by all mathematicians. To wit, if the function $f(x)$ becomes infinitely large as the argument approaches a particular value c in the interval (a, b), then obviously the sum S, whatever degree of smallness we attribute to the δ's, can obtain any value whatsoever. Thus it has no limiting value, and $\int_a^b f(x)dx$, as above, would have no meaning. But, should

$$\int_a^{c-\alpha_1} f(x)dx + \int_{c+\alpha_2}^b f(x)dx,$$

when α_1 and α_2 become infinitely small, approach a fixed limit, $\int_a^b f(x)dx$ is understood to be this limit value.

Other statements by Cauchy about the concept of the definite integral in cases where, under the basic concept, one does not exist may be useful for certain classes of investigations. However, they were not introduced at a general level and, given their great arbitrariness, are scarcely appropriate.

§ 5

Second, let us now look at the scope of validity of this concept, or at the question: in which cases is a function integrable and in which not?

We shall first look at the concept of integral in the strict sense, i.e., we assume that the sum S converges when all the δ's become infinitely small. If we call the greatest variation of the function between a and x_1, i.e., the difference between its largest and smallest values in this interval, D_1, that between x_1 and x_2, $D_2, \ldots,$ that between x_{n-1} and b, D_n, then

$$\delta_1 D_1 + \delta_2 D_2 + \cdots + \delta_n D_n$$

must become infinitely small with the quantities δ. We further assume that, as long as all δ's remain smaller than d, the greatest value that this sum can have is Δ; Δ will then be a function of d that is always decreasing along with d and becomes infinitely small with this quantity. Now if the total size of the intervals in which the variations are greater than σ is $= s$, the contribution these intervals make to the sum $\delta_1 D_1 + \delta_2 D_2 + \cdots + \delta_n D_n$ is evidently $\geq \sigma s$. One therefore has

$$\sigma s \leq \delta_1 D_1 + \delta_2 D_2 + \cdots + \delta_n D_n \leq \Delta,$$

consequently $s \leq \frac{\Delta}{\sigma}$.

Now $\frac{\Delta}{\sigma}$, when σ is given, can always be made infinitely small by a suitable choice of d; hence the same is true of s, and this results in:

For the sum S to converge as all δ's become infinitely small, not only does the function $f(x)$ need to be finite, but also the total size of the intervals in which the variations are $> \sigma$ (whatever σ may be) must be able to be made arbitrarily small by a suitable choice of d.

This theorem can also be inverted:

If the function $f(x)$ is always finite and, as all the quantities δ decrease the total size s of those intervals where the variations of function $f(x)$ are larger than a given value σ continually becomes infinitely small, then the sum S will converge as all δ's become infinitely small.

For, those intervals where the variations are $> \sigma$ contribute an amount smaller than s to the sum $\delta_1 D_1 + \delta_2 D_2 + \cdots + \delta_n D_n$ multiplied by the greatest variation of the function between a and b, which (by hypothesis) is finite. The remaining intervals contribute an amount $< \sigma(b - a)$. Evidently one can first assume σ to be arbitrarily small and then always (by hypothesis) so determine the size of the intervals that s also becomes arbitrarily small, whence the sum $\delta_1 D_1 + \delta_2 D_2 + \cdots + \delta_n D_n$ can be given any arbitrary smallness one wishes, and consequently the sum S can be contained within arbitrarily narrow limits.

Thus we have found conditions that are necessary and sufficient for the sum S to converge as the values of δ infinitely decrease and so we can speak in the strict sense of an integral of the function $f(x)$ between a and b.

If the concept of the integral is now expanded as above, it is evident that for integration always to be possible, the latter of the two conditions we found is still necessary; but in place of the condition that the function always be finite comes the condition that the function only becomes infinite as the argument approaches particular values, and that a determinate limiting value emerge as the integration limits are approach infinitely close to this value.

A.5 Schwarz on Squaring the Circle

From H.A. Schwarz, Ueber einige Abbildungsaufgaben, *Journal für die reine und angewandte Mathematik* vol. 70 (1869), pp. 105–120; in *Gesammelte Mathematische Abhandlung* vol. 2 (1890), pp. 65–83, this extract pp. 66–70.[8]

[8]This translation is adapted from the one in *A Source Book in Classical Analysis*, G. Birkhoff and U. Merzbach (eds.) 1973, pp. 56–59.

[Schwarz first commented on the striking contrast between Riemann's general mapping theorem and the absence of specific formulas for mapping triangles (say) on the unit circle. He then began to solve that problem as follows.]

The following fruitful theorem leads to the solution of this and many other mapping problems: If to a continuous sequence of real values of the complex argument of an analytic function there corresponds a continuous sequence of real values of the function, then to any two conjugate values of the argument there correspond conjugate values of the function. In the u-plane, whose points represent the values of a complex variable u geometrically, let U' be a bounded, simply connected domain whose boundary contains a finite segment ℓ of the real axis.

Let a single-valued analytic function t of the complex variable u, $t = f(u)$ behave like an entire function for all values of u in the interior of U'. That is, if u_0 denotes an arbitrary value of u in the interior of the domain U' then the function $f(u)$ can be expanded in a power series in $u - u_0$ for the values of u lying in a neighbourhood of u_0 that converges for all sufficiently small absolute values of $u - u_0$. We shall assume that t remains bounded as u approaches the boundary ℓ and is real for all points of the line ℓ, and that for all values of the argument u in the interior and the boundary of the domain U' the function $t = f(u)$ varies continuously with the argument u.

To the domain U' there corresponds a domain U'' whose points lie symmetrically with the points of U' with respect to the real axis.

At all points of the domain U'' an analytic function t is defined that in the domains U' and U'' takes conjugate values at conjugate values of u in the domains. If one thinks of the two domains U' and U'' as joined along the segment ℓ, then one obtains a simply connected domain $U' + U''$. For all values of the argument u in the interior of $U' + U''$, the value of t is uniquely defined; and indeed in the interior of U' and the interior of U'' as an analytic function of this argument which behaves like an entire function. On crossing the line ℓ, and along that line, the value of t varies continuously. It follows from this that the function t determined for the domain U'' is an analytic continuation of the function determined for the domain U', and is indeed a continuation extending beyond the line ℓ. The proof of the validity of this statement can be carried out as follows if, as may be assumed, the domain $U' + U''$ everywhere covers the u-plane only simply.

If u_0 denotes a value of u in the interior of U' then, by a theorem of Cauchy, the integral

$$\frac{1}{2\pi i}\int \frac{f(u)}{u - u_0}du$$

taken in a positive sense around the boundary of the domain U' or that of U'', has the value $f(u_0)$ in the first case and the value 0 in the second. On adding these two integrals, the integrations along the line ℓ, carried out twice and in opposite senses, cancel out; and the following equation holds

$$f(u_0) = \frac{1}{2\pi i}\int \frac{f(u)}{u - u_0}du,$$

where the integral is taken along the boundary of the domain $U' + U''$, and for all values of the quantity u_0 that are among the points represented geometrically in the interior of these domains, represents a continuous function of this argument whose values coincide everywhere with the values of the function $t = f(u)$.

It follows that the function so determined also behaves like an entire function for all values in the interior of the segment ℓ. Thus, under the assumptions made, conjugate values of the argument correspond to conjugate values of the function, or, expressed geometrically, the conformal mapping of the (u)-plane onto the (t)-plane, whose points represent the values of the complex quantity t, is symmetric with respect to the real axis for both planes: to symmetric points there correspond symmetric images.

If one analytically continues the function $t = f(u)$, symmetrically on both sides of the real axis in the (u)-plane, one is led to the result that singular points of some kind lie either singly on the real axis or in pairwise symmetry on both sides of it. This theorem can immediately be extended to an analytic function which maps a straight line segment in the domain of the argument or on its boundary onto a straight line segment in the plane whose points represent the value of the analytic function geometrically.

For the special problem of mapping the surface of a square onto the surface of a disk, one easily guesses that, to make the centre of the square correspond to the centre of the disk, the four straight lines that are symmetry axes of the square should in any case be mapped onto straight lines. This comparison locates the four singular points on the circumference of the circle that correspond to the four corners of the square under this special assumption.

Now the solution of the given problem could evidently be simplified by replacing the disk by a half- plane, which can be achieved by the transformation by reciprocal radii [inversion, JJG]. Indeed, the resulting simplification lies in the circumstance that the boundaries of the two regions to be mapped onto each other are now straight lines. By the general law given above, the mapping function can be continued analytically outside of the square in which it had been originally defined. If the centre of the transformation is taken to be one of the singular points on the circumference of the circle, it follows that the points $t = \infty, t = -1, t = 0, t = +1$ on the real axis can be taken as singular points, while the half-plane lying on the positive side of the real axis is a conformal image of the disk.

If the position of a point inside the given square is determined by the complex number u then the problem requires that for all [such] z, the complex variable t be an analytic function for all values of the argument u that correspond to points lying in the interior of the given square, with the property that all values corresponding to points u on the periphery of the square shall have real values. The domain of the argument u can now be extended by the above principle, first to the interior of four squares symmetrically placed adjacent to the given square, then by repeated application to an arbitrarily large domain of the (u)-plane.

It follows that the [extended] function t must, for all finite values of its argument u in the extension of its domain, be a single-valued function and indeed a doubly periodic function of u, the ratio of whose fundamental periods is $\sqrt{-1}$. This already indicates the lemniscatic function.

The boundary of the square has singular points at its four corners. Under the requirement that the perimeter of the square is to be mapped onto the perimeter of the circle or onto the perimeter of the half-plane so that they are equal in the smallest parts [infinitesimally conformal], these points must be excepted; otherwise the given problem would contain an impossible condition.

Each piece of the surface of the square near a corner, being a right-angled sector near the vertex, must be mapped onto a straight angle by the function describing the given mapping.

This leads to the problem of finding the most general function that maps the sector subtending an angle $\alpha\pi$ lying near the vertex $u = 0$ in the (u)-plane,

$$u = re^{\phi i}; \quad 0 \leq \phi \leq \alpha\pi; \quad 0 < r < r_0$$

conformally onto the half-plane

$$t = \rho e^{\psi i}, \quad 0 \leq \psi \leq \pi$$

so that inside the given boundaries each point $u = re^{\phi i}$ corresponds continuously to a point $t = \rho e^{\psi i}$, while the values

$$r = 0, \rho = 0; \phi = 0, \psi = 0; \phi = \alpha\pi, \psi = \pi$$

correspond to each other. The simplest function providing such a mapping is the function $v = u^{1/\alpha}$. Every other function t of the argument u that also provides a mapping with the stated properties considered as a function of the variable v has, by the above theorem, the character of an entire function for the value $v = 0$ and in the vicinity of this value.

Likewise conversely the quantity v can be expressed as an analytic function of the argument t, which behaves like an entire function for all values of the complex quantity t in the vicinity of and including $t = 0$. Hence one obtains the following analytic representations, valid in the vicinities of the values $v = 0$ and $t = 0$:

$$v = u^{1/\alpha}; \quad t = Cv(1 + a_1 v + a_2 v^2 + \cdots); \, v = \frac{1}{C} t(1 + b_1 t + b_2 t^2 + \cdots);$$

$$u = v^{\alpha}; \quad u = \frac{1}{C^{\alpha}} t^{\alpha}(1 + c_1 t + c_2 t^2 + \cdots).$$

The constant C is positive and different from 0; the coefficients a, b, c all have real values; the latter follows from the fact that to all sufficiently small positive values of u with respect to the quantity v positive values of t must again correspond.

[Schwarz continued and eventually obtained what became known as the 'Schwarz–Christoffel' transformation

$$u' = \int_0^t \frac{dt}{\sqrt{4t(1-t^2)}}$$

which is a lemniscatic integral that represents the interior of two half-planes divided by the real axis onto the interior of a square with sides $\int_0^1 \frac{dt}{\sqrt{4t(1-t^2)}}$. The substitution $s = \frac{t-i}{t+i}$ takes one from the half-plane lying on the positive side of the real axis in the (t)-plane to the surface of the circle lying in the (s)-plane with radius 1 and centre the point $s = 0$.]

Appendix B
Series of Functions

Very often, functions are defined as convergent infinite sums of other functions, and when this is done we can ask if the properties of the individual terms of the sum (integrable, continuous, differentiable, ...) also belong to the limit function. The over-simplified general philosophy is that they do if the convergence is uniform, otherwise not. The troublesome case is differentiability: the terms may be differentiable and the convergence uniform, without the limit function being differentiable. When the relevant property is assured, the question of term by term integration or differentiation arises.

Two classes of infinite series are generally encountered: power series and Fourier series. A power series has a radius of convergence, and on any smaller interval (or, in the complex case, disc) convergence is uniform, so with that restriction on the domain of definition, the theorems stated below do apply.

Fourier series are much more complicated. The 'standard hypotheses' on a function are that it is periodic (say with period 2π), regulated on $[-\pi, \pi]$, and has the 'right' Fourier coefficients:

$$a_n = \frac{1}{\pi}\int_{-\pi}^{\pi} f(x)\cos nx dx, \qquad b_n = \frac{1}{\pi}\int_{-\pi}^{\pi} f(x)\sin nx dx.$$

A regulated function is defined as the uniform limit of step functions. Among the regulated functions are functions continuous on a closed interval, say $[a, b]$, and they, like all regulated functions, are integrable. A regulated function, f and its indefinite integral, $F(t) := \int_a^t f(x)dx$, satisfy the fundamental theorem of the calculus:

$$F'(t) = f(t).$$

Even when a function f satisfies the standard hypotheses, sufficient hypotheses for the Fourier series of the function f to convergence pointwise to f are that the function f be differentiable, and sufficient hypotheses for a Fourier series of a function f to

J. Gray, *The Real and the Complex: A History of Analysis in the 19th Century*,
Springer Undergraduate Mathematics Series, DOI 10.1007/978-3-319-23715-2

converge uniformly to the function are that the function f be C^2 (twice continuously differentiable).

Uniform Convergence and Continuity

A sequence of functions $f_n(x)$ defined on a domain D tends to a function $f(x)$ on that domain if for each $x \in D$ and every $\varepsilon > 0$ there is an $N = N(\varepsilon, x)$ such that $n > N$ implies that $|f_n(x) - f(x)| < \varepsilon$. If N can be chosen for all $x \in D$ the convergence is said to be uniform, otherwise it is said to be pointwise.

If a sequence of functions $\{f_n\}$, each continuous at a point $x_0 \in (a, b)$, tends uniformly to a function f then the function f is continuous at the point x_0.

If a series of functions $\sum_n f_n$, each continuous at a point x_0, tends uniformly to a function f then the function f is continuous at the point x_0.

Uniform Convergence and Integrability

If a series of functions $\sum_n f_n$, each continuous and Riemann-integrable on an interval (a, b), tends uniformly to a function f then the function f is Riemann-integrable and the integral of the sum is sum of the integrals; i.e. term by term integration is valid.

Uniform Convergence and Differentiability

Here, uniformity alone is not enough. Indeed, it is possible for a sequence of differentiable functions $\{f_n\}$ to converge uniformly on an open interval and for the sequence $\{f_n'\}$ to fail to converge pointwise on that interval, as we shall see below. However, the following more delicate results hold.

If $\{f_n\}$ is a sequence of functions, each with a finite derivative at every point of the open interval (a, b), and:
If there is at least one point x_0 in (a, b) such that the sequence $\{f_n(x_0)\}$ converges, and
If there is a function g on (a, b) such that the sequence of derived functions $\{f_n'\}$ converges uniformly to the function g on (a, b),
Then there is a function f such that the sequence $\{f_n\}$ converges uniformly to f on (a, b), and for each x_0 in (a, b) the derivative $f'(x)$ exists and $f'(x) = g(x)$.

If $\sum_n f_n$ is a series of functions, each with a finite derivative at every point of the open interval (a, b), and:
If there is at least one point x_0 in (a, b) such that the series $\sum_n f_n(x_0)$ converges, and
If there is a function g on (a, b) such that the series of derived functions $\sum_n f_n'$ converges uniformly to the function g on (a, b),
Then there is a function f such that the series $\sum_n f_n$ converges uniformly to f on (a, b), and for each x_0 in (a, b) the derivative $f'(x)$ exists and $\sum_n f'(x) = g(x)$.

Non-uniform Convergence of Smooth Functions

Define $f(x)$ by:

$$f(x) = \begin{cases} -1 & x \leq -\frac{1}{n} \\ \sin\left(\frac{n\pi x}{2}\right) & -\frac{1}{n} \leq x \leq \frac{1}{n} \\ 1 & x \geq \frac{1}{n} \end{cases}$$

All these functions are differentiable, but the limit is -1 when $x < 0$, 0 when $x = 0$, and $+1$ when $x > 0$.

Indeed, worse can happen. For example, a sequence of functions, (f_n), can tend uniformly to a smooth function, f, and yet the sequence of derived functions (f_n') can fail to converge even pointwise, and a fortiori therefore there is no question of the sequence converging to some limit f'. This is illustrated by the sequence $\frac{\sin(nx)}{\sqrt{n}}$.

Suppose, however, that a sequence of functions, (f_n) defined on the interval (a, b) tends pointwise (not necessarily uniformly) to a smooth function, f, that each function f_n is differentiable and has a bounded derivative on (a, b) and the sequence of derived functions (f_n') converges uniformly to some limit g. Then the conditions of the above theorem apply, and we can deduce that the sequence of derived functions (f_n') converges uniformly to some limit f', the derivative of f—in other words, term by term differentiation is permitted.

The subtleties are well illustrated by Abel's series. Recall that $f_n(x) := \sum_{k=1}^{n}(-1)^{k-1}\frac{\sin(kx)}{k}$. In the interval $(-1, 1)$, say, these functions do a remarkable job of approximating the function $x/2$ (recall Fig. 4.3). Indeed, they do so well that you might imagine that the derived functions

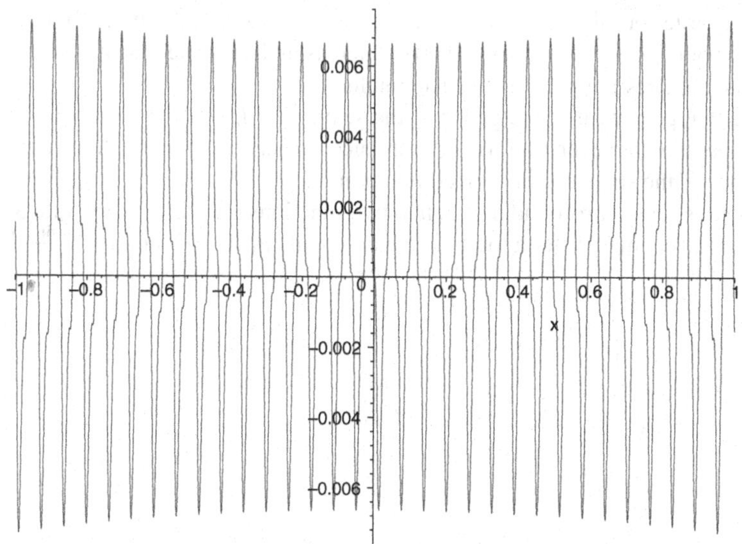

Fig. B.1 Much of the tail of Abel's series

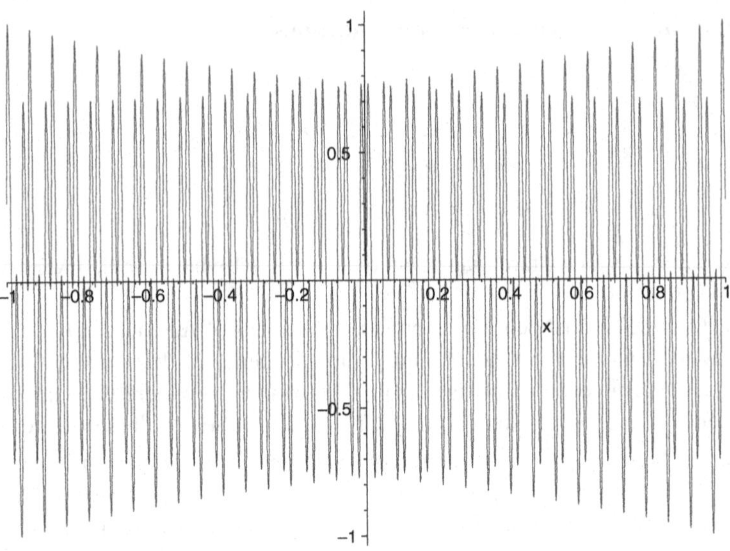

Fig. B.2 Much of the tail of the derivative of Abel's series

$$f_n'(x) := \Sigma_k^n (-1)^{k-1} \cos(kx)$$

approximate the constant function $1/2$ equally well—but they do not (recall Fig. 4.4). Figure B.1 shows the contribution of terms 101 to 300 to $f_n(x) - x/2$ and Fig. B.2 shows their contribution to $f_n'(x) - 1/2$; they make very visible the fact that the convergence of the $f_n(x)$ to $x/2$ is uniform, but that the derived functions do not even converge for all x. You need to look carefully to see that in the first case the uniform bound is already down to 0.007, whereas in the second case it is still 1. This suggests, but does not prove the stated results.

Note that pointwise convergence of the sequence (f_n) would have been enough for the theorem, but even uniform convergence, as here, is not enough to control the behaviour of the sequence of derived functions.

This illustrates a profound moral, that two functions may differ very little and yet their derivatives may differ a lot, as is the case with $f_n(x) = \Sigma_k^n (-1)^{k-1} \frac{\sin(kx)}{k}$ and $x/2$.

Appendix C
Potential Theory: A Mathematical Summary

The next two subsections revise the mathematics of what was discussed historically in Chap. 13. The first covers vector methods (**div**, **grad**, and **curl**). The second looks ahead to the connection between potential theory and the theory of harmonic functions, which became very important for Riemann's pioneering development of complex function theory.

Vector Methods

The calculus of vector fields is concerned with three operators on functions or vector fields: **grad**, **div**, and **curl**. To define them we need some notation. We write a function u of n variables $x_1, x_2, \ldots, x_n$ as $u(x_1, x_2, \ldots, x_n)$, and the ith partial derivative of u as either $\frac{\partial u}{\partial x_i}$, or as u_{x_i}. We write a vector field as $\mathbf{v} = (v_1, v_2, \ldots, v_n)$, where each component v_i is a function of the variables $x_1, x_2, \ldots, x_n$.

First, **grad** applies to a function u and produces a vector:

$$\mathbf{grad}\ u = \nabla u := \left(u_{x_1}, u_{x_2}, \ldots, u_{x_n}\right).$$

Next, **div** applied to a vector field **v** produces a scalar:

$$\mathbf{div}(\mathbf{v}) = \nabla \cdot \mathbf{v} = v_{x_1} + v_{x_2} + \cdots + v_{x_n}.$$

Finally, **curl** applied to a vector field produces a vector field, but the definition we give here (the usual one) only makes sense when there are 3 variables ($n = 3$)—be careful. In this setting, the ith component of the vector field $\mathbf{curl}\ \mathbf{v} = \nabla \times \mathbf{v}$ is $\frac{\partial v_j}{\partial x_k} - \frac{\partial v_k}{\partial x_j}$, where the variables are taken in cyclic order i, j, k.

A major feature of the study of vector fields in the plane or in space is the use of line and surface integrals. The main theorems about them, Gauss's divergence theorem and Stokes's theorem, generalise the fundamental theorem of the one variable calculus. Both concern an arbitrary vector field **v** in space.

J. Gray, *The Real and the Complex: A History of Analysis in the 19th Century*,
Springer Undergraduate Mathematics Series, DOI 10.1007/978-3-319-23715-2

Gauss's divergence theorem concerns a region U of $\mathbb{R}^3$ inside a closed surface S in space, and relates the normal component of $\mathbf{v}$ on the boundary S to the integral of the divergence of the vector field inside U:

$$\int_{surf} \mathbf{v}.\mathbf{n} = \int_{vol} \nabla.\mathbf{v}.$$

It is about surface and volume integrals.

Stokes's theorem concerns a surface in S in $\mathbb{R}^3$ bounded by a closed curve γ, and relates the integral of the tangential component of $\mathbf{v}$ along the boundary to the integral of the normal component of the curl of the vector field over the surface S:

$$\int_{curve} \mathbf{v}.ds = \int_{surf} \nabla \times \mathbf{v}.\mathbf{n}.$$

It is about line and surface integrals.

A **potential function** associated to a vector field $\mathbf{v}$ is a function u such that $\nabla u = \mathbf{v}$. It is an immediate consequence of Stokes's theorem that in many circumstances the line integral of a potential functions depends only on the end points, not the path. It is interesting to note that Stokes first stated his theorem as a problem in the Smith's Prize examination at Cambridge in 1854. The Smith's Prize was an annual event designed to test the best recent Cambridge graduates and to give then a chance to develop an interest in original research. Technical mastery of the material, which is all the proof requires, was regarded a necessary pre-requisite for any kind of further study.

The Gauss–Green Theorem

A good way to think of the mathematics involved in Gauss's and Green's theorems is to realise that it all derives from one theorem, which I shall call the Gauss–Green theorem. This theorem refers to a region U of $\mathbb{R}^n$ with a boundary ∂U on which there is a unit outward normal $^\circ = (\nu^1, \nu^2, \ldots, \nu^n)$. Write the gradient of a function u with continuous first derivatives on U as $\nabla u = \left(\frac{\partial u}{\partial x_1}, \frac{\partial u}{\partial x_2}, \ldots, \frac{\partial u}{\partial x_n}\right)$, and the (outward) normal derivative as $\frac{\partial u}{\partial \nu} = {}^\circ.\nabla u$

The Gauss–Green theorem states that

$$\int_U u_{x_i} dx = \int_{\partial U} u\nu^i dS, \quad i = 1 \ldots n.$$

It immediately implies, on integrating uv by parts, the corollary that

$$\int_U u_{x_i} v dx = -\int_U u v_{x_i} dx + \int_{\partial U} uv\nu^i dS, \quad i = 1 \ldots n.$$

We now upgrade the functions u, v to have continuous second derivatives, and obtain **Green's formulas**:

$$1. \quad \int_U \Delta u dx = \int_{\partial U} \frac{\partial u}{\partial \nu} dS$$

$$2. \quad \int_U \nabla v.\nabla u dx = -\int_U u\Delta v + \int_{\partial U} u\frac{\partial v}{\partial \nu} dS$$

$$3. \quad \int_U (u\Delta v - v\Delta u)\, dx = \int_{\partial U} \left(u\frac{\partial v}{\partial \nu} - v\frac{\partial u}{\partial \nu}\right) dS.$$

Formula 1 follows from the Corollary on replacing u with u_{x_i} and v with 1.
Formula 2 follows from the Corollary on replacing v with v_{x_i}.
Formula 3 follows from Formula 2 by interchanging u and v and subtracting.

Potential and Harmonic Functions

There is a close connection between potential and harmonic functions, as the next result indicates. The function $f(r) = \frac{1}{r}$ is the one appropriate to problems involving mass, magnetism or electricity. Write this as

$$f(x, y, z) = (x^2 + y^2 + z^2)^{-1/2}$$

and set $R = (x^2 + y^2 + z^2)^{-1/2}$. We note that $\frac{\partial R}{\partial x} = -xR^3$ with similar formulae for the other partial derivatives, so

$$\nabla f = (xR^3, yR^3, zR^3).$$

We now compute the Laplacian of f, $\Delta f = \nabla \cdot \nabla f$. We differentiate the first component, xR^3 with respect to x and find

$$\frac{\partial}{\partial x}(xR^3) = R^3 - 3x^2R^5.$$

So the Laplacian turns out to be

$$\begin{aligned} &R^3 - 3x^2R^5 + R^3 - 3y^2R^5 + R^3 - 3z^2R^5 \\ &= 3R^3 - 3(x^2 + y^2 + z^2)R^5 = 3R^3 - 3R^3 = 0. \end{aligned}$$

This computation is valid in any region in which $f \neq \infty$, that is, in any region in which there is no mass, and we deduce that in any region containing no mass the potential function $f(r) = \frac{1}{r}$ is harmonic—it satisfies Laplace's equation.

This turns out to be generally true: typically a potential function in a region where it is never infinite is harmonic. This makes for a very close connection between potential theory and the theory of harmonic functions.

In fact, although it is a matter of vector algebra that $\nabla \cdot \nabla \times \mathbf{F} = 0$ for any smooth vector field $\mathbf{F}$, and that $\nabla \times \nabla f = 0$ for any smooth function f, both these results have non-trivial consequences. In each case, we consider only nice regions of $\mathbb{R}^3$—for example star-shaped regions—in which nothing becomes infinite. In physical terms there are no sources present. In the first case, we can say that $\nabla \cdot \mathbf{F} = 0$ if and only if there is a smooth vector field $\mathbf{G}$ such that $\mathbf{F} = \nabla \times \mathbf{G}$. In the second case, we can say that $\nabla \times \mathbf{F} = 0$ if and only if there is a smooth function f such that $\mathbf{F} = \nabla f$. We will prove these claims in a moment.

Using these results, consider a vector field $\mathbf{F}$ which admits a potential function f, so $\mathbf{F} = \nabla f$. Now we calculate the Laplacian of the function f, and we find

$$\Delta f = \nabla \cdot \nabla f = \nabla \cdot \mathbf{F},$$

and so by the first result $\Delta f = 0$ if and only if f is the potential function of a vector field whose divergence vanishes, which happens if and only if the vector field is a curl, that is, it is of the form $\nabla \times G$ for some other vector field $\mathbf{G}$.

We deduce that every potential function, in regions where it is not infinite (no sources are present) is a harmonic function.

We shall now see that in any simply connected domain, if a vector field $\mathbf{G} = 0$ then there is vector field $\mathbf{F}$ such that $\mathbf{G} = \nabla \times \mathbf{F}$. Let us first fix some notation: $\mathbf{F} = (F_1, F_2, F_3)$ and $\mathbf{G} = (G_1, G_2, G_3)$.

Now we use the fact that for any smooth function f, $\nabla \times \nabla f = 0$ to simplify the problem. We find a function f such that $\frac{\partial f}{\partial z} = -F_3$, which is easy: define f by $f(\mathbf{r}) = -\int_0^{\mathbf{r}} f_3 dz +$ any arbitrary function $g(x, y)$. The integration is done treating f as a function of z alone. The arbitrary function g can be set to zero or used to keep the expressions symmetric when working on a specific problem.

We are reduced to solving the problem of finding a vector field $\mathbf{F}$ of the form $\mathbf{F} = (F_1, F_2, 0)$. The equations we have to solve are therefore these three

$$\begin{aligned} F_{2_z} &= G_1, \\ -F_{1_z} &= G_2, \\ F_{1_y} - F_{2_x} &= G_3. \end{aligned}$$

This is three equations for two unknowns, so the system is said to be over-determined and perhaps cannot be solved. Notice that they imply that $G_{1_x} + G_{2_y} + G_{3_z} = 0$, which is just the condition that $\nabla \cdot G = 0$, so the given condition is necessary; we are about to see that it is sufficient.

We integrate the first equation with respect to z and find

$$F_2(\mathbf{r}) = \int_0^{\mathbf{r}} G_1 dz + h(x, y),$$

where $h(x, y)$ is an arbitrary function of x and y. We substitute this into the third for equation, and get these two equations for the remaining component F_1 of the vector field $\mathbf{F}$:

$$F_{1_y}(\mathbf{r}) = \int_0^{\mathbf{r}} G_1 dz + G_3 + h_x(x, y),$$

$$F_{1_z} = -G_2.$$

This is two equations for F_1. We solve the second one by a simple integration, to get

$$F_1(\mathbf{r}) = -\int_0^{\mathbf{r}} G_2 dz + k(x, y),$$

where $k(x, y)$ is another arbitrary function of x and y. Then we differentiate this expression for F_1 with respect to y and substitute that in the other equation, to get

$$F_{1_y}(\mathbf{r}) = -\int_0^{\mathbf{r}} G_{2_y} dz + k_y(x, y) = \int_0^{\mathbf{r}} G_1 dz + G_3 + h_x(x, y).$$

This rearranges to become

$$k_y(x, y) - h_x(x, y) = \int_0^{\mathbf{r}} \left(G_{2_y} - G_1\right) - G_3.$$

To finish off we can set one of h or k to zero and integrate. In any given problem it is likely that one or other of these methods will make the integration easy or the expression for F_1 attractive. But in any case we have shown that if a vector field $\mathbf{G} = 0$ then there is vector field $\mathbf{F}$ such that $\mathbf{G} = \nabla \times \mathbf{F}$. The only condition is that the path integrals make sense, and this is guaranteed in $\mathbb{R}^3$ or any other simply connected space.

Next, let us be more precise about when a vector field $\mathbf{F}$ is the gradient of a (potential) function. Notice that this is the case if and only if the line integral of the field is independent of the path. 'Only if' goes this way: we have $\int_\gamma \mathbf{F}.d\mathbf{r} = \int_\gamma \nabla f.d\mathbf{r}$ and this works out, by the chain rule, to be the integral of df along the path, which is plainly the difference in the values of f at the endpoints whatever path is chosen. The 'if' part is little more than writing down $f(\mathbf{x}) := \int_{\mathbf{p}}^{\mathbf{x}} \mathbf{F}.d\mathbf{r}$ for an arbitrary $\mathbf{p}$ and checking that it makes sense—which it does because the integral is assumed to be independent of the path.

So a vector field is the gradient of a function if and only if integrals along it are independent of the path, which is the same—by Stokes's theorem—as saying if and only if integrals around closed loops always enclose a surface within which the curl $\nabla \times \mathbf{F}$ vanishes. And we know that the vanishing of the curl of $\mathbf{F}$ is a necessary condition for $\mathbf{F}$ to be given by a potential function. So a vector field is given by a potential function if and only if any closed loop in the regions outside the sources of the field contains a surface. This is usually true in questions about gravitation, but not in electro-magnetic theory where circuits flow along closed loops of wire or, it is sometimes supposed, along infinitely long wires. Later mathematicians (and indeed Gauss) were to see this as the start of the study of the topology of spaces of the form $\mathbb{R}^3 - stuff$.

References

Abel, N.H.: Ueber einige bestimmte Integrale. J. für die Reine und Angewandte Mathematik **2**, 22–30 (1828a). (tr. as Sur quelques intégrales définies, in Oeuvres, 2nd edn. (no. 15) I, 251–262)

Abel, N.H.: Solution d'un problème général concernant la transformation des fonctions elliptiques. Astr. Nachr. vol. VI, col. 365–388. (Addition au Mémoire précédent. Astr. Nachr. VII, 147, in Oeuvres complètes 2nd edn. (nos. 19, 20) I, 103–428 and 429–443)

Abel, N.H.: In: Sylow, L., Lie, S. (eds.) Oeuvres complètes de Niels Hendrik Abel, 2nd edn. Christiania (Oslo) (1881)

Abel, N.H.: Correspondance d'Abel comprenant ses lettres et celles qui lui ont été adressées. Lettres relatives à Abel. In: Holst, E., Størmer, C., Sylow, L. (eds.) Niels Henrik Abel: Mémorial publié à l'occasion du centenaire de sa naissance, pp. 111–135 (1902). (J. Dybwald, Kristiania; Gauthier-Villars, Paris; Williams and Norgate, London; Teubner, Leipzig)

Alexander, J.W.: An example of a simply connected surface bounding a region which is not simply connected. Proc. Natl Acad. Sci. USA **10**, 8–10 (1924)

Ampère, A.M.: Recherches sur quelques points de la théorie des functions dérivés, etc. J. École Polytech. **6**, 148–181 (1806)

Argand, J.R.: In: Hoüel, J. (ed.) Essai sur une manière de représenter les quantités imaginaires dans les constructions géometriques, 2nd edn. (1806). 1874

Argand, J.R.: Réflexions sur la nouvelle théorie des imaginaires, suivie d'une application à la démonstration d'un théorème d'analyse. Ann. de Mathématiques **5**, 197–209 (1814)

Arthur, R.T.W.: Leibniz's syncategorematic infinitesimals. Arch. Hist. Exact Sci. **67**, 553–593 (2013)

Belhoste, B.: Augustin-Louis Cauchy: A Biography. Springer, New York (1991)

Belhoste, B.: Autour d'un mémoire inédit: la contribution d'Hermite au développement de la théorie des fonctions elliptiques. Revue d'Histoire des Mathématiques **2**, 1–66 (1996)

Belhoste, B., Lützen, J.: Joseph Liouville et le Collège de France. Revue d'Histoire des Sci. **37**, 255–304 (1984)

Bertrand, J.: Notice sur les travaux du Commandant Laurent, pp. 389–393. Eloges académiques Hachette, Paris (1890)

Bertrand, J.: La vie et les travaux du baron Cauchy, par C. A. Valson. J. des Savants 205–215 (1869)

Biermann, K.-R.: Die Mathematik und ihre Dozenten an der Berliner Universitaät 1810–1920: Stationen auf dem Wege eines mathematischen Zentrums von Weltgeltung. Akademie-Verlag, Berlin (1973)

Birkhoff, G., Merzbach, U.: A Source Book in Classical Analysis. Harvard U.P., Cambridge (1973)

J. Gray, *The Real and the Complex: A History of Analysis in the 19th Century*,
Springer Undergraduate Mathematics Series, DOI 10.1007/978-3-319-23715-2

Bjerknes, C.A.: Niels-Henrik Abel. Tableau de sa vie et son action scientifique. Bordeaux Mémoires (3) **1**, 1–365 (1885)
Björling, E.G.: Doctrinae serierum infinitarum exercitationes. Nova acta Regiae Societatis Scientiarum Upsaliensis **13**(61–87), 143–187 (1846)
Björling, E.G.: Sur une classe remarquable de séries infinies. J. de Mathématiques Pures et Appliquées **17**, 454–472 (1852)
Björling, E.G.: Om oändliga serier, hvilkas termer äro continuerliga functioner af en reel variabel mellan ett par gränser, mellan hvilka serierna äro convergerande. Öfversigt af Kongl. Vetenskaps-Akademiens Forhandlingar **10**, 147–159 (1853)
Borel, É.: Leçons sur la Théorie des fonctions. Gauthier-Villars, Paris (1914)
Bos, H.J.M., Kers, C., Oort, F., Raven, D.W.: Poncelet's closure theorem. Expositiones Mathematicae **5**, 289–364 (1987)
Bottazzini, U.: Riemann's Einfluss auf E. Betti und F. Casorati. Arch. Hist. Exact Sci. **18**, 27–37 (1977)
Bottazzini, U.: Il calcolo sublime: storia dell' analisi matematica da Euler a Weierstrass. Editore Boringhieri, Torino (1981). (English translation The Higher Calculus, Springer, New York (1986))
Bottazzini, U.: Geometrical rigour and 'modern' analysis. An introduction to Cauchy's. Cours d'analyse, pp. xi–clxvii (1990). (Bottazzini's 1992 edition of (Cauchy 1821))
Bottazzini, U.: 'Algebraic Truths' vs. 'Geometric Fantasies': Weierstrass's response to Riemann. In: Proceedings of the International Congress of Mathematicians, Beijing 2002, vol. 3, pp. 923–934. Higher Education Press, Beijing (2002)
Bottazzini, U., Gray, J.J.: Hidden Harmony-Geometric Fantasies: The Rise of Complex Function Theory. Springer, New York (2013)
Bråting, K.: A new look at E.G. Björling and the Cauchy sum theorem. Arch. Hist. Exact Sci. **61**, 519–535 (2007)
Bressoud, D.M.: A Radical Approach to Real Analysis. Mathematical Association of America and Cambridge U.P., Cambridge (1994)
Bressoud, D.M.: A Radical Approach to Lebesgue's Theory of Integration. Mathematical Association of America and Cambridge U.P., Cambridge (2008)
Brill, A., Noether, M.: Ueber die algebraischen Functionen und ihre Anwendung in der Geometrie. Mathematische Annalen **7**, 269–316 (1874)
Briot, C., Bouquet, J.: Étude des fonctions d'une variable imaginaire. J. de l'École Polytech. **21**, 85–132 (1856a)
Briot, C., Bouquet, J.: Recherches sur les propriétes des fonctions définies par des équations différentielles. J. de l'École Polytech. **21**, 133–198 (1856b)
Briot, C., Bouquet, J.: Mémoire sur l'intégration des équations différentielles au moyen des fonctions elliptiques. J. de l'École Polytech. **21**, 199–254 (1856c)
Briot, C., Bouquet, J.: Théorie des fonctions doublement périodiques et, en particulier, des fonctions elliptiques. Paris (1859)
Buée, M.: Mémoire sur les quantités imaginaires. Philos. Trans. R. Soc. 13–88 (1806)
Buchwald, J.Z., Feingold, M.: Newton and the Origin of Civilization. Princeton U.P., Princeton (2012)
Cannell, M.: George Green, Mathematician and Physicist, 1793–1841. The Background to His Life and Work. The Athlone Press (1993)
Cantor, G.: Ueber die Ausdehnung eines Satzes aus der Theorie der trigonometrischen Reihen. Mathematische Annalen **5**, 123–133 (1872). (in Gesammelte Abhandlungen, 92–102)
Cantor, G.: Fernere Bemerkung über trigonometrische Reihen. Mathematische Annalen **16**, 267–269 (1880). (in Gesammelte Abhandlungen, 104–106)
Cantor, G.: Ueber unendliche lineare Punktmannigfaltigkeiten. Mathematische Annalen **21**, 545–591 (1882). (in Gesammelte Abhandlungen, 139–246)
Cantor, G.: Über eine elementare Frage der Mannigfaltigkeitslehre. Jahresbericht der Deutschen Mathematiker-Vereinigung **1**, 75–78 (1891). (in Gesammelte Abhandlungen, 278–280)

Cantor, G.: In: Zermelo, E. (ed.) Gesammelte Abhandlungen mathematischen und philosophischen Inhalts. Springer, Berlin (1932)

Carathéodory, C.: Elementarer Beweis für den Fundamentalsatz der konformen Abbildungen. H.A. Schwarz-Festschrift. Mathematische Abhandlungen, Hermann Amandus Schwarz zu seinem fünfzigjährigen Doktorjubiläum am 6. August 1914 gewidmet von Freunden und Schülern, 19–14

Casorati, F.: Teorica delle funzioni di variabili complesse. Pavia (1868)

Cauchy, A.L.: Mémoire sur les intégrales définies. Mém. Div. Sav. Inst. Fr. (1814). ((2), 1, (1827), 601-799 in Oeuvres (1) 1, 319-506)

Cauchy, A.L.: Cours d'analyse de l'Ecole Royale Polytechnique. 1re partie. Analyse algébrique. Paris (1821). (in Oeuvres (2) 3, rep. U. Bottazzini (ed.) CLUEB, Bologna, 1992)

Cauchy, A.L.: Résumé des leçons donnés à l'école royale polytechnique sur le calcul infinitesimal. Tome premier. Paris (1823). (in Oeuvres (2) 4, 5–261)

Cauchy, A.L.: Mémoire sur les intégrals définies, prises entre des limites imaginaries. (1825) (in Oeuvres (2) 15, 41-89)

Cauchy, A.L.: Leçons sur le calcul différentiel. Paris (1829). (in Oeuvres (2) 4, 263–409)

Cauchy, A.-L.: Sui metodi analitici, Bibl. Ital. **60**, 202–219; **61**, 321–334; **62**, 373–386 (1830–1831). (Rep. as Dei metodi analitici. Tipografia delle Belle Arti, Roma 1843 in Oeuvres (2) 15, 149–181)

Cauchy, A.L.: Extrait du mémoire présenté à l'Académie de Turin le 11 Octobre 1831 (1831). (in Oeuvres (2), 15, 262–411)

Cauchy, A.L.: Sulla meccanica celeste e sopra un nuovo calcolo chiamato calcolo dei limiti. In: Frisiani, P., Piola, G. (eds.) Opuscoli matem. fis. **2**, 1–84, 133–202, 261–316 (1834)

Cauchy, A.-L.: Première lettre sur la détermination complète de toutes les racines des équations de degré quelconque. Comptes Rendus de l'Académie des Sci. **4**, 773–783 (1837a). (in Oeuvres (1) 4, 48–60)

Cauchy, A.-L.: Deuxième lettre sur la résolution des équations de degré quelconque. Comptes Rendus de l'Académie des Sci. **4**, 805–821 (1837b). (in Oeuvres (1) 4, 61–80)

Cauchy, A.L.: Mémoire sur l'intégration des équations différentielles des mouvements planetaires. Comptes Rendus de l'Académie des Sci. **9**, 184–190 (1839). (in Oeuvres (1) 4, 483–491)

Cauchy, A.L.: Considérations nouvelles sur la théorie des suites et sur les lois de leur convergence. Exercises d'Analyse et de Physique Mathématique, **1**, 269–287 (1840). (in Oeuvres (2) 11, 331–353)

Cauchy, A.L.: Note sur le développement des fonctions en séries. Comptes Rendus de l'Académie des Sci. **13**, 910–914 (1841). (in Oeuvres (1) 6, 359–365)

Cauchy, A.L.: Mémoire sur l'emploi du nouveau calcul, appelé calcul des limites, dans l'intégration d'un système d'équations différentielles. Comptes Rendus de l'Académie des Sci. **15**, 14–25 (1842). (in Oeuvres (1) 7, 5–17)

Cauchy, A.L.: Note sur le développement des fonctions en séries ordonnées suivant les puissances entiéres positives et négatives des variables. Comptes Rendus de l'Académie des Sci. **17**, 193–198 (1843a). (in Oeuvres (1) 8, 5–10)

Cauchy, A.L.: Note. Comptes Rendus de l'Académie des Sci. **17**, 370 (1843b). (in Oeuvres (1) 8, 17–18)

Cauchy, A.L.: Note sur le développement des fonctions en séries convergentes ordonnées suivant les puissances entières des variables. Comptes Rendus de l'Académie des Sci. **17**, 940–942 (1843c). (in Oeuvres (1) 8, 117–120)

Cauchy, A.L.: Sur les intégrales qui s'étendent à tous les points d'une courbe fermée. Comptes Rendus de l'Académie des Sci. **23**, 251–255 (1846a). (in Oeuvres (1) 10, 70–74)

Cauchy, A.L.: Mémoire sur les intégrales dans lesquelles la fonction sous le signe $\int$ change brusquement de valeur. Comptes Rendus de l'Académie des Sci. **23**, 557–563. (in Oeuvres (1) 10, 135–143)

Cauchy, A.L.: Mémoire sur les intégrales imaginaires des équations différentielles, etc. Comptes Rendus de l'Académie des Sci. **23**, 563–569 (1846c). (in Oeuvres (1) 10, 143–150)

Cauchy, A.L.: Mémoire sur une nouvelle théorie des imaginaires, et sur les racines symboliques des équations et des équivalences. Comptes Rendus de l'Académie des Sci. **24**, 1120–1130 (1847). (in Oeuvres (1) 10, 312–323)

Cauchy, A.L.: Mémoire sur les quantités géométriques. Exercises d'Analyse et de Physique Mathématique **4**, 157–180 (1849). (in Oeuvres (2) 14, 175–202)

Cauchy, A.L.: Note sur les séries convergentes dont les divers termes sont des functions continues d'une variable réelle ou imaginaire, entre des limites données. Comptes Rendus de l'Académie des Sci. **36**, 454–459 (1853). (in Oeuvres (1) 12, 30–36)

Cauchy, A.L.: Considerations nouvelles sur la théorie des suites, etc. Exercises d'Analyse, **2** (1841). (in Oeuvres (2) 11, 331–353)

Cavaillès, J., Noether, E. (eds.): Briefwechsel Cantor–Dedekind. Paris

Cayley, A.: On the geometrical representation of imaginary variables by a real correspondence of two planes. Proc. Lond. Math. Soc. **9**, 31–39 (1878). (in The Collected Mathematical Papers of Arthur Cayley 10, 316–323)

Clebsch, R.F.A.: Ueber die Anwendung der Abelschen Functionen in der Geometrie. J. für die Reine und Angewandte Mathematik **63**, 189–243 (1864)

Clebsch, R.F.A., Gordan, P.: Theorie der Abelschen Functionen. Teubner, Leipzig (1866)

Clifford, W.K.: On the canonical form and dissection of a Riemann's surface. Proc. Lond. Math. Soc. **8**, 292–304 (1877). (in Mathematical Papers, 241–254)

Cogliati, A.: On Jacobi's transformation theory of elliptic functions. Arch. Hist. Exact Sci. **68**, 529–545 (2014)

Corry, L.: Axiomatics, Empiricism, and Anschauung in Hilbert's conception of geometry: between arithmetic and general relativity. In: Ferreirós, J., Gray, J.J. (eds.) The Architecture of Modern Mathematics, pp. 133–156. Oxford U.P., Oxford (2006)

Cox, D.A.: The arithmetic–geometric mean of Gauss. L'Enseignement Mathématique **30**(2), 275–330 (1984)

le Rond D'Alembert, J.: Essai d'une nouvelle théorie de la résistance des fluides (1752)

Darboux, G.: Sur les functions discontinues. Annales Scientifiques de l'École Normale Supérieure **4**(2), 57–112 (1875)

Darboux, G.: Les origines, les méthodes et les problèmes de la géométrie infinitésimale. In: Atti del IV Congresso Internazionale dei Matematici, Roma 1908, vol. 1, pp. 105–122 (1909). (Tipografia Accademia dei Lincei, Rome, G. Castelnuovo (ed.))

Darrigol, O.: Electrodynamics from Ampère to Einstein. Oxford U.P., Oxford (2000)

Dauben, J.W.: Georg Cantor: His Mathematics and Philosophy of the Infinite. Harvard U.P., Cambridge (1979)

Dedekind, R.: Stetigkeit und irrationale Zahlen, Vieweg (1872), tr. W.W. Beman as Continuity and irrational numbers, Dover (1963)

Dedekind, R.: Gesammelte mathematische Werke, I. New York (1969)

Dhombres, J.: French mathematical textbooks from Bézout to Cauchy. Hist. Sci. **28**, 91–137 (1985)

Dieudonné, J.: History of Functional Analysis. North-Holland Mathematics Studies, no. 49 (1981)

Dini, U.: Lezioni di analisi infinitesimale, vol. 1. Pisa (1877)

Dini, U.: Fondamenti per la teorica delle funzioni di variabili reali. Pisa (1878)

Dirichlet, P.G.L.: Sur la convergence des séries trigonométriques. J. für die Reine und Angewandte Mathematik **4**, 157–169 (1829). (in Werke, I, 117–132)

Dirichlet, P.G.L.: Vorlesungen über die im umgekehrten Verhältniss des Quadrats der Entfernung wirkenden Kräfte. In: F. Grube (ed.) Teubner, Leipzig (1876)

Dirichlet, P.G.L.: In: Fuchs, L., Kronecker, L. (eds.) Gesammelte Werke, 2 vols. Berlin (1889, 1897)

Dirksen, E.H.: A. L. Cauchy's Lehrbuch der algebraischen Analysis. Aus dem Französischen übersetzt von C.L.B. Huzler, Königsberg 1828. Jahrbücher für Wissenschaftliche Kritik **2**, 211–222 (1829)

Du Bois-Reymond, P.: Versuch einer Classification der willkürlichen Functionen etc. J. für die Reine und Angewandte Mathematik **79**, 21–37 (1875)

Dugac, P.: Éléments d'analyse de Karl Weierstrass. Arch. Hist. Exact Sci. **10**, 41–176 (1973)

Dunnington, G.W.: Carl Friedrich Gauss, Titan of Science: A Study of His Life and Work. Exposition Press, New York (1955). (Re-edition with a new introduction and appendices by J.J. Gray. Mathematical Association of America, Washington, D.C. 2004)

Durège, H.: Elemente der Theorie der Functionen einer complexen veränderlichen Grösse. Mit besonderer Berücksichtigung der Schöpfungen Riemanns, [etc.]. Teubner, Leipzig (1864)

Enneper, A.: In: Müller, F. (ed.) Elliptische Functionen. Theorie und Geschichte, 2nd edn. (1890). Halle a. S. L. Nebert

Euler, L.: Introductio in analysin infinitorum 1, Opera Omnia (1) **8** (1748). (tr. Blanton, J.: Introduction to Analysis of the Infinite, Book I, Springer, 1988, E 101)

Euler, L.: Observationes analyticae variae de combinationibus. Commentarii Academiae Scientiarum Petropolitanae **13**, 64–93 (1751). (in Opera Omnia (1) 2, 163–193, E 158)

Euler, L.: Institutiones calculi differentialis cum eius usu in analysi finitorum ac doctrina serierum. Opera Omnia (1) **10**, E 212 (1755)

Euler, L.: De integratione aequationis differentialis $\frac{mdx}{\sqrt{(1-x^4)}} = \frac{ndy}{\sqrt{(1-y^4)}}$. Novi Commentarii Academiae Scientiarum Petropolitanae **6**, 37–57 (1761a). (in Opera Omnia (1) 20, 58–79, E 251)

Euler, L.: Observationes de comparatione arcuum curvarum irrectificibilium. Novi Commentarii Academiae Scientiarum Petropolitanae **6**, 58–84 (1761b). (in Opera Omnia (1) **20**, 80–107, E 252)

Euler, L.: De miris proprietatibus curvae elasticae sub aequatione $y = \int (xxdx)/\sqrt{(1-x^4)}$ contentae. Acta Academiae Scientarum Imperialis Petropolitinae 34–61 (1786). (in Opera Omnia (1) 21, 91–118, E 605)

Fagnano, G.: Produzioni matematiche, 2 vols. Stamperia Gavelliana, Pesaro in Opere matematiche (1750) (V. Volterra, G. Loria, and D. Gambioli (eds.) 3 vols. Dante Alighieri, Milano, 1911)

Fauvel, J., Gray, J.J.: The History of Mathematics—A Reader. Macmillan (1987)

Ferraro, G., Panza, M.: Lagrange's theory of analytical functions and his ideal of purity of method. Arch. Hist. Exact Sci. **66**, 95–197 (2012)

Ferreirós, J.: Labyrinth of Thought: A History of Set Theory and its Role in Modern Mathematics. Birkhäuser, Basel (1999). (2nd edn. 2007)

Feymnan, R.: Lectures in Physics. Addison-Wesley, Reading (1964)

Fourier, J.: Théorie analytique de la chaleur (1822). (in Oeuvres 1, reprinted Gabay, Paris, 1988, tr. as The analytical theory of heat tr. A. Freeman, Cambridge 1878, Dover reprint 1950)

Freudenthal, H.: Augustin-Louis Cauchy. Dict. Sci. Biogr. **3**, 131–148 (1971)

Freudenthal, H.: Bernhard Riemann. Dict. Sci. Biogr. **11**, 447–456 (1975)

Fuchs, L.I.: Zur Theorie der linearen Differentialgleichungen mit veränderlichen Coefficienten, Jahresberichte der Gewerbeschule Berlin in Ges. Math. Werke **1**, 111–158 (1865)

Fuchs, L.I.: Zur Theorie der linearen Differentialgleichungen mit veränderlichen Coefficienten. J. für die reine und angewandte Mathematik **66**, 121–160 (1866). (in Ges. Math. Werke **1**, 159–204)

Fuss, P.H.: Correspondance mathématique et physique de quelques célèbres géomètres du XVIIIème siècle, 2 vols (1845)

Gandon, S., Perrin, Y.: Le probléme de la définition de l'aire d'une surface gauche: Peano et Lebesgue. Arch. Hist. Exact Sci. **63**, 665–704 (2009)

Gauss, C.F.: Disquisitiones arithmeticae, G. Fleischer, Leipzig (1801). (in Werke I. English translation W.C. Waterhouse, A.A. Clarke, Springer (1986))

Gauss, C.F.: Disquisitiones generales circa seriem infinitam, Pars prior. Comm. Soc. Reg. Gött. II. (1812a). (in Werke III, 123–162)

Gauss, C.F.: Determinatio seriei nostrae per aequationem differentialem secundi ordinis, Ms. (1812b). (in Werke, pp. 207–230)

Gauss, C.F.: Determinatio attractionis quam in punctum quodvis positionis datae exerceret, etc. Comm. Soc. Reg. Göttingen **4**, 21–48 (1818). (in Werke 3, 331–356)

Gauss, C.F.: Allgemeine Auflösung der Aufgabe die Theile einer gegebenen Fläche auf einer andern gegebenen Fläche so abzubilden, dass die Abbildung dem Abgebildeten in den kleinisten Theilen ähnlich wird. Astronomische Abhandlungen 3, 1825. Gauss (1822). (Werke IV 1880, 189–216)

Gauss, C.F.: Theoria residuorum biquadraticorum, Commentatio secunda. Göttingische Gelehrte Anzeigen. 625–638 (1831). (in Werke 2, 169–178)

Gauss, C.F.: Allgemeine Lehrsätze in Beziehung auf die im verkehrten Verhältnisse des Quadrats der Entfernung wirkenden Anziehungs-und Abstossungs-Kräfte. Leipzig (1840)

Gauss, C.F.: Werke, vols. I–V, K. Ges. Wiss. Göttingen (1863–1867)

Gauss, C.F.: Briefwechsel zwischen Gauss und Bessel. Engelmann, Leipzig (1880)

Geppert, H.: Bestimmung der Anziehung eines elliptischen Ringes. Nachlass zur Theorie des arithmetisch geometrischen Mittels und der Modulfunktion von C. F. Gauss. Teubner, Leipzig (1927)

Gilain, Chr.: Le théorème fondamental de l'algèbre et la théorie géométrique des nombres complexes au XIXe visages, 51–73, D. Flament (ed.) Maison des Sciences de l'Homme, Paris (1997)

Gispert, H.: Sur les fondements de l'analyse en France, à partir des lettres inédits de G. Darboux. Arch. Hist. Exact Sci. **28**, 37–106 (1983)

Giusti, E.: Cauchy's "errors" and the foundations of analysis. (Italian). Boll. Storia Sci. Mat. **4**, 24–54 (1984)

Grabiner, J.V.: The Origins of Cauchy's Rigorous Calculus. MIT Press, Cambridge (1981)

Grattan-Guinness, I.: The Cauchy-Stokes-Seidel story on uniform convergence again: was there a fourth man? Bulletin de la Société Mathématique de Belgique **38**, 225–235 (1986)

Gray, J.J.: On the history of the Riemann mapping problem. Supplemento ai Rendiconti del Circolo Matematico di Palermo **34**(2), 47–94 (1994)

Gray, J.J.: Linear Differential Equations and Group Theory from Riemann to Poincaré, 2nd edn. Birkhäuser, Boston (2000a)

Gray, J.J.: The Hilbert Challenge. Oxford U.P., Oxford (2000b)

Gray, J.J.: Worlds Out of Nothing; A Course on the History of Geometry in the 19th Century. Springer, New York (2011)

Green, G.: An Essay on the application of mathematical analysis to the theories of electricity and magnetism. Nottingham. Mathematical Papers, pp. 356–374 (1828)

Green, G.: Mathematical Papers. Macmillan, Cambridge (1871)

Griffiths, P., Harris, J.: On Cayley's explicit solution to Poncelet's porism. L'Enseignement mathématique **24**(2), 31–40 (1978)

Gudermann, Chr.: Theorie der Modular-Functionen. J. für die Reine und Angewandte Mathematik **18**, 1–54, 142–258, 303–364 (1818)

Hales, T.C.: Jordan's proof of the Jordan curve theorem. Stud. Log. Gramm. Rhetor. **10**(23), 45–60 (2007)

Hamilton, W.R.: Theory of conjugate functions, or algebraic couples; with a preliminary and elementary essay on algebra as the science of pure time. Trans. R. Irish Acad. **17**, 293– 422 (1837). (in Mathematical Papers 3, 3–96)

Hankel, H.: Untersuchungen über die unendlich oft oszillierenden und unstetigen Functionen, Gratulationsprogramm der Tübinger Universität (1870). (rep. Mathematische Annalen **20**, 63–112 (1882))

Hankins, T.L.: Jean d'Alembert; Science and the Enlightenment. Oxford U.P., Oxford (1970)

Hardy, G.H.: Weierstrass's non-differentiable function. Trans. Am. Math. Soc. **17**, 301–325 (1916)

Hardy, G.H.: Sir George stokes and the concept of uniform convergence. Collect. Pap. **7**, 505–513 (1918)

Harnack, A.: Grundlagen der Theorie des logarithmischen Potentiales, etc. Teubner, Leipzig (1887)

Hausdorff, F.: Grundzüge der Mengenlehre. Teubner, Leipzig (1914)

Hausdorff, F.: Bemerkung über den Inhalt von Punktmengen. Mathematische Annalen **75**, 428–433 (1915). (in Werke IV, 3–18)

Hawkins, T.: Lebesgue's Theory of Integration; Its Origins and Development. Chelsea, New York (1975). (rep, American Mathematical Society. Providence, Rhode Island (1999))

Heine, E.: Über trigonometrische Reihen. J. für die Reine und Angewandte Mathematik **71**, 353–365 (1870)

Heine, E.: Die Elemente der Functionenlehre. J. für die Reine und Angewandte Mathematik **74**, 172–188 (1872a)

Hermite, Ch.: Sur les fonctions algébriques. Comptes Rendus de l'Académie des Sci. **32**, 358–361 (1851). (in Oeuvres 1, 276–280)
Hermite, C.: Oeuvres, 4 vols. Paris (1905–1917)
Hilbert, D.: Ueber die stetige Abbildung einer Linie auf ein Flächenstück. Mathematische Annalen **38**, 459–460 (1891). (in Gesammelte Abhandlungen 3, 1–2)
Hilbert, D.: Die Theorie der algebraischen Zahlkörper. Jahrsbericht den Deutschen mathematiker Vereinigung **4**, 175–546 (1897). (in Gesammelte Abhandlungen, 1, 63–363, English edition, The Theory of Algebraic Number Fields, F. Lemmermeyer and N. Schappacher (trs. and eds.). Springer)
Hilbert, D.: Grundlagen der Geometrie, (Festschrift zur Einweihung des Göttinger Gauss-Weber Denkmals). Leipzig (1899). (Revised 2nd edn. 1903, English translation. Foundations of Geometry, numerous subsequent editions and translations)
Hilbert, D.: Über das Dirichletsche Princip. Jahresbericht der Deutschen Mathematiker-Vereinigung **8**, 184–187 (1900). (in Gesammelte Abhandlungen, 3, 15–37)
Hilbert, D.: Über das Dirichletsche Prinzip. Mathematische Annalen **59**, 161–186 (1904). (in Gesammelte Abhandlungen, 3, 1–2)
Holzmüller, G.: Einführung in die Theorie der isogonalen Verwandtschaften und der conformen Abbildungen, verbunden mit Anwendungen auf mathematische Physik. Teubner, Leipzig (1882)
Houzel, Ch.: The work of Niels Henrik Abel. In: Laudal, O.A., Piene, R. (eds.) The Legacy of Niels Henrik Abel: The Abel Bicentennial. Springer, Oslo (2004)
Jacobi, C.G.J.: Demonstratio theorematis ad theoriam functionum ellipticarum spectantis. Astronomische Nachrichten **6** (1827). (in Gesammelte Werke 1, (2nd edn.) 37–48)
Jacobi, C.G.J. Fundamenta Nova Theoriae Functionum Ellipticarum (1829). (in Gesammelte Werke 1, (2nd ed.) 49–239)
Jacobi, C.G.J.: Zur Geschichte der elliptischen und Abelschen Transcendenten. Ms. Gesammelte Werke 2, 516–521 (1847)
Jacobi, C.G.J.: Gesammelte Werke, 8 vols., 2nd edn. Chelsea
Jacobi, C.G.J.: Vorlesungen über analytische Mechanik. Berlin (1996). (1847/48, Pulte, H. (ed.) Vieweg)
Jesseph, D.M.: Leibniz on the elimination of infinitesimals. In: Goethe, N.B., Beeley, P., Rabouin, D. (eds.) G.W. Leibniz, Interrelations Between Mathematics and Philosophy, pp. 189–205. Springer, New York (2015)
Jordan, C.: Cours d'analyse, 1st edn. Gauthier-Villars, Paris (1887)
Jourdain, P.: Introduction to Georg Cantor (1915). (Contributions to the founding of the Theory of transfinite Numbers, Dover edition, 1955)
Killing, W.: Karl Weierstrass. Rede, gehalten beim Antritt des Rectorats an der Kgl. Akademie zu Münster am 15 October 1897. Natur und Offenbarung, vol. 43, pp. 705–725 and Aschensdorff'sche Buchhdl., Münster (1897)
Klein, C.F.: Riemann und seine Bedeutung für die Entwicklung der modernen Mathematik. Jahresbericht der Deutschen Mathematiker-Vereinigung **4**, 71–87 (1894–95). (in Ges. Math. Abh. 3, 482–497)
Klein, C.F.: In: Fricke, R., Ostrowski, A.M., Vermeil, H., Bessel-Hagen, E. (eds.) Gesammelte mathematische Abhandlungen, 3 vols. Springer, Berlin (1921–1923)
Klein, C.F.: Vorlesungen über die Entwicklung der Mathematik im 19. Jahrhundert (1926–1927). (R. Courant and O. Neugebauer (eds.), 2 vols. Springer, rep. Chelsea, New York 1967)
Kline, M.: Mathematical Thought from Ancient to Modern Times. Oxford U.P., Oxford (1972)
Koenigsberger, L.: Vorlesungen über die Theorie der elliptischen Functionen, nebst einer Einleitung in die allgemeine Functionenlehre, 2 vols. Teubner, Leipzig (1874)
Koenigsberger, L.: Weierstrass' erste Vorlesung über die Theorie der elliptischen Funktionen. Jahresbericht der Deutschen Mathematiker-Vereinigung **25**, 393–424 (1917)
Krantz, S.G., Parks, H.R.: The Implicit Function Theorem: History, Theory, and Applications. Birkhäuser, Boston (2002)

Krazer, A.: Zur Geschichte des Umkehrproblems der Integrale. Jahresbericht der Deutschen Mathematiker-Vereinigung **18**, 44–75 (1909)

Kummer, E.E.: Über die hypergeometrische Reihe, etc. J. für die Reine und Angewandte Mathematik **15**, 39–83, 127–172 (1836). (in Coll. Papers 2, 75–166)

Lagrange, J.-L.: Nouvelle méthode pour résoudre les équations littérales par le moyen des séries. Hist. Acad. Sci. Berlin **24**, 251–326 (1770). (in Oeuvres 3, 5–78)

Lagrange, J.-L.: Sur le problème de Kepler. Hist. Acad. Sci. Berlin **25**, 204–233 (1771). (in Oeuvres 3, 113–138)

Lagrange, J.-L.: Théorie des fonctions analytiques contenant les principes du calcul différentiel, dégagé de toute considération d'infiniment petits ou d'évanouissans, de limites ou de fluxions, et réduits à l'analyse algébrique des quantités finis, L'Imprimerie de la République, Paris (1797). (in Serret, J.-A. (ed.) Oeuvres 9. Gauthiers-Villars, Paris (1881))

Lagrange, J.-L.: Traité de la résolution des équations numériques de tous les degrés. Paris (1st edn. 1798, 3rd edn. 1826) (1808). (in Oeuvres 8, J.-A. Serret (ed.) Paris, Gauthiers-Villars, 1881)

Laplace P.S.: Mémoire sur l'usage du calcul aux différences partielles dans la théorie des suites. Hist. Acad. Sci. Paris **1777**, 99–122 (1780). (in O. C. 9, 313–335)

Laugwitz, D.: Bernhard Riemann, tr A. Shenitzer. Birkhäuser, Basel (2000)

Laurent, P.A.: Extension du théorème de M. Cauchy, relatif á la convergence du développement d'une fonction suivant les puissances ascendantes de la variable (1843). (in J. Peiffer, Les premiers exposés globaux de la théorie des fonctions de Cauchy, 1840–1860, Thesis, Paris)

Lebesgue, H.: Intégrale, longeur, aire. Annali di Matamatica Pure et Applicata **7**(3), 231–259 (1902). (in Oeuvres scientifiques I, 203–391 Geneva)

Lebesgue, H.: Leçons sur l'intégration et la recherche des fonctions primitives. Gauthier-Villars, Paris (1904)

Lebesgue, H.: Sur la non-applicabilité de deux domaines appartenant à des espaces à n et $n + p$ dimensions. Mathematische Annalen **70**, 166–168 (1911)

Legendre, A.-M.: Mémoire sur les integrations par les arcs d'ellipse. Hist. Acad. Sci. Paris **1786**, 616–643 (1788a)

Legendre, A.-M.: Seconde mémoire sur les intégrations par d'arcs d'ellipse et sur la comparison de ces arcs. Hist. Acad. Sci. Paris **1786**, 644–683 (1788b)

Legendre, A.-M.: Mémoire sur les transcendantes elliptiques, où l'on donne des méthodes faciles pour comparer et évaluer ces transcendantes [etc.]. Du Pont & Firmin–Didot, Paris (1792). (English translation in Leybourn, T. New Series of the Mathematical Repository 2 (1809) 1–45)

Legendre, A.-M.: Recherches sur diverses sortes d'intégrales définies. Mém. Inst. Fr. **9**, 416–509 (1809)

Legendre, A.-M.: Exercises de calcul intégral, vol. II, 3 vols. Paris (1814)

Legendre, A.-M.: Traité des fonctions elliptiques et des Intégrales Euleriennes, 3 vols. Paris (1825/27)

Legendre, A.-M., Jacobi, C.G.J.: Correspondance mathématique entre Legendre et Jacobi, J. für die Reine und Angewandte Mathematik **80**, 205–279 (1875). (Jacobi. Gesammelte Werke **1**, 385–461 (1875))

Liouville, J.: Leçons sur les fonctions doublement périodiques faites en 1847. J. für die Reine und Angewandte Mathematik **80**, 277–310 (1880)

Lipschitz, R.: De explicatione per series trigonometicas. J. für die Reine und Angewandte Mathematik **63**, 296–308 (1864)

Lützen, J.: Joseph Liouville, 1809–1882. Master of Pure and Applied Mathematics. Springer, New York (1990)

Lützen, J.: The foundation of analysis in the 19th century. In: Jahnke, H.N. (ed.) A History of Analysis, pp. 155–196. American and London Mathematical Societies, HMath 24, Providence (2003)

Mittag-Leffler, G.: Die ersten 40 Jahre des Lebens von Weierstrass. Acta Mathematica **39**, 1–57 (1923)

Moigno, F.: Leçons de calcul différentiel et de calcul intégral, 2 vols. Bachelier, Paris (1840–1844)

Moore, E.H.: On certain crinkly curves. Trans. Am. Math. Soc. **1**, 72–90 (1900)
Moore, G.H.: Zermelo's Axiom of Choice: Its Origins. Development and Influence. Springer, New York (1982)
Neuenschwander, E.: Studies in the history of complex function theory. The Casorati-Weierstrass theorem. Historia Mathematica **5**, 139–166 (1978a)
Neuenschwander, E.: Der Nachlass von Casorati (1835–1890) in Pavia. Arch. Hist. Exact Sci. **19**, 1–89 (1978b)
Neumann C.A.: Vorlesungen über Riemann's Theorie der Abel'schen Integrale. Teubner, Leipzig (1865). (2nd revised edition 1884)
Ore, O.: Niels Henrik Abel: Mathematician extraordinary, University of Minnesota Press (1957). (rep. Chelsea, New York (1974))
Osgood, W.F.: A Jordan curve of positive area. Trans. Am. Math. Soc. **4**, 107–112 (1903)
Painlevé, P.: Sur la théorie de la représentation conforme. Comptes Rendus de l'Académie des Sci. **112**, 653–657 (1891)
Peano, G.: Sur une courbe, qui remplit toute une aire plane. Mathematische Annalen **36**, 157–160 (1890)
Petrova, S.S.: Sur l'histoire des démonstrations analytiques du théorème fondamental de l'algèbre. Historia Mathematica **1**, 255–261 (1974)
Pincherle, S.: Gli elementi della teoria delle funzioni analitiche. Zanichelli, Bologna (1922)
Poincaré, H.: L'œuvre mathématique de Weierstrass. Acta **22**, 1–18 (1899). (Not in Oeuvres)
Poincaré H.: Science et méthode. Paris (1908)
Poincaré H.: Oeuvres, 11 vols. Paris (1916–1954)
Poincaré, H.: Analyse des travaux scientifiques de Henri Poincaré faite par lui-même. Acta Mathematica **38**, 1–135 (1921)
Poisson, S.D.: Rapport sur l'ouvrage de M. Jacobi intitulé Fundamenta nova theoriae functionum ellipticarum. Mém. Acad. Sci. Paris 10, 73–117 (1831)
Poncelet, J.V.: Traité des propriétés projectives des figures. Bachelier, Paris (1822)
Prym, F.E.: Zur Integration der Differentialgleichung $\frac{\partial^2 u}{\partial x^2} + \frac{\partial^2 u}{\partial y^2} = 0$. J. für die Reine und Angewandte Mathematik **73**, 340–364 (1871)
Puiseux, V.: Recherches sur les fonctions algébriques. J. de Mathématiques Pures et Appliquées **15**, 365–480 (1850)
Puiseux, V.: Recherches sur les fonctions algébriques. Suite. J. de Mathématiques Pures et Appliquées **16**, 240–288 (1851)
Remmert, R.: Theory of Complex Functions. Springer, New York (1991)
Riemann, B.: Grundlagen für eine allgemeine Theorie der Functionen einer veränderlichen complexen Grösse (Inaugural dissertation), Göttingen (1851). (in Werke 3–45)
Riemann, B.: Über die Darstellbarkeit einer Function durch einer trigonometrische Reihe. K. Ges. Wiss. Göttingen **13**, 87–132 (1854a). (in Werke, 227–271)
Riemann, B.: Ueber die Hypothesen welche der Geometrie zu Grunde liegen. K. Ges. Wiss. Göttingen **13**, 1–20 (1854b). (in Werke, 272–287)
Riemann, B.: Beiträge zur Theorie der durch Gauss'sche Reihe $F(\alpha, \beta, \gamma, x)$ darstellbaren Functionen. K. Ges. Wiss. Göttingen (1857a). (in Werke, 67–83)
Riemann, B.: Selbstanzeige der vorstehenden Abhandlung. Göttingen Nachr. no. 1 (1857b). (in Werke, 84–87)
Riemann, B.: Theorie der Abelschen Functionen. J. für die reine und angewandte Mathematik **54**, 115–155 (1857c). (in Werke, 88–144)
Riemann, B.: Ueber die Anzahl der Primzahlen unter einer gegebene Grösse. Monatsberichte Berlin Akademie, pp. 671–680 (1859). (in Werke, 145–153)
Riemann, B.: Bernhard Riemann's Gesammelte Mathematische Werke und Wissenschaftliche Nachlass. In: Dedekind, R., Weber, H. (eds.) with Nachträge, ed. M. Noether and W. Wirtinger. 3rd ed. R. Narasimhan, Springer (1990). (English translation Collected Papers, trans. R. Baker, C. Christenson, and H. Orde, Kendrick Press)

Roch, G.: Ueber Functionen complexer Grössen. Zeitschrift für Mathematik und Physik **8**(12–26), 183–203 (1863)
Roch, G.: Ueber die Anzahl der willkürlichen Constanten in algebraischen Functionen. J. für die Reine und Angewandte Mathematik **64**, 372–376 (1865)
Rouché, E.: Mémoire sur la série de Lagrange. J. de l'École Polytechnique **22**, 193–224 (1862)
Rüdenberg, L., Zassenhaus, H. (eds.): Hermann Minkowski – Briefe an David Hilbert. Springer, New York (1973)
Russ, S.: The Mathematical Works of Bernhard Bolzano. Oxford U.P., Oxford (2004)
Sagan, H.: Space-Filling Curves. Springer, New York (1994)
Scharlau, W., Opolka, H.: From Fermat to Minkowski. Springer, New York (1985)
Scheeffer, L.: Allgemeine Untersuchungen über Rectification der Curven. Acta Mathematica **5**, 49–82 (1884)
Schlesinger, L.: Über Gauss's Arbeiten zur Funktionentheorie. Göttingen Nachrichten (Beiheft) in Gauss Werke **10**(2), 1–222 (1912). (Separate pagination)
Schlömilch, O.: Vorlesungen über einzelne Theile der höheren Analysis gehalten an der K.S. Polytechnischen Schule zu Dresden. Vieweg & Sohn, Braunschweig (1866)
Schoenflies, A.: Beiträge zur Theorie der Punktmengen III. Mathematische Annalen **62**, 286–328 (1906)
Schottky, F.: Ueber die conforme Abbildung mehrfach zusammenhängender ebener Flächen. J. für die Reine und Angewandte Mathematik **83**, 300–351 (1877)
Schubring, G.: Conflicts Between Generalization, Rigor, and Intuition: Number Concepts Underlying the Development of Analysis in 17–19th Century France and Germany. Sources and Studies in the History of Mathematics and Physical Sciences. Springer (2005)
Schubring, G.: Lettres de mathématiques français à Weierstrass – documents de sa réception en France. In: Suzanne, F. (ed.) Aventures de l'analyse de Fermat à Borel: Mélanges en l'honneur de Christian Gilain, pp. 567–594. Éditions Universitaires de Lorraine (2012)
Schwarz, H.A.: Ueber einige Abbildungsaufgaben. J. für die Reine und Angewandte Mathematik **70**, 105–120 (1869a). (in Abhandlungen II, 65–83)
Schwarz, H.A.: Zur Theorie der Abbildung. Programm ETH Zürich (1869c). (Abhandlungen II, 108–132)
Schwarz, H.A.: Ueber einen Grenzübergang durch altenirendes Verfahren. Vierteljahrschrift Natur. Gesellschaft Zürich **15**, 272–286 (1870a). (in Abhandlungen II, 133–143)
Schwarz, H.A.: Zur integration der partiellen Differentialgleichung $\Delta u = 0$ unter vorgeschriebenen Grenz und Unstetigkeits bedingungen. Monatsber. K. A. der Wiss.Berlin, 767–795 (1870b). (in Abhandlungen II, 144–171)
Schwarz, H.A.: Beispiel einer stetigen nicht differentiirbaren Function. Verhandlungen der Schweizerischen Naturforschenden Gesellschaft, 252–258 (1878). (in Abhandlungen II, 269–274)
Schwarz, H.A.: Sur une définition erronée de l'aire d'une surface courbe. Cours de M. Hermite, 35–36 (18781–82). (Paris, 1883, in Abhandlungen II, 309–311)
Schwarz, H.A.: Gesammelte Mathematische Abhandlungen, 2 vols. (1st ed.) Berlin (1890). ((2nd ed.) rep. in 1 vol. Chelsea, 1972)
Seidel, P.L.: Note über eine Eigenschaft der Reihen, welche discontinuirliche Funktionen darstellen. Abhandlungen Bayerische Akademie der Wissenschaften **5**, 381–394 (1847). (rep. in Ostwald's Klassiker, H. Liebmann (ed.) 116, 35–45, Leipzig, 1900)
Smith, D.E.: A Source Book in Mathematics, 1st edn. Dover, New York (1929)
Smith, H.J.S.: On the integration of discontinuous functions. Proc. Lond. Math. Soc. **6**, 140–153 (1875). (in Collected Mathematical Papers 1, 86–100)
Smith H.J.S.: Collected Mathematical Papers. Oxford U.P., Oxford (1894). (2 vols, Chelsea reprint 1965)
Sørensen, H.K.: Exceptions and counterexamples: understanding Abel's comment on Cauchy's theorem. Historia Mathematica **32**, 453–480 (2005)
Stahl, H.: Theorie der Abel'schen Functionen. Teubner, Leipzig (1896)
Stahl, H.: Elliptischen Functionen: Vorlesungen von B. Riemann, Teubner, Leipzig (1899)

Stirling, J.: Methodus differentialis: sive tractatus de summatione et interpolatione serierum infinitarum. G. Bowyer, London (1730)
Stokes, Sir G.: On the critical values of the sums of periodic series. Trans. Camb. Philos. Soc. **8**, 533–583 (1849)
Struik, D.: A Source Book in Mathematics, 1200–1800. Harvard U.P., Cambridge (1969)
Stubhaug, A.: Niels Henrik Abel and His Times. Called Too Soon by Flames Afar. Springer, New York (2000)
Sturm, C.: Mémoire sur une classe d'équations à differences partielles. J. de Mathématiques Pures at Appliquées **1**, 373–444 (1836)
Thomae, J.: Einleitung in die Theorie der bestimmten Integrale. Halle, Nebert (1875)
Thomae, J.: Elementare Theorie der analytischen Functionen einer complexen Veränderlichen. Nebert, Halle (1880). (2nd. edn. Nebert, Halle 1898)
Tobies, R.: Felix Klein. Teubner, Leipzig (1981)
Viertel, K.: Geschichte der gleichmäßigen Konvergenz: Ursprünge und Entwicklungen des Begriffs in der Analysis des 19. Jahrhunderts. Springer (2014)
Von der Mühll, K., et al.: Clebsch Rudolf Friedrich Alfred - Versuch einer Darlegung und Würdigung seiner wissenschaftlichen Leistungen von einigen seiner Freunde. Mathematische Annalen **7**, 1–55 (1874)
Von Koch, H.: Une méthode giéométrique élémentaire pour l'étude de certaines questions de la théorie des courbes planes. Acta Mathematica **30**, 145–174 (1906)
Weierstrass K.T.W.: Darstellung einer analytischen Function einer complexen Veränderlichen, deren absolute Betrag zwischen zwei gegebenen Grenzen liegt, ms (1841a). (in Werke 1, 51–66)
Weierstrass K.T.W.: Zur Theorie der Potenzreihen, ms (1841b). (in Werke, I, 67–74)
Weierstrass K.T.W.: Definition analytischer Functionen einer Veränderlichen vermittelst algebraischer Differentialgleichungen, ms (1842). (in Werke I, 75–84)
Weierstrass K.T.W.: Zur Theorie der Abel'schen functionen. J. für die reine und angewandte Mathematik **47**, 289–306 (1854). (in Werke I, 133–152)
Weierstrass K.T.W.: Über die Theorie der analytischen Facultäten. J. für die reine und angewandte Mathematik **51**, 1–60 (1856a). (in Werke, I, 153–221)
Weierstrass K.T.W.: Theorie der Abel'schen functionen, J. für die reine und angewandte Mathematik 52, 285–339 (1856b). (in Werke, I, 297–355)
Weierstrass, K.T.W.: Über die sogenannte Dirichlet'sche Princip (1870). (in Werke, vol. 2, pp. 49–54)
Weierstrass, K.T.W.: Über continuerliche Functionen eines reellen Arguments, die für keinen Werth des Letzeren einen bestimmten Differentialquotient besitzen, read to the Königlichen Akademie der Wissenschaften, Berlin (1872) 18 July 1872. (in Werke, II, pp. 71–74))
Weierstrass, K.T.W.: Zur Funktionenlehre. Monatsberichte Berlin, pp. 719–743 (1880). (Nachtrag, Monatsberichte Berlin, 1881, 228–230, rep. in (Weierstrass 1886, 67–101, 102–104), in Werke 2, 201–233)
Weierstrass, K.T.W.: Über die analytische Darstellbarkeit sogenannter willkürlicher Funktionen reeller Argumente. Berlin Berichte, pp. 633–639, 789–805 (1885). (in Werke 3, 1–37)
Weierstrass, K.T.W.: Abhandlungen aus der Funktionenlehre. Springer, Berlin (1886)
Weierstrass, K.T.W.: Werke. 7 vols. Olms, Hildesheim (1894–1927)
Weierstrass, K.T.W.: Vorlesungen über die Theorie der Abelschen Transcendenten (1902). (in Werke, IV)
Weierstrass, K.T.W.: Einfürung in die Theorie der analytischen Funktionen, nach einer Vorlesungmitschrift von Wilhelm Killing aus dem Jahr 1868. In: Scharlau, W. (ed.) Drucktechnische Zentralstelle Universität Münster, Münster (1968)
Weierstrass, K.T.W.: In: Ullrich, P. (ed.) Einleitung in die Theorie der analytischen Funktionen. Vorlesung Berlin 1878 in einer Mitschrift von Adolf Hurwitz. Vieweg & Sohn, Braunschweig (1988)
Weil, A.: Number Theory: An Approach Through History from Hammurapi to Legendre. Birkhäuser, Boston (1984)

Whittaker, E.T., Watson, G.N.: A Course of Modern Analysis, 4th edn. Cambridge U.P., Cambridge (1927)
Yandell, B.H.: The Honors Class. A.K. Peters (2002)
Zaremba, S.: Sur l'équation aux dérivées partielles $\Delta u + \lambda u + f = 0$ et sur les fonctions harmoniques. Annales Scientifiques de l'École Normale Supérieure **16**(3), 427–464 (1899)

Index

J. Gray, *The Real and the Complex: A History of Analysis in the 19th Century*, Springer Undergraduate Mathematics Series, DOI 10.1007/978-3-319-23715-2

Printed in the United States
By Bookmasters

Printed in the United States
By Bookmasters